Optics and Vision

Leno S. Pedrotti

Center for Occupational Research
and Development
Waco, Texas

Emeritus Professor of Physics
Air Force Institute of Technology
Dayton, Ohio

Frank L. Pedrotti, S.J.

Rockhurst College
Kansas City, Missouri

PRENTICE HALL
Upper Saddle River, New Jersey 07458

Library of Congress Cataloging-in-Publication Data

Pedrotti, Leno S.,
    Optics and vision / Leno S. Pedrotti, Frank L. Pedrotti.
      p. cm.
    Includes bibliographical references and index.
    ISBN 0-13-242223-9 (hardcover)
    1. Physiological optics. 2. Optics. I. Pedrotti, Frank L.
  II. Title.
    [DNLM: 1. Vision--physiology. 2. Optics. 3. Refraction, Ocular-
-physiology. 4. Light. 5. Lenses. 6. Eye--physiology. WW 103
P372o 1998]
QP475.P33 1998
612.8'4--dc21
DNLM/DLC
for Library of Congress   97-26179
                                               CIP

Executive editor: Alison Reeves
Assistant editor: Wendy Rivers
Total concept coordinator: Kimberly P. Karpovich
Production: ETP Harrison
Copy editor: Jane Loftus
Cover designer: Patricia H. Wosczyk
Cover photo: Eve Ritscher, Assoc/SPL/Photo Researchers, Inc.
Manufacturing manager: Trudy Pisciotti

 © 1998 by Prentice-Hall, Inc.
Simon & Schuster / A Viacom Company
Upper Saddle River, New Jersey 07458

Printed in the United States of America

10 9 8 7 6 5 4 3 2 1

ISBN 0-13-242223-9

Prentice-Hall International (UK) Limited, *London*
Prentice-Hall of Australia Pty. Limited, *Sydney*
Prentice-Hall Canada Inc., *Toronto*
Prentice-Hall Hispanoamericana, S.A., *Mexico*
Prentice-Hall of India Private Limited, *New Delhi*
Prentice-Hall of Japan, Inc., *Tokyo*
Simon & Schuster Asia Pte Ltd., *Singapore*
Editora Prentice-Hall do Brasil, Ltda., *Rio de Janeiro*

*This book is dedicated
to Jean and the rest of our support team:
Daro, Michael, Sandra, Laura, Catherine,
Leno M., Mary, and John.*

Contents

List of Tables

Physical Constants

Speed of light	$c = 2.998 \times 10^8$ m/s
Electron charge	$e = 1.602 \times 10^{-19}$ C
Electron rest mass	$m = 9.109 \times 10^{-31}$ kg
Planck constant	$h = 6.626 \times 10^{-34}$ Js
Boltzmann constant	$k = 1.3805 \times 10^{-23}$ J/K
Permittivity of vacuum	$\varepsilon_0 = 8.854 \times 10^{-12}$ C^2/N-m^2
Permeability of vacuum	$\mu_0 = 4\pi \times 10^{-7}$ T-m/A

Preface

This textbook, *Optics and Vision*, is an adaptation of *Introduction to Optics*, a textbook by the same authors, and as such, borrows heavily from it. *Optics and Vision* was prepared with two major goals in mind: first, to provide a text less demanding in mathematical preparation and second, a text more applicable to students of optics and vision. Thus, students without training in calculus should not be at a disadvantage. Calculus does not appear explicitly, and in the few places where it is essential, the results are stated without proof. A number of the more advanced chapters that appear in *Introduction to Optics* are omitted. In addition, several new chapters are added and others expanded to deal more closely with the eye and vision. Accordingly, this textbook should meet the needs of college optics courses in which calculus cannot be presumed, as well as introductory optics courses for students of optometry and related sciences.

Organization of the material follows essentially traditional lines. The first 10 chapters deal with geometrical optics and the remainder deal primarily with wave optics and its applications. Additionally, the particle (or photon) model of light is introduced in the first chapter and used especially in the chapter on "Lasers and the Eye."

Chapter 1 presents an historical review of attempts to understand the process of vision, the development of spectacles, and the gradual clarification of the nature of light itself. It is hoped that the concentration of this material, rather difficult to find in one place, will be helpful in providing an overview of the development of our knowledge of light and vision.

In Chapter 2, we describe a variety of common sources and detectors of light, as well as the radiometric and photometric units of measurement that are used throughout the book. In this chapter, and in the remainder of the text, the MKS system of units is employed.

Chapter 3 reviews the geometrical optics covered by introductory physics courses, deriving the usual reflection and refraction relations for mirrors and lenses. The method of vergence of wave fronts is introduced and applied in this chapter as an alternative to the ray description of object-image formation. The method of vergence is ideally suited to the optics of the eye and topics of direct interest to students of optometry and ophthalmology. Chapter 4 presents both analytical and ray-tracing methods of handling thick lenses, when the modification of the wave front of light passing through a refracting medium is not negligible as it is usually assumed to be in a thin lens. The *cardinal points* explained here are also applicable to more complicated optical systems.

In Chapter 5, the method of vergences is applied to vision and to the correction of refractive defects in vision. Chapter 6 describes cylindrical lenses, a topic usually omitted in standard optics courses, but central to dealing with the correction of astigmatic defects of the eye. Chapter 7 discusses a qualitative treatment of third-order aberrations as they apply to spherical surfaces and, in particular, to the eye. In Chapter 8, we present the rather demanding subject of stops, pupils, and windows, important in controlling image brightness and field of view in an optical system. Chapter 9 deals with the principles of geometrical optics and aberration theory as applied to a number of optical instruments, including prisms, cameras, microscopes, and telescopes. Chapter 10 takes up more realistic models of the eye, biologically and optically. Common ophthalmic instruments used in assessing defects in vision are presented and explained.

The second part of the text, dealing primarily with the consequences of the wave nature of light, begins with Chapter 11, in which the mathematical description of waves and their superposition is presented and applied to electromagnetic waves. Chapter 12 appeals to the photon model of light to explain the essential elements and basic physics of lasers, reviews laser types and applications, and describes laser therapy for selected ocular defects. The phenomena of interference, evident when light passes through small apertures or interacts with thin films, is presented in Chapter 13, followed by selected applications in Chapter 14, where a qualitative description of holography is included. The physical mechanisms responsible for the polarization of light are treated in Chapter 15. In Chapter 16, we provide a standard treatment of far-field, or Fraunhofer, diffraction, ending with the diffraction grating. In the final Chapter 17, we present an introduction to the field of fiber optics.

The authors, both having served as classroom teachers for many years, are aware that professors generally adapt a particular text to fit the special needs of their students, as well as to allow for the inevitable constraints of available time. Nevertheless, the authors feel that the following guidelines may be helpful to teachers in adapting this text to fit special situations.

For students of vision—pursuing medical careers as opticians, optometrists, and ophthalmologists—the material presented in the full text is all generally applicable to their field of study, with particular emphasis on the subject matter developed especially for them in:

Chapter 5, "Vergence and Vision"
Chapter 6, "Cylindrical Lenses and Astigmatism"

Chapter 10, "Optics of the Eye," and
Chapter 12, "Lasers and the Eye."

For students of physics and engineering in a standard, one-semester, college optics course, minimum coverage would normally include:

Chapter 1, "Light and Vision—An Historical Perspective"
Chapter 3, "Geometrical Optics"
Chapter 9, "Optical Instrumentation"
Chapter 11, "Light as Waves"
Chapter 13, "Interference Phenomena"
Chapter 15, "Polarized Light"
Chapter 16, "Fraunhofer Diffraction"

In a two-semester course for general students of physics and engineering, coverage would also include the other chapters in the text, with the possible exception of those chapters or sections of chapters having special emphasis on the science of vision and its defects.

Several features have been included to make the text more useful to students and teachers alike. Each chapter contains a comprehensive set of problems designed to illustrate the concepts developed in the text material. Answers to these problems are provided for the student's convenience at the end of the book. Also, a *Teacher's Manual*, with detailed solutions, is available from the publisher for optional use by the instructor. Finally, for those interested in collateral reading on related subject matter, a list of bibliographical references is included at the end of each chapter, referred to as needed in the text using the author's name and the publication date.

We wish to thank those teachers of optics who responded generously to our questionnaire and so influenced our selection of topics for students of optometry and related sciences. They are: Drs. Robert Gilman (NOVA Southeastern University), Richard P. Hemenger (Southern California College of Optometry), Michael P. Keating (Ferris State University), Stanley A. Klein (University of California Berkeley), Alan L. Lewis (Ferris State University), Roger W. West (Northeastern State University), and Graeme S. Wilson (University of Alabama at Birmingham). We are also grateful to Michael Pedrotti, O. D., for his review and suggestions in topics dealing with the eye. We wish to thank Judy Lawson for her sketch of Einstein that graces the first chapter.

We would also like to thank Copy Editor Jane Loftus and Project Manager Gina Gordon of ETP Harrison. Finally, we express our gratitude to the editorial and production staff of Prentice Hall. In particular, we are indebted to Physics Executive Editor, Alison Reeves; Director of Marketing, Kelly McDonald; and Assistant Editor, Wendy Rivers.

Leno S. Pedrotti
Frank L. Pedrotti, S.J.

1

Light and Vision— A Historical Perspective

INTRODUCTION

Gradual development of our understanding of the process of vision and of the nature of light is characterized by the interplay of parallel lines of progress. Knowledge of the structure and functioning of the eye, including artificial aids to vision provided by spectacles, increased together with the adoption of more accurate models of light itself. In what follows, for purposes of clarity, these lines of investigation will be separated into three parts: the history of vision, the history of spectacles, and the history of the physics of light. As presented here, these histories share in the uncertainties of the historical record and are not necessarily complete. There will always be uncertainties in knowledge of the origins of the subject, and there may be injustices in attributing to one person a particular advancement in knowledge, when discoveries, both theoretical and experimental, are concurrently pursued by a number of investigators. Once an investigation is undertaken, the apparent inevitability of scientific progress often produces the same discovery by independent investigators. It is not always a simple matter to credit the most worthy individuals for their achievements. Finally, in what follows we limit ourselves to a thumbnail sketch of histories that, in their completeness, would require several volumes.

Efforts to understand the structure and functioning of the eye in human vision date back to ancient Greek times. Whatever may have been accomplished in pre-classical civilization is unknown. During the time of the Greek philosopher, Aristotle (384–322 B.C.), the prevalent theory, which we shall refer to as the "tentacle theory," held that *visual rays* reached out from the eyes and touched the objects that were seen. These visual rays were imagined to be some type of subtle threads or tentacles. When these slender beams, which approximate our notion of rays, met with an external object, the act of vision was accomplished. This theory was accepted by Aristotle's teacher, Plato, but was rejected by Aristotle himself. Aristotle could not accept the notion that rays emanated from the eye, pointing out that in this case, the eye should be able to see in the dark. Aristotle's theory, a more reasonable explanation, was that the object seen generates a motion that acts on the senses. In his writings, Aristotle also contrasted long and short sight, probably a generalization of his own and others' experience. In this regard, Aristotle also noted that the ability to focus varied with age.

In the Alexandrian period, which bridged the Hellenistic and Roman periods and in which both Euclid and Ptolemy lived, many structural elements of the eye were identified and named. Among them, the optic nerve was described as a "hollow channel." Around 300 B.C., the Greek mathematician, Euclid, described the tentacles as light rays traveling in straight lines, but accepted the prevalent theory that these visual rays were emitted by the eye. Claudius Ptolemy, a Greco-Egyptian mathematician and astronomer, best known for his geocentric world system, wrote on optics around A.D. 150. Ptolemy also used the concept of straight-line rays to study refraction at a plane surface, and also accepted the idea that rays emanated from the eye. Around the same time, the Greek anatomist, Galen, produced writings that included the most complete description of the eye to date, though it was primarily a summary of what had been learned by others. Galen's work was considered the authoritative account of medicine in Europe until the time of the 15th century Renaissance. Galen continued the misleading notions of the tentacle theory and mistakenly identified the crystalline lens as the essential organ of vision. Modifications in the lens, according to Galen, were perceived by the brain in the act of vision.

During the early Middle Ages, Arabian civilization, in its turn, borrowed much of Greek learning. The greatest mathematician and physicist among the Arabs was Ibn al-Haitham, known in the West as Alhazen (A.D. 956–1038). After long periods of neglect, Alhazen succeeded in reviving interest in physiological optics through his treatise, *Book of Optics*, written in Cairo. In this work, Alhazen contributed much that was new. He rejected the tentacle theory of vision, proving that rays travel in straight lines from each point of a luminous object to the eye. However, Alhazen held that only the one ray from each point that struck the sensitive front surface of the lens perpendicularly contributed to the image as it passed on to the optic nerve. He proved experimentally the law of reflection and studied refraction, but, without a more complete knowledge of refraction, he was unable to appreciate the action of the cornea and lens in focusing the rays of light entering the eye. In particular, he did not realize the retina's function as the screen receiving the visual image. Alhazen was aware of the formation of inverted images in a *camera obscura* (the lensless pinhole camera), and considered it a model of the eye. Thus, although he believed a visual image was formed within the eye, he did not understand where and how it was formed. Still, Alhazen's influence in reviving the science, as well as his own contributions, merited for him in some quarters the title of "father of physiological optics."

Little advancement in visual science occurred after Alhazen, until the work of Roger Bacon (1214–1292), English philosopher and scientist. In his treatise on optics, the Franciscan monk described the principles of the *camera obscura*. The ability of the pinhole camera to form inverted images would later be greatly enhanced by placing a convex lens at the pinhole, an innovation introduced by the Neapolitan, Giambattista della Porta (1535–1615). This step improved the analogy between the camera and the eye, but della Porta believed that, in the eye, the lens itself was the photoreceptive element and the retina acted as a concave reflector. The correct assignment of the photoreceptive function was later made by Felix Plater (1536–1614), who proved experimentally that the crystalline lens of the eye acted simply as a spectacle lens.

It is interesting to note that the great Italian genius, Leonardo da Vinci (1452–1519), aware of the existence of a visual image in the eye, believed he had to discover a way in which the eye reversed the inverted image a second time to present an erect image on which the mind could act. Because of this belief, he could not succeed in tracing the actual path of light rays in the eye. Da Vinci recognized that the essential organ of vision was the retina, not the lens. Among other contributions, he drew attention to differences in peripheral vision and discussed the nature of depth perception.

Probably the earliest explanation of myopia and hyperopia was due to Francis Maurolycus (1494–1577), though he attributed these defects in vision to curvature of the lens rather than to the cornea or the axial length of the eyeball.

The formation of a retinal image in the eye—real, inverted, and demagnified—became generally accepted, especially due to the work of Johannes Kepler (1571–1630), the great astronomer and mathematician. He is perhaps best known for his laws of planetary motion. Kepler made other notable contributions to the science of vision. For example, he defined the "far point" of vision using ray tracings. He explained the effects of convex and concave lenses on light entering the eye, and described the improvement of vision possible by reshaping the curvature of the crystalline lens. Kepler also pointed out that sharpest vision took place at the central portion of the retina.

It was left to the Jesuit, Christoph Scheiner (1575–1650), to actually observe the inverted retinal image visually, an experiment that had to be repeated by Descartes and others before it convinced the scientific community. Scheiner also introduced several techniques for studying the refraction of the eye that became the basis for the invention of the ophthalmoscope. He was the first to measure the refractive index of the various media of the eye. Also, he was probably the first to discover that accommodation of the eye, when focusing on objects at different distances, did not occur when the lens was absent, thus pointing out the indispensable role of the lens in accommodation.

The Dutch naturalist, Antonie van Leeuwenhoek (1632–1723) used the microscope he constructed to perform detailed investigations of the retina and other parts of the eye. The British physician and professor of physics in London, Thomas Young (1773–1829), is credited with showing that the eye changes focus or accommodates due to changes in the lens. In his paper of 1799, *Sound and Light*, Young propounded the wave theory of light and suggested experimental ways to demonstrate the wave theory. Among these is the double-slit interference experiment, which still goes by his name. Interest in color blindness was given a boost when John Dalton (1766–1844), English chemist and physicist and pioneer in atomic theory, noticed color blindness in himself and gave a good description of this defect. Dalton also described the nature of astigmatism in his work. A Czech physiologist, Johannes Purkinje (1787–1840), determined how the lens and the muscles attached to the lens work together to bring about accommodation. He also studied reflected images from the cornea and lens (Purkinje images), using these images to estimate the powers of the cornea and lens. In England, the astronomer, George Airy (1801–1892), fabricated spherocylindrical lenses and used them to correct astigmatism. Around the same time, the British astronomer and lensmaker, Charles Babbage (1792–1871) invented the optical instrument, the ophthalmoscope. Around the middle of the century, the pioneer in physiological optics, Hermann von Helmholtz (1821–1894), produced his treatise, *Physiological Optics*. With this work, the famous German physicist and physiologist is credited with laying the foundation of modern visual science. His many contributions include formalizing the theory of color perception, in which the three color components, red, green, and blue, account for the vision of color. He also discussed the mechanism of accommodation and gave a clear explanation of how the two eyes work together to produce single vision. The photoreceptive role of the rods and cones was demonstrated experimentally by Heinrich Müller (1820–1864), a contemporary of Helmholtz. A few decades later, Max Schultze laid the foundations for understanding the twofold action of the retina—that cones are active under daylight conditions and sensitive to color, while rods respond to weaker stimuli and are insensitive to color.

1-2 HISTORY OF SPECTACLES

It is not hard to imagine how, looking through odd pieces of glass, even primitive people could become aware of the magnifying power of a crude lens. When held in sunlight, the focusing power of crude lenses must also have been noticed. The Athenian playwright, Aristophenes, in his comedy, *The Clouds*, is credited with one of the earliest references to the use of a convex lens as a *burning glass*. It is believed that as early as the 10th century the Chinese were using magnifying glasses set in frames.

Until around the end of the 13th century, lenses were fashioned from naturally occurring quartz and rock crystal. The use of glass lenses had to await the accidental discovery and development of glass. The discovery of glass is believed to have occurred in Syria, although Egypt soon became the center of the early glass industry. When sufficiently clear glass was later produced, bottles could be made that, when filled with water, would act as burning glasses and refract light to show magnification of small detail. With irregular pieces of glass, the prism effect of producing colors would also have been observed. It is interesting to note that, as far back as the first century A.D., the Roman stoic philosopher, Seneca, described the prism effect as well as the magnification of minute lettering that was rendered larger and clearer when viewed through a glass bulb filled with water.

The real art of glass-making was developed in northern Italy where, in the 11th century, the industry flourished in places like Pisa, Venice, and the island of Murano. The birth of the optical industry probably occurred there, with the invention of corrective spectacles around 1285. Spectacles had certainly been invented by the time of the arrival of the printing press in 1440—a technological development that ensured the future widespread need for spectacles. Clearly, the invention and development of spectacles was tied to parallel discoveries and developments in optics, as well as to the nature of the process of vision itself.

The earliest recorded reference (1268) to the use of lenses for optical purposes is credited to the Franciscan monk, Roger Bacon, mentioned earlier. Bacon alludes to magnifying glasses used for reading and to the usefulness of convex lenses in correcting hyperopia. Around that time, a manuscript in Italy mentions "glasses known as spectacles for reading." In 1305, Giordana da Rivalto, a monk in Pisa, writes that it was "not yet 20 years since the invention of spectacles." There is also a memorial bust, dated 1317, of Salvino d'Armato in Florence, which reads, "inventor of spectacles." Dating around the same time, a Heidelberg ballad refers to the use of a lens to assist in reading. Soon thereafter other evidences of the use of spectacles appear, such as Tommaso da Modena's portrait (1352) of Hugh of Provence, showing glasses, and a painting (1480) of the Christian scholar, St. Jerome, shown with glasses dangling from his desk.

Early spectacles were all convex, worn only when needed to magnify small print or other detail. Real motivation to improve the quality of optical glass came with the invention and early use of optical instruments, like the telescope (Galileo, 1608) and the microscope (Leeuwenhoek, 1684). The earliest evidence of concave lenses used to help myopes is in a book by Nicholas Causanus (1401–1464), and the earliest picture of them is Raphael's portrait (1517) of Pope Leo X, in which the concavity of the lens is evident from the reflected image.

The first appearance of bifocals is due to the American statesman, Benjamin Franklin, about 1760, who directed their construction for his own use. The first bifocals were simply the halves of two lenses of different power for near and far vision, their straight edges joined together along a horizontal. Then in 1804, a British chemist and physicist, William Wollaston (1766–1828), obtained a patent on a meniscus lens, designed to improve vision when the line of sight was off the center of the lens. George Airy (1801–1892), the English astronomer noted for the *Airy disc* in diffraction theory, receives credit for initiating work on spherocylindrical lenses to correct for the defect of astigmatism.

Extremely important in the development of lenses was the advent of the plastic industry, which grew rapidly in the last decades of the 19th century. Since the end of World War II, hard, or relatively scratch-resistant plastics were produced. To make plastic lenses for vision, it was necessary to develop clear optical plastic with a surface hard enough to serve for spectacles. The advantages of plastic lenses were their light weight and resistance to fracture. Since the decade of the 1950s, their use has grown steadily.

Another benefit of the plastic lens is its use in contact lenses. Elementary forms of the contact lens had been described as far back as Leonardo da Vinci and René Descartes, but the credit for the invention of the modern contact lens goes to the English astronomer, Sir John F. W. Herschel, who in 1827 proposed a glass lens with an inner surface fitted to the irregular cornea of an eye, with the help of a transparent, animal jelly. Then A. E. Fick, a university professor in Zurich, made a small glass lens that could float on the cornea, rest under the eye lids, and be held by the capillary action of tear fluid. When clear optical plastic became available in the United States in 1936, the use of the light-weight and shatter-proof material for contact lenses became widespread. In the 1970s, plastics were developed that were both thinner and softer, with the additional advantage of being water absorbent and permeable to the supply of oxygen to the cornea.

Asking questions about the structure and functioning of parts of the eye, or questions about how light behaves on reflection and refraction, was not nearly as sophisticated as asking about the nature of light itself. To make progress in answering this question, much more refined observations were required. Before 1600, much progress was made in dealing with the mechanisms of vision and the behavior of light rays as they move from one medium to another. Simple laws of optics were formulated, the usefulness of lenses was explored, and appropriate spectacles to counter visual defects of the eye were developed. Very little was done to answer the deeper question: what is the nature of light itself? After 1600, this was among the central questions dealt with by the scientific community.

The evolution in our understanding of the physical nature of light forms one of the most fascinating accounts in the history of science. Since the dawn of modern science in the 16th and 17th centuries, light has been pictured either as particles or as waves—incompatible models—each of which enjoyed a period of prominence among the scientific community. In the 20th century, it became clear that somehow light was both wave and particle, yet it was precisely neither. For some time this perplexing state of affairs, referred to as the *wave-particle duality*, motivated the greatest scientific minds of our age to find a resolution to apparently contradictory models of light. The solution was achieved through the creation of *quantum electrodynamics*, one of the most successful theoretical structures in the annals of physics.

In what follows, we will be content to sketch briefly a few of the high points of this developing understanding.[1] Certain areas of physics once considered to be disciplines apart from optics—electricity and magnetism, and atomic physics— are very much involved in this account. This alone suggests that the resolution achieved also constitutes one of the great unifications in our understanding of the physical world. The final result is that light and subatomic particles, like electrons, are both considered to be manifestations of matter or energy under the same set of formal principles.

The 17th and 18th centuries produced partial answers to this fundamental question. The gifted Descartes presented a theory of the nature of light that served as the beginning of modern theories. The notion of an ether permeating all space may have originated with Descartes when he described light as a kind of pressure in the ether, transmitted from object to eye, suggestive of the wave theory of light. Descartes' theory, however, required that the speed of light increase in a denser medium. This conclusion was challenged by the French mathematician, Pierre de Fermat (1601–1665), who correctly argued that the speed should be less, consistent with Snell's law. Fermat also introduced a generalization of Hero of Alexander's principle, which held that light rays travel from one point to another by taking the shortest path. To be consistent with the experimental fact of refraction, in which the shortest path is *not* the one taken, Fermat claimed that the path taken is rather the path of least time.

Robert Hooke (1635–1703), English physicist and mathematician, who constructed the first Gregorian telescope, also objected to Descartes' "pressure-in-the-ether" theory and held that light propagates in the ether by longitudinal oscillations of the particles of the ether. The Italian Jesuit, Francesco Grimaldi (1618–1663), discovered and named the phenomenon of diffraction, describing

---

[1]A more in-depth historical account may be found, for example, in Ronchi (1979).

the wavelike spreading of light around obstacles that prevented the formation of perfectly sharp shadows.

Christiaan Huygens (1629–1695), a Dutch scientist contemporary with Newton, championed the view (in his *Treatise on Light*) that light is a wave motion, spreading out from a light source in all directions and propagating through an all-pervasive elastic medium called the ether. He was impressed, for example, by the experimental fact that when two beams of light intersected, they emerged unmodified, just as in the case of two water or sound waves. Adopting a wave theory, Huygens showed how the progress of a wave front could be determined from a given initial position (Huygens' principle). Using this principle, he was then able to derive the laws of reflection and refraction and to explain double-refraction in calcite as well.

In the 17th century, the most prominent advocate of a particle theory of light was Isaac Newton (1642–1727), the same creative genius who had erected a complex science of mechanics and gravity. He is perhaps best known in optics for his experiments with prisms, showing that white light is composed of all colors that can be separated out (and recombined) by the prism's dispersive action. In his treatise *Optics*, Newton clearly regarded rays of light as streams of very small particles emitted from a source of light and traveling in straight lines. Although Newton often argued forcefully against positing hypotheses that were not derived directly from observation and experiment, here he adopted a particle hypothesis, believing it to be adequately justified by the phenomena he had observed. Important in his considerations was the observation that light can cast sharp shadows of objects, in contrast to water and sound waves, which bend around obstacles in their paths. At the same time, Newton was aware of the phenomenon now referred to as *Newton's rings*. Such light patterns are not easily explained by viewing light as a stream of particles traveling in straight lines. Newton maintained his basic particle hypothesis, however, and accounted for the phenomenon by endowing the particles themselves with what he called "fits of easy reflection and easy transmission," a kind of periodic motion due to the attractive and repulsive forces imposed by material obstacles. Newton's eminence as a scientist was such that his point of view dominated the century that followed his work.

Within two years of the centenary of the publication of Newton's *Optics*, a British professor of physics in London, Thomas Young (1773–1829), performed a decisive experiment that seemed to demand a wave interpretation, turning the tide of support to the wave theory of light. It was Young's double-slit experiment, in which an opaque screen with two small, closely spaced openings was illuminated by monochromatic light from a small source. The "shadows" observed formed a complex interference pattern like those produced with water waves.

Victories for the wave theory continued up to the twentieth century. In the mood of scientific confidence that characterized the latter part of the nineteenth century, there was little doubt in the scientific community that light, like most classical areas of physics, was well understood. We mention a few of the more significant confirmations.

In 1821, the French physicist and engineer, Augustin Fresnel (1788–1827), published results of his experiments and analysis, which required that light be a transverse wave. On this basis, double refraction in calcite could be understood as a phenomenon involving polarized light. It had been assumed that light waves in an ether were necessarily longitudinal, like sound waves in a fluid, which cannot support transverse vibrations. For each of the two components of polarized light, Fresnel developed the *Fresnel equations*, which gave the amplitude of light reflected and transmitted from a plane interface separating two optical media. From his theoretical assumptions, Fresnel was able to describe a method of producing circularly polarized light.

Working in the field of electricity and magnetism, the Scottish theoretical physicist, James Clerk Maxwell (1831–1879), synthesized known principles in his set of four *Maxwell equations*. The equations yielded a prediction for the speed of an electromagnetic wave in the ether that turned out to be the measured speed of light, suggesting its electromagnetic character. From then on, light was viewed as a particular region of the electromagnetic spectrum of radiation. The experiment (1887) of Albert Michelson and Edward Morley, which attempted to detect optically Earth's motion through the ether, and the special theory of relativity (1905) of Albert Einstein (1879–1955) were of monumental importance. Together they led inevitably to the conclusion that the assumption of an ether was superfluous. The problems associated with transverse vibrations of a wave in a fluid thus vanished.

If the 19th century served to place the wave theory of light on a firm foundation, this foundation was to crumble as the century came to an end. The wave-particle controversy was to resume with vigor. Again, we mention only briefly some of the key events along the way. Difficulties in the wave theory seemed to show up in situations that involved the interaction of light with matter. In 1900, at the very dawn of the 20th century, Max Planck announced at a meeting of the German Physical Society that he was able to derive the correct blackbody radiation spectrum only by making the curious assumption that atoms emit light in discrete energy packets rather than in a continuous manner. Thus *quanta* and *quantum mechanics* were born. According to Planck, the energy E of a quantum of electromagnetic radiation is proportional to the frequency f of the radiation,

$$E = hf \tag{1-1}$$

where the constant of proportionality, *Planck's constant*, has the very small value of 6.63×10^{-34} J-s. Five years later, in the same year that he published his theory of special relativity, Albert Einstein offered an explanation of the photoelectric effect, the emission of electrons from a metal surface when irradiated by light. Central to his explanation was the conception of light as a stream of photons whose energy is related to frequency by Planck's equation, Eq. (1-1). Then in 1913, the Danish physicist, Niels Bohr, once more incorporated the quantum of radiation in his explanation of the emission and absorption process of the hydrogen atom, providing a physical basis for understanding the hydrogen spectrum. Again in 1922, the photon model of light came to the rescue for Arthur Compton, who explained the scattering of X-rays from electrons as particle-like collisions between photons and electrons in which both energy and momentum are conserved.

All such victories for the photon or particle model of light indicated that light could be treated in some cases like a particular kind of matter, possessing both energy and momentum. It was Luis de Broglie who saw the other side of the picture. In 1924 he published his speculations that subatomic particles are endowed with wave properties. He suggested, in fact, that a particle with momentum p had an associated wavelength of

$$\lambda = \frac{h}{p} \tag{1-2}$$

where h was, again, Planck's constant. Experimental confirmation of de Broglie's hypothesis appeared during the years 1927–1928 when Clinton Davisson and Lester Germer (in the United States) and Sir George Thomson (in England) performed experiments that could only be interpreted as the diffraction of a beam of electrons.

Thus the wave-particle duality came full circle. Light behaved like waves in its propagation and in the phenomena of interference and diffraction; it could, however, also behave as particles in its interaction with matter, as in the photoelectric effect. On the other hand, electrons usually behaved like particles, as observed in the pointlike scintillations of a phosphor exposed to a beam of electrons; in other situations they were found to behave like waves, as in the diffraction produced by an electron microscope.

Photons and electrons that behaved both as particles and as waves seemed at first an impossible contradiction, since particles and waves are very different entities indeed. Gradually it became clear, to a large extent through the reflections of Niels Bohr and especially in his *principle of complementarity*, that photons and electrons were neither waves nor particles, but something more complex than either.

In attempting to explain physical phenomena, it is natural to appeal to well-known physical models like waves and particles. As it turns out, however, the full intelligibility of a photon or an electron is not exhausted by either model. In certain situations, wavelike attributes may predominate; in other situations, particle-like attributes stand out. We can appeal to no simpler physical model that is adequate to handle all cases.

Quantum mechanics, or wave mechanics, as it is often called, deals with all particles more or less localized in space and so describes both light and matter. Combined with special relativity, the momentum, wavelength, and speed for both material particles and photons are given by the same general equations. A crucial difference between particles like electrons and neutrons and particles like photons is that the latter have zero rest mass. The equations show that while nonzero rest mass particles like electrons have a limiting speed c (the speed of light in vacuum), zero rest-mass particles like photons *must* travel in vacuum with the constant speed c. The energy of a photon is not a function of its speed but of its frequency, as expressed in Eq. (1-1).

Another important distinction between electrons and photons is that electrons obey statistical laws called *Fermi statistics*, whereas photons obey other statistics, called *Bose statistics*. A consequence of Fermi statistics is the restriction that no two electrons in the same interacting system be in the same *state*, that is, have precisely the same physical properties. Bose statistics impose no such prohibition so that identical photons with the same energy and momentum can occur together in large numbers. Because light beams can possess so many similar photons in proximity, the granular structure of the beam is not ordinarily experienced, and the beam can be adequately represented by a continuous electromagnetic wave. From this point of view, electromagnetic fields appear as a special manifestation of photons.

A profound consequence of the wave nature of particles is embodied in the *Heisenberg principle* of indeterminacy. As a result of this principle, particles do not obey deterministic laws of motion. Rather, the theory predicts only probabilities. Wave functions are associated with the particles through the fundamental wave equation of quantum mechanics. The wave amplitudes, or better, the square of the wave amplitudes assigned to these particles, provide a means of expressing the probability that a particle will be found within a region of space during an interval of time. Thus the *irradiance* (power/area) of these waves at some intercepting surface, also proportional to the square of the wave amplitudes, provides a measure of this probability. When large numbers of particles are involved, probabilities approach certainties, so that the irradiance E_e of light at a location is proportional to the number of photons passing through the location per second:

$$n(\text{photons/m}^2\text{-s}) = \frac{E_e}{hf} \tag{1-3}$$

In this way, the interference and diffraction patterns previously explained by waves can be interpreted as manifestations of particles. Particle wave amplitudes predict the probabilities of their locations in the same patterns.

In the modern theory that was developed, called *quantum electrodynamics*, which combines the principles of quantum mechanics with those of special relativity, photons are assumed to interact only with charges. An electron, for example, is capable of both absorbing and emitting a photon, with a probability that is proportional to the square of the charge. There is no conservation law for photons as there is for the charge associated with particles. In this theory the wave-particle duality becomes reconciled. Essential distinctions between photons and electrons are removed. Both are considered subject to the same general principles. Through this unification, light is viewed as basically just another form of matter. Nevertheless, the complementary aspects of particle and wave descriptions of light remain, justifying our use of one or the other description when appropriate. The wave description of light will be found adequate to describe most, but not all, of the optical phenomena treated in this text.

PROBLEMS

1-1. Show that **(a)** $\lambda = \dfrac{h}{\sqrt{2mE}}$ **(b)** $p = \dfrac{E}{c}$

(Recall that momentum $p = mv$, kinetic energy $E = \frac{1}{2}mv^2$, and one electron volt (eV) of energy $= 1.602 \times 10^{-19}$ Joules of energy.)

1-2. Calculate the de Broglie wavelength of **(a)** a golf ball of mass 50 g moving at 20 m/s and **(b)** an electron with kinetic energy of 10 eV.

1-3. The threshold of sensitivity of the human eye is about 100 photons per second. The eye is most sensitive at a wavelength of around 550 nm. For this wavelength, determine the threshold in watts of power.
(Recall: 1 nm of length $= 1 \times 10^{-9}$ meter, and wavespeed $c = f\lambda$.)

1-4. What is the energy, in electron volts (eV), of light photons at the ends of the visible spectrum, that is, at wavelengths of 400 and 700 nm?
(See Problem 1-1 for eV equivalence.)

1-5. Solar radiation is incident at Earth's surface at an average of 1353 watts/m^2 on a surface normal to the rays. For a mean wavelength of 550 nm, calculate the number of photons falling on 1 cm^2 of the surface each second.

1-6. Two parallel beams of electromagnetic radiation with different wavelengths deliver the same power to equivalent surface areas normal to the beams. Show that the numbers of photons striking the surfaces per second for the two beams are in the same ratio as their wavelengths.

REFERENCES

BERTHOLD-GEORG, ENGLERT, MARLAN O. SCULLY, and HERBERT WALTHES. 1994. "The Duality in Matter and Light," in *Scientific American* (December): 86.

CANTOR, G. N. 1983. *Optics after Newton*. Dover, N.H.: Manchester University Press.

CROMBIE, A. C. 1964. "Early Concepts of the Senses and the Mind," in *Scientific American* (5): 108–16.

FEINBERG, GERALD. 1969. "Light," in *Lasers and Light*, pp. 4–13. San Francisco: W. H. Freeman and Company, Publishers.

GREGG, JAMES R. 1965. *The Story of Optometry*. New York: The Ronald Press Company.

GUILLEMIN, VICTOR. 1968. *The Story of Quantum Mechanics.* New York: Scribner's Sons.

HIRSCH, MONROE J., and RALPH E. WICK. 1968. *The Optometric Profession*. New York: Chilton Book Company.

HOFFMANN, BANESH. 1959. *The Strange Story of the Quantum*. New York: Dover Publications.

POLYAK, STEPHEN L. et al. 1943. "History of Vision," in *The Human Eye in Anatomical Transparencies*, Rochester, N.Y.: Bausch & Lomb Press.

RONCHI, VASCO. 1979. *The Nature of Light: An Historical Survey*. New York: Oxford University Press.

WALDMAN, GARY. 1983. *Introduction to Light*. Englewood Cliffs, N.J.: Prentice Hall.

ZEKI, SEMIR. 1992. "The Visual Image in Mind and Brain," in *Scientific American* (September): 68.

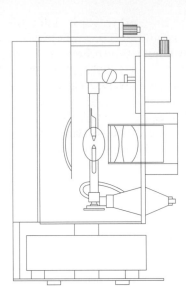

2

Production and Measurement of Light

INTRODUCTION

Electromagnetic radiation may vary in wavelength (or frequency) and in "strength." Classification due to variation in wavelength is summarized in the *electromagnetic spectrum*. Variations in strength are described in more precise physical terms, which have developed in the areas called *radiometry* and *photometry*. *Sources* and *detectors* of electromagnetic radiation can be classified on the basis of their spectral range and the strength of signal produced or detected, respectively. These considerations are essential to the production and measurement of electromagnetic radiation.

In the broad field of optics, and in the science of vision, the particular regions of the electromagnetic spectrum of most interest to us are denoted the *visible* (light), *infrared,* and *ultraviolet* regions. In this chapter we concentrate on the production and measurement of electromagnetic radiation in these regions.

2-1 ELECTROMAGNETIC SPECTRUM

An electromagnetic disturbance that propagates through space as a wave may be *monochromatic*, that is, characterized for practical purposes by a single wavelength, or *polychromatic*, in which case it is represented by many wavelengths, either discrete or in a continuum. The distribution of energy among the various constituent waves is called the *spectrum* of the radiation, and the adjective *spectral* implies a dependence on wavelength. Various regions of the *electromagnetic spectrum* are referred to by particular names—such as radio waves, microwaves, cosmic rays, light, and ultraviolet radiation—because of differences in the way they are produced or detected. Most of the common descriptions are given in Figure 2-1, in which the electromagnetic spectrum is displayed in terms of both wavelength (λ) and frequency (f). The two quantities are related, as with all wave motion, to the wave speed (c), as given in Eq. (2-1).

$$c = \lambda f \qquad\qquad (2\text{-}1)$$

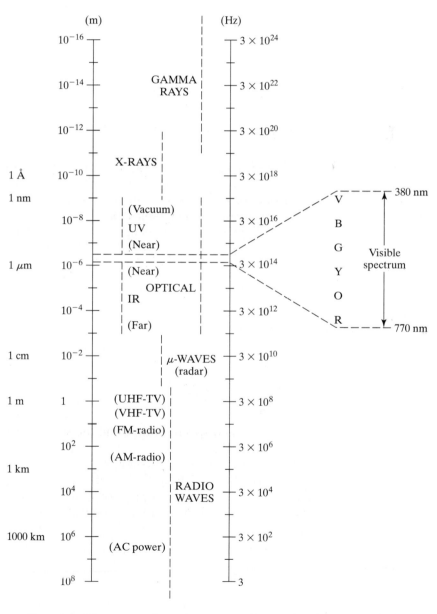

Figure 2-1 Electromagnetic spectrum, arranged by wavelength in meters and frequency in hertz. The narrow portion occupied by the visible spectrum is highlighted, varying from V (violet) to R (red).

The radiation described in Figure 2-1 is assumed to propagate in free space, for which, approximately, $c = 3 \times 10^8$ m/s. Common units for wavelength shown are the *angstrom* (1 Å = 10^{-10} m), the *nanometer* (1 nm = 10^{-9} m), and the *micrometer* (1 μm = 10^{-6} m). The regions ascribed to various types of waves, as shown, are not precisely bounded. Regions may overlap, as in the case of the continuum from X-rays to gamma rays. The choice of label will depend on the manner in which the radiation is either produced or used. The narrow range of electromagnetic waves from approximately 380 to 770 nm is capable of producing a visual sensation in the human eye and is properly referred to as "light." This visible region of the spectrum, which includes the spectrum of colors from red (long wavelength end) to violet (short-wavelength end) is bounded by the invisible ultraviolet and infrared regions, as shown. The three regions taken together make up the *optical spectrum*, that region of the electromagnetic spectrum of special interest in a textbook on optics.

2-2 RADIOMETRY

Radiometry is the science of measurement of electromagnetic radiation. In the discussion we present the *radiometric quantities* or physical terms used to describe the energy and power content of radiation. Later we discuss briefly some of the more common principles used in the instruments designed to measure radiation. Many radiometric terms have been introduced and used; however, we include here only approved International System (SI) units. These terms and their units are summarized in Table 2-1.[1]

Radiometric quantities appear either without subscripts or with the subscript *e* (*electromagnetic*) to distinguish them from similar photometric terms, to

TABLE 2-1   RADIOMETRIC TERMS

Term	Symbol	Description	Defining equation	Units of measure
Radiant energy	Q_e	Energy	——	J
Radiant energy density	w_e	Energy per unit volume	$w_e = \dfrac{\Delta Q_e}{\Delta V}$	J/m³
Radiant flux	Φ_e	Energy per unit time (power)	$\Phi_e = \dfrac{\Delta Q_e}{\Delta t}$	J/s or W
Radiant exitance	M_e	Power emitted per unit area of source	$M_e = \dfrac{\Delta \Phi_e}{\Delta A}$	W/m²
Irradiance	E_e	Power falling on unit area of target	$E_e = \dfrac{\Delta \Phi_e}{\Delta A}$	W/m²
Radiant intensity	I_e	Source power radiated per unit solid angle	$I_e = \dfrac{\Delta \Phi_e}{\Delta \omega}$	W/sr
Radiance	L_e	Source power radiated per unit area per unit solid angle	$L_e = \dfrac{\Delta \Phi_e}{\Delta \omega \Delta A} = \dfrac{I_e}{\Delta A}$	W/m²-sr

Abbreviations: J, joule; W, watt; m, meter; s, second; sr, steradian.

[1]The introduction "in the abstract" of so many new units, some rarely used and others misused, is not very palatable pedagogically. Table 2-1 and the following Table 2-2 are meant to serve as convenient summaries that can be referred to when needed.

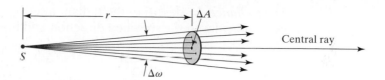

Figure 2-2 The radiant intensity is the flux through the cross section ΔA per unit of solid angle. Here the solid angle $\Delta\omega = \Delta A/r^2$.

be described later. The terms *radiant energy, Qe* (J = joules), *radiant energy density, w_e* (J/m³), and *radiant flux, Φ_e* (W = watts = J/s), need no further explanation. *Radiant flux density* at a surface, measured in watts per square meter, may be either emitted (scattered, reflected) from a surface, in which case it is called *radiant exitance, M_e*, or incident onto a surface, in which case it is called *radiant incidence* or simply *irradiance, E_e*. The radiant flux (Φ_e) emitted per unit of solid angle (ω) by a point source in a given direction (Figure 2-2) is called the *radiant intensity, I_e*. This quantity, often confused with *irradiance*, is given by

$$I_e = \frac{\Delta\Phi_e}{\Delta\omega} \ (\text{W/sr}) \tag{2-2}$$

where $\Delta\Phi_e$ is the radiant flux (power) radiated by the source through solid angle $\Delta\omega$, and where sr = steradian. The radiant intensity I_e from a sphere radiating Φ_e watts of power uniformly in all directions, for example, is ($\Phi_e/4\pi$) W/sr, since the total surrounding solid angle is 4π sr. The radiant intensity I_e from a hemisphere radiating Φ watts of power would be ($\Phi_e/2\pi$) W/sr, and so on.

The familiar inverse-square law of radiation from a point source, illustrated in Figure 2-3, now becomes apparent by calculating the irradiance E_e of a point source on a spherical surface surrounding the point, of solid angle 4π sr and surface area $4\pi r^2$. Thus

$$E_e = \frac{\Phi_e}{A} = \frac{4\pi I_e}{4\pi r^2} = \frac{I_e}{r^2} \tag{2-3}$$

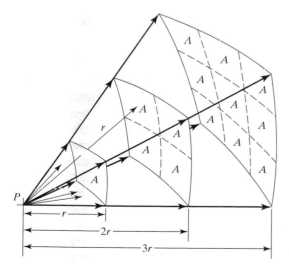

Figure 2-3 Illustration of the inverse square law. The flux leaving a point source within any solid angle is distributed over increasingly larger areas, producing an irradiance that decreases inversely with the square of the distance.

The *radiance*, L_e, describes the radiant intensity, I_e, per unit of *projected* area, $A_\perp$, perpendicular to the specified direction and is given by

$$Le = \frac{\Delta I_e}{\Delta A_\perp} = \frac{\Delta\Phi_e}{(\Delta\omega)(\Delta A \cos\theta)} \text{ W/(m}^2\text{-sr)} \tag{2-4}$$

The importance of the radiance is suggested in the following considerations. Suppose a plane radiator or reflector is perfectly *diffuse*, by which we mean that it radiates or reflects uniformly in all directions. The radiant intensity is measured for a fixed solid angle defined by the fixed aperture A_p at some distance r from the radiating surface, as shown in Figure 2-4. The aperture might be the input aperture of a detecting instrument measuring all the flux that so enters. When viewed at $\theta = 0°$, along the normal to the surface, a certain maximum intensity $I(0)$ is observed. As the aperture is moved along the circle of radius r, thereby increasing the angle θ, the cross section of radiation presented by the surface decreases in such a way that

$$I(\theta) = I(0) \cos\theta \tag{2-5}$$

a relation called *Lambert's cosine law*. If the radiance is determined at each angle θ, it is found to be constant, because the intensity must be divided by the projected area $A \cos\theta$ such that the cosine dependence cancels:

$$L_e = \frac{I(\theta)}{A \cos\theta} = \frac{I(0)\cos\theta}{A \cos\theta} = \frac{I(0)}{A} = \text{constant} \tag{2-6}$$

Thus when a radiating (or reflecting) surface has a radiance that is independent of the viewing angle, the surface is said to be perfectly diffuse or a *Lambertian surface*.

The following example illustrates the use of the radiometric terms and defining equations listed in Table 2-1.

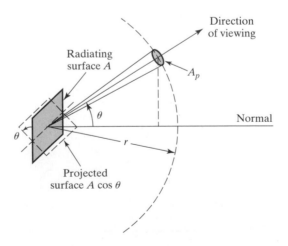

Figure 2-4 Radiant flux collected along a direction making an angle θ with the normal to the radiating surface. The projected area of the surface is shown by the dashed rectangle.

❏ Example

A small Philips Krypton, PR 102, flashlight bulb is rated at 2.4 volts and 0.7 amperes. It is roughly spherical in shape with a diameter of about 1.0 centimeter. For the calculations outlined below, ignore the metal base and assume the bulb radiates over its entire spherical surface.

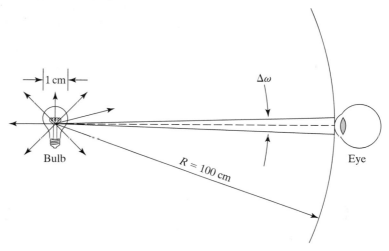

Find:

(a) Radiant flux (Φ_e) emitted by the bulb, in watts (W).

(b) Radiant intensity (I_e) of the bulb, in W/sr.

(c) Radiant exitance (M_e) of the bulb, in W/cm².

(d) Radiance (L_e) of the bulb in W/cm²-sr.

(e) The amount of energy, in joules, radiated by the bulb in 5 minutes.

(f) Irradiance (E_e) received by an eye 100 cm from the bulb, if the pupil of the eye is open to a diameter of 6.0 millimeters.

Solution

(a) The radiant flux Φ_e of the bulb is equal to its radiant power. Assuming that all of the electrical power is converted to useful, visible power (not a valid assumption, since some electrical energy is lost as heat) we can write

$$\Phi_e = P_{\text{electrical}} = \text{volts} \times \text{amps} = 2.4 \text{ V} \times 0.7 \text{ A} = 1.68 \text{ V-A} = 1.68 \text{ W}$$

since 1 volt-amp is equal to 1 watt of electrical power.

(b) The radiant intensity I_e of the bulb is equal to the radiated power Φ_e divided by the total solid angle into which the bulb radiates. Since the bulb is assumed to be spherical and radiates uniformly into the surrounding space of $\Delta\omega = 4\pi$ steradians, we have

$$I_e = \frac{\Delta\Phi_e}{\Delta\omega} = \frac{1.68 \text{ W}}{4\pi \text{ sr}} = 0.134 \text{ W/sr}$$

(c) The radiant exitance M_e is equal to the power per unit surface area radiated by the bulb. The bulb is a sphere of surface area $4\pi r^2$, where r, the radius of the bulb, is equal to 0.5 cm. Thus

$$M_e = \frac{\Delta\Phi_e}{\Delta A_{\text{bulb}}} = \frac{1.68 \text{ W}}{4\pi(0.5 \text{ cm})^2} = 0.54 \text{ W/cm}^2$$

(d) The radiance L_e of the bulb is equal to the power it radiates per unit area of surface, per unit solid angle.

$$L_e = \frac{\Delta\Phi_e}{(\Delta A_{\text{bulb}})(\Delta\omega)} = \frac{1.68 \text{ W}}{4\pi(0.5 \text{ cm})^2 \, (4\pi \text{ sr})} = 0.043 \, \frac{\text{W}}{\text{cm}^2\text{-r}}$$

(e) The amount of energy Q_e radiated by the bulb is given by

$$Q_e = \Phi_e(\Delta t) = (1.68 \text{ W})(5 \text{ min})(60 \text{ sec/min}) = 504 \text{ W-s}$$

or, since 1 watt-s equals 1 joule, $Q_e = 504$ joules.

(f) The definition for irradiance tells us that we need to know how much power falls on the target area (the pupil of the eye) at 100 cm from the bulb. At this distance, the bulb may be considered a point source. The pupil subtends a solid angle $\Delta\omega$ given by

$$\frac{\Delta A_{\text{pupil}}}{R^2} = \frac{\pi r^2}{R^2} = \frac{\pi(0.6 \text{ cm})}{(100 \text{ cm})^2} = 1.13 \times 10^{-4} \text{ sr}$$

Since the bulb radiates 0.134 W/sr (part b), we know that a power of $(0.134 \text{ W/sr} \times 1.13 \times 10^{-4} \text{ sr}) = 1.51 \times 10^{-5}$ W reaches the pupil. Since the area of the pupil is $\pi(0.6 \text{ cm})^2$, we then have for the irradiance,

$$E_e = \frac{\Delta\Phi_{\text{eye}}}{\Delta A_{\text{eye}}} = \frac{1.51 \times 10^{-5} \text{ W}}{1.13 \text{ cm}^2} = 1.34 \times 10^{-5} \, \frac{\text{W}}{\text{cm}^2}$$

or about 13 microwatts/cm². Thus, even though the bulb emits a total of 1.68 watts of power, at 1 meter distance, a normal eye receives a very small fraction of the available power. ∎

2-3 PHOTOMETRY

Radiometry applies to the measurement of all electromagnetic radiation. *Photometry*, on the other hand, applies only to the visible portion of the optical spectrum. Whereas radiometry involves purely physical measurement, photometry takes into account the response of the human eye to radiant energy at various wavelengths and so involves psycho-physical measurements. The distinction rests on the fact that the human eye, as a detector, does not have a "flat" spectral response; that is, it does not respond with equal sensitivity at all wavelengths. If three sources of light of equal radiant power (but radiating blue, yellow, and red light, respectively) are observed visually, the yellow source will appear to be far brighter than the others. When we use photometric quantities, then, we are measuring the properties of visible radiation as they appear to the normal eye, rather than as they appear to an unbiased detector. Since not all human eyes are identical, a standard response has been determined by the international Commission on Illumination (CIE) and is reproduced in Figure 2-5. The relative response or sensation of brightness for the eye is plotted versus wavelength, showing that peak sensitivity occurs at the yellow-green wavelength of 555 nm. Actually the curve shown is the luminous efficiency of the eye for *photopic vision*, that is, when adapted for day vision. For lower levels of illumination, when adapted for night or *scotopic vision*, the curve shifts toward the green, peaking at 510 nm. It is interesting to note that human color sensation is a function of illumination and is

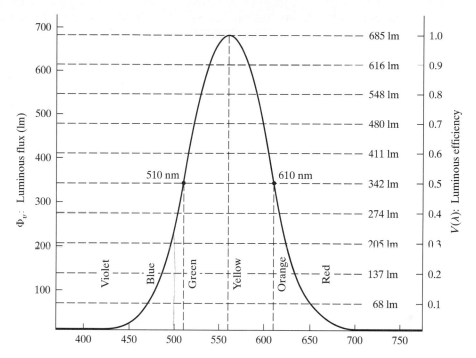

Figure 2-5 CIE luminous efficiency curve. The luminous flux Φ_v corresponding to 1 W of radiant power at any wavelength λ is given by the product of 685 lm and the luminous efficiency $V(\lambda)$ at the same wavelength: $\Phi_v(\lambda) = 685\ V(\lambda)$ for each watt of radiant power.

almost totally absent at lower levels of illumination. One way to confirm this is to compare the color of stars, as they appear visually, to their photographic images made on color film using a suitable time exposure. The variety of colors among the stars, not so evident with the naked eye, are much more evident on film. Another, dramatic way to demonstrate human color dependence on illumination is to project a 35-mm color slide of a scene onto a screen with a low current in the projector bulb. At sufficiently low currents, the scene appears black and white. As the current is increased, the full colors in the scene gradually emerge. On the other hand, very intense radiation may be visible even beyond the limits of the CIE curve. The reflection from a target surface of an intense laser beam of 694.3 nm, produced by a ruby laser, is easily seen. Even the infrared radiation around 900 nm from a gallium-arsenide semiconductor laser can be seen as a deep red.

Radiometric quantities are related to photometric quantities through the luminous efficiency curve of Figure 2-5 in the following way: Corresponding to a radiant flux of 1 W at the peak wavelength of 555 nm, where the luminous efficiency of the eye is maximum, the *luminous flux* or *power* is defined to be 685 lumens (lm). Then, for example, near $\lambda = 610$ nm in the orange region where the luminous efficiency is 0.5 or 50%, 1 W of radiant flux would produce only 0.5×685 or 342 lm of luminous flux. The curve shows that again at $\lambda = 510$ nm, in the blue-green region, the brightness has dropped to 50%.

Photometric units, in terms of their definitions, parallel radiometric units. This is amply demonstrated in Table 2-2. In general, analogous units are related by the following equation:

$$\text{photometric unit} = K(\lambda) \times \text{radiometric unit}$$

TABLE 2-2 PHOTOMETRIC TERMS

Term	Symbol	Description	Defining equation	Units of measure	Analogous radiometric term
Luminous energy	Q_v	Luminous energy in visible spectrum	——	talbot	joule
Luminous power	Φ_v	Luminous power per unit time (power)	$\Phi_v = \dfrac{\Delta Q_v}{\Delta t}$	lumen (talbot/s)	watt (joule/s)
Luminous exitance	M_v	Luminous power per unit area of source	$M_v = \dfrac{\Delta \Phi_v}{\Delta A}$	lumen/m² (lux)	watt/m²
Illuminance	E_v	Luminous power per unit area of target	$E_v = \dfrac{\Delta \Phi_v}{\Delta A}$	lumen/m² (lux)	watt/m²
Luminous intensity	I_v	Luminous power emitted per unit solid angle	$I_v = \dfrac{\Delta \Phi_v}{\Delta \omega}$	lumen/sr (cd)	watt/sr
Luminance	L_v	Luminous power per unit solid angle per unit area of source	$L_v = \dfrac{\Delta \Phi_v}{\Delta \omega \, \Delta A}$	lumen/m²-sr	watt/m²-sr

Abbreviations: sr, steradian; lm, lumen; lx, lux; cd, candela (candle); ft-cd, foot-candle.

Equivalences: lx = lm/m²; cd = lm/sr; ft-cd = lm/ft² = 10.76 lm/m².

where $K(\lambda)$ is called the *luminous efficacy*. If $V(\lambda)$ is the *luminous efficiency*, as given on the CIE curve, then

$$K(\lambda) = 685 \, V(\lambda)$$

Photometric terms are preceded by the word *luminous* and the corresponding units are subscripted with the letter *v* (*visual*); otherwise the symbols are the same. Notice that the SI unit of luminous energy is the *talbot*, the unit of luminous exitance and illuminance is the *lux* (lx), and the unit of luminous intensity is the *candela* (cd). Notice also the distinction between the analogous terms *irradiance* (radiometric) and *illuminance* (photometric).

❏ **Example**

A light bulb emitting 100 W of radiant power Φ_e is positioned 2 m from a surface. The surface is oriented perpendicular to a line from the bulb to the surface. Calculate the irradiance E_e at the surface. If all 100 W is emitted from a red bulb at $\lambda = 650$ nm, calculate also the illuminance E_v at the surface.

Solution The irradiance

$$E_e = \Delta\Phi_e/\Delta A = 100 \text{ W}/4\pi(2\text{m})^2 \cong 2 \text{ W/m}^2$$

From the CIE curve, $V(650 \text{ nm}) = 0.1$. Thus illuminance

$$E_v = K(\lambda) \times \text{irradiance} = 685 \, V(\lambda) \times E_e$$

$$E_v = 685 \times 0.1 \times 2 = 137 \text{ lm/m}^2 \text{ or lux}$$

Thus, whereas a *radiometer* with aperture at the surface measures 2 W/m², a *photometer* in the same position would be calibrated to read 137 lx. ■

When the radiation consists of a spread of wavelengths, the radiometric and photometric terms may be functions of the wavelength. This dependence is noted by preceding the term with the word *spectral* and by using a subscript λ or adding the λ in parentheses. For example, *spectral radiant flux* is denoted by $\Phi_{e\lambda}$ or Φ_e (λ). The total radiant flux is then determined by summing the spectral radiant flux in each narrow wavelength region over the entire wavelength region of interest.

2-4 COMPARING LUMINOUS INTENSITIES: THE PHOTOMETER

A photometer is an optical instrument used to compare the luminous intensity of light sources. If one of the sources is a standard, then the photometer can be used to determine the luminous intensity of an unknown source. The laboratory instrument is, in itself, a simple device, as shown schematically in Figure 2-6. It is made up of a long, calibrated bar, usually about 1 meter long. The two light sources are situated at opposite ends of the bar. The light sources are generally small and regarded as point sources. Between the two sources, free to slide along the calibrated bar, is a double-faced, rough-surfaced screen. Each side of the screen is a Lambertian surface, perfectly diffuse, generally with a surface reflectivity equal to one so that all incident light is reflected.

In operation, the double-faced screen is moved between the two light sources until both sides appear to the observer to be equally illuminated. When a match of the two sources is achieved, the illuminance E_S due to the standard source equals the illuminance E_X due to the unknown source. Then it follows from Eq. (2-3) that one can write the equality

$$\frac{I_S}{r_S^2} = \frac{I_X}{r_X^2}$$

Knowing I_S, r_S, and r_X, one can solve for r_X.

❑ **Example**

Using a photometer, an unknown light source is matched against a known source of luminous intensity I_S of 40 lm/sr (also called 40 candelas). At the position of the match, the photometer shows that $r_S = 35$ cm and $r_X = 65$ cm. What is the luminous intensity I_X of the unknown?

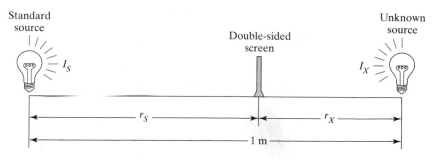

Figure 2-6 Measuring luminous intensity with a photometer.

Solution   Using the relation $\dfrac{I_S}{r_S^2} = \dfrac{I_X}{r_X^2}$ and substituting known values, one obtains

$$I_X = 40\,\frac{(65)^2}{(35)^2} = 137.9 \text{ lm/sr, or about 138 candelas}$$

The photometer shows that the unknown source has a luminous intensity about 3.5 times as high as the standard source. ∎

In the practice of illumination engineering, the value of the total flux emitted by a lamp is more important than its luminous intensity—that is, how many lumens it emits per unit solid angle in a given direction. Consequently, with the help of a photometer, the total flux is measured as follows: The lamp in question is placed in an *integrating sphere*, a large sphere with a diffuse, perfectly reflecting interior surface. The flux emerging from a small hole in the sphere is measured with a photometer near the hole. Then, from knowledge of the surface areas of the sphere and hole and the photometer reading, one can calculate the total flux emitted by the lamp.

2-5   BLACKBODY RADIATION

A *blackbody* is an ideal absorber: All radiation falling on a blackbody, regardless of wavelength or angle of incidence, is completely absorbed. It follows that a blackbody in equilibrium at a particular temperature is also a perfect emitter or radiator. No body at the same temperature as a blackbody can emit more radiation at any wavelength or into any direction than a blackbody. Blackbodies are approached in practice by blackened surfaces and by tiny apertures in radiating cavities. An excellent example of a blackbody is the surface formed by the series of sharp edges in a stack of razor blades. The array of blade edges effectively traps the incident light, resulting in almost perfect absorption.

The spectral radiant exitance M_λ of a blackbody can be derived on theoretical grounds. It was first so derived by Max Planck, who found it necessary to postulate quantization in the process of radiation and absorption by the blackbody. The result of this calculation (Resnick 1991) is given by

$$M_\lambda = \frac{3.745 \times 10^8}{\lambda^5}\left(\frac{1}{e^{14388/\lambda T} - 1}\right) \text{ W/(m}^2\text{-}\mu\text{m)} \tag{2-9}$$

where λ is in micrometers and T is in Kelvin. The spectral radiant exitance M_λ of Eq. (2-9) is plotted versus wavelength λ in Figure 2-7 for different temperatures. Examination of Figure 2-7 shows that M_λ increases with absolute temperature at each wavelength. Note also that, as the varying area under the separate curves indicates, the total energy radiated increases rapidly with the temperature. The energy radiated at 6000 K, for example, is far more than twice the energy radiated at 3000 K. Additionally, the peak value of M_λ shifts toward shorter wavelengths with increasing temperature, falling into the visible spectrum (between dashed vertical lines) at $T = 5000$ and 6000 K. The variation of $\lambda_{\max}$, the wavelength at which M_λ peaks, is given by the *Wien displacement law*,[2]

$$\lambda_{\max} T = \frac{hc}{5k} = 2.88 \times 10^3 \ (\mu\text{m-K}) \tag{2-10}$$

[2] Although Wien's law is often found written in this form, the number 5 is an approximation to 4.965, which, when used, gives a more accurate Wien constant of $2.898 \times 10^3 \ \mu$m-K.

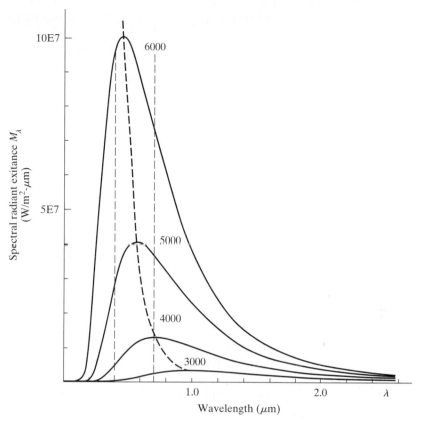

Figure 2-7 Blackbody radiation spectral distribution at four Kelvin temperatures. The vertical dashed lines mark the region of the visible spectrum, and the dashed curve connecting the peaks of the four curves illustrates the Wien displacement law. (The notation 5E7, for example, stands for 5×10^7.)

and is indicated in Figure 2-7 by the dashed curve. If, on the other hand, the spectral exitance M_λ of Eq. (2-9) is summed over all wavelengths, the total radiant exitance or area under the blackbody radiation curve at temperature T is

$$M = \sigma T^4 \tag{2-11}$$

known as the *Stefan-Boltzmann law*, with σ as the Stefan-Boltzmann constant, equal to 5.67×10^{-8} W/m²-K⁴.

The radiation from real surfaces is always less than that of the blackbody or *Planckian source* and is accounted for quantitatively by the *emissivity ε*. Distinguishing now between the radiant exitance M of a measured specimen and that of a blackbody M_{bb} at the same temperature, we define the emissivity $\varepsilon(T)$ as

$$\varepsilon(T) = \frac{M}{M_{bb}} \tag{2-12}$$

If the radiant exitance of the blackbody and the specimen are compared in various narrow wavelength intervals, a spectral emissivity is calculated, which is not, in general, a constant. In those special cases where the emissivity is independent of wavelength, the specimen is said to be a *graybody*. In this instance the spectral exitance of the specimen is proportional to that of the blackbody and their curves are the same except for a constant factor. The spectral radiation from a heated tungsten wire, for example, is close to that of a graybody with $\varepsilon = 0.4$ to 0.5.

Blackbody radiation is used to establish a color scale in terms of absolute temperature alone. The *color temperature* of a specimen of light is then the temperature of the blackbody with the closest spectral energy distribution. In this way, a candle flame can be said to have a color temperature of 1900 K, whereas the sun has a typical color temperature of 5500 K.

❏ **Example**

The human body, at a temperature of about 98.6° F, emits a continuous stream of electromagnetic radiation. Estimate the peak wavelength of the electromagnetic radiation and identify the region of the electromagnetic spectrum where this radiation lies.

Solution   To estimate the peak wavelength λ_{max}, treat the body as a blackbody radiator and use the Wien displacement law

$$\lambda_{max} = \frac{2.88 \times 10^3 \ \mu\text{m-K}}{T_K}$$

To find T_K in degrees Kelvin, use the formula $T_C = 5/9 \ (T_F - 32)$ to convert the Fahrenheit temperature of 98.6° F to a Celsius temperature of 37° C. Then use the formula $T_K = T_C + 273.15$ to obtain a value of $T_K = 310.15$ K or approximately 310 K. Then

$$\lambda_{max} = \frac{2.88 \times 10^3 \ \mu\text{m-K}}{310 \ \text{K}} \cong 9.3 \ \mu\text{m}$$

Referring to Figure 2-1, we see that a wavelength of 9.3 μm lies within the infrared region. Not surprisingly then, the human body glows with radiation and is easily seen by infrared cameras and snooper scopes, especially at night.   ∎

❏ **Example**

The human eye is most sensitive to light of wavelength $\lambda = 555$ nm. To what temperature should a blackbody be heated to radiate most strongly at this wavelength?

Solution   Express the maximum wavelength in μm:

$$\lambda_{max} = 555 \times 10^{-9} \ \text{m} = 0.555 \times 10^{-6} \ \text{m} = 0.555 \ \mu\text{m}$$

Then use the Wien displacement law to determine T_K:

$$T_K = \frac{2880 \ \mu\text{m-K}}{\lambda_{max}} = \frac{2880 \ \mu\text{m-K}}{0.555 \ \mu\text{m}} = 5189 \ \text{K}$$ ∎

❏ **Example**

Assume that the typical color temperature for the sun is 5500 K. If it is considered as a blackbody, how many watts of radiant power does it emit from each square meter of its surface?

Solution Use the Stefan-Boltzmann law to find M, the power radiated per square meter of surface, at all radiated wavelengths.

$$M = \sigma T^4 = \left(5.67 \times 10^{-8} \frac{\text{W}}{\text{m}^2\text{-K}^4}\right) T^4$$

$$M = (5.67 \times 10^{-8})(5500)^4 \cong 5.2 \times 10^7 \text{ W/m}^2$$

Thus the sun radiates into space about 52 million watts of radiant power for each square meter of its surface, in the visible, ultraviolet, and infrared. ∎

2-6 SOURCES OF OPTICAL RADIATION

Sources of light may be natural, as in the case of sunlight and skylight, or artificial, as in the case of incandescent or discharge lamps. Light from various sources may also be classified as monochromatic, spectral line, or continuous. The way in which energy is distributed in the radiation determines the color of the light and, consequently, the color of surfaces seen under the light. Anyone who has used a camera is aware that the actual color response of film depends on the type of light used to illuminate the subject.

The following brief survey of sources of light cannot hope to be comprehensive; rather it is intended to direct attention to an extensive area of practical information. For the purposes of this limited survey, we classify a number of sources as follows:

- A. Sunlight, skylight
- B. Incandescent sources
  - 1. Blackbody sources
  - 2. Nernst glowbar
  - 3. Tungsten filament
- C. Discharge lamps
  - 1. Monochromatic and spectral sources
  - 2. High-intensity sources
    - a. Carbon arc
    - b. Compact short arc
    - c. Flash
    - d. Concentrated zirconium arc
  - 3. Fluorescent lamps
- D. Semiconductor light-emitting diodes (LEDs)
- E. Coherent source-laser

Sunlight and skylight. Daylight is a combination of sunlight and skylight. Direct light from the sun has a spectral distribution that is clearly different from that of skylight, which has a predominantly blue hue. A plot of spectral solar irradiance M_λ is given in Figure 2-8. Extraterrestrial solar radiation indicates that the sun behaves approximately as a blackbody with a temperature of 6000 K at its center and 5000 K at its edge, but the radiation received at the earth's surface is modified by absorption in the earth's atmosphere. The annual average of total irradiance just outside the earth's atmosphere is the *solar constant*, 1350 W/m². Although solar radiation is not routinely used as a light source in the

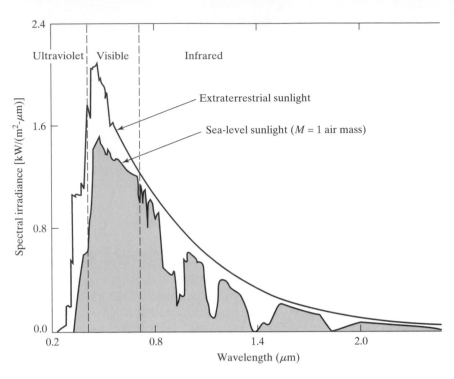

Figure 2-8 Solar spectral irradiation above the atmosphere and on a horizontal surface at sea level on a clear day with the sun at zenith.

laboratory, high-pressure xenon lamps, with appropriate filters, provide an excellent artificial source for solar simulation and are available commercially.

Incandescent sources. Artificial optical sources that use light produced by a solid material heated to incandescence by an electric current are called *incandescent lamps*. Radiation arises from the de-excitation of the atoms or molecules of the material after they have been thermally excited. The energy is emitted over a broad continuum of wavelengths. Commercially available *blackbody sources* consist of cavities equipped with a small hole. Radiation from the small hole has an emissivity that is essentially constant and equal to unity. Such sources are available at operating temperatures from that of liquid nitrogen ($-196°$ C) to $3000°$ C. Incandescent sources particularly useful in the infrared include the *Nernst glower*. This source is a cylindrical tube or rod of refractory material (zirconia, yttria, thoria), heated by an electric current and useful from the visible to around 30 μm. The Nernst glower behaves like a graybody with an emissivity greater than 0.75. When the material is a rod of bonded silicon carbide, the source is called a *globar*, approximating a graybody with an average emissivity of 0.88 (see Figure 2-9).

The tungsten filament lamp is the most popular source for optical instrumentation designed to use a continuous radiation in the visible and into the infrared region. The lamp is available in a wide variety of filament configurations and bulb and base shapes. The filament is in coil or ribbon form, the ribbon providing a more uniform radiating surface. The bulb is usually a glass envelope, although quartz is used for higher-temperature operation. Radiation over the visible spectrum approximates that of a graybody, with emissivities approaching unity for tightly coiled filaments. Lumen output depends both on the filament temperature and the electrical power input (wattage). During operation, tung-

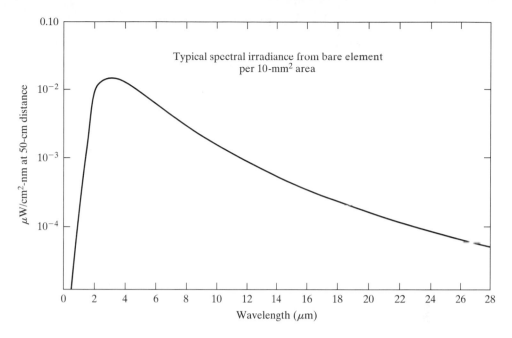

Figure 2-9 Globar infrared source, providing continuous usable emission from 1 μm to over 25 μm at a temperature variable up to 1000 K. The source is a 6.2-mm diameter silicon carbide resistor. (Oriel Corp., General Catalogue, Stratford, Conn. Used by permission.)

sten gradually evaporates from the filament and deposits on the inner bulb surface, leaving a dark film that can decrease the flux output by as much as 18% during the life of the lamp. This process also weakens the filament and increases its electrical resistance. The presence of an inert gas, usually nitrogen or argon, introduced at around 0.8 atmospheric pressure, helps to slow down the evaporation. More recently this problem has been minimized by the addition of a halogen vapor (iodine, bromine) to the gas in the *quartz-halogen* or *tungsten-halogen* lamp. The halogen vapor functions in a regenerative cycle to keep the bulb free of tungsten. Iodine reacts with the deposited tungsten to form the gas tungsten iodide, which then dissociates at the hot filament to redeposit the tungsten and free the iodine for repeated operation. A typical spectral irradiance curve for a 100-W quartz-halogen filament source is given in Figure 2-10. The lamp approximates a 3000-K graybody source, providing a useful continuum from 0.3 to 2.5 μm. In *tungsten arc lamps*, an arc discharge between two tungsten electrodes heats the electrodes to incandescence in an atmosphere of argon, providing a spectral distribution of radiation like that of tungsten lamps at 3100 K.

Discharge lamps. Lamps that depend for their radiation output on the dynamics of an electrical discharge in a gas are called *discharge lamps*. A current is passed through the ionized gas between two electrodes, sealed in a glass or quartz tube. (Glass envelopes absorb ultraviolet radiation below about 300 nm, whereas quartz transmits down to about 180 nm.) An electric field accelerates electrons sufficiently to ionize the vapor atoms. The source of the electrons may be a heated cathode (*thermionic emission*), a strong field applied at the cathode (*field emission*), or the impact of positive ions on the cathode (*secondary emission*). De-excitation of the excited vapor atoms provides a release of energy in the form of photons of radiation. High-pressure and high-current operation of

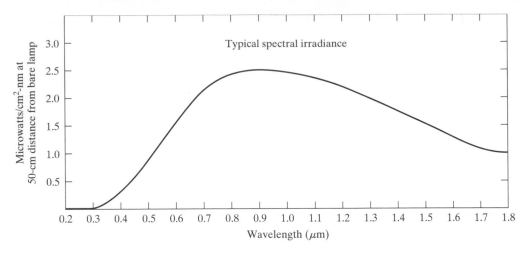

Figure 2-10 Spectral irradiance from a 100-W quartz halogen lamp, providing continuous radiation from 0.3 μm to 2.5 μm. (Oriel Corp., General Catalogue, Stratford, Conn. Used by permission.)

discharge lamps generally results in a continuous spectral output, in addition to spectral lines characteristic of the vapor. At lower pressure and current, sharper spectral lines appear and the background continuum is minimal. When sharp spectral lines are desired, as in *monochromatic sources*, the lamp is designed to operate at low temperature, pressure, and current. The *sodium arc lamp*, for example, provides radiation almost completely confined to a narrow "yellow" band due to the spectral lines at 589.0 and 589.6 nm. The low-pressure *mercury discharge tube* is often used to provide, with the help of isolating filters, strong monochromatic radiation at wavelengths of 404.7 and 435.8 nm (violet), 546.1 nm (green), and 577.0 and 579.1 nm (yellow). Other gases or vapors may be used to provide spectral lines of other desired wavelengths. For the highest spectral purity, particular isotopes of the gas are used.

When high intensity rather than spectral purity is desired, other designs become available. Perhaps the oldest source of this kind is the *carbon arc*, still widely used in searchlights and motion picture projectors. The high-current arc is formed between two carbon rods in air. A 200-A carbon arc lamp may have a peak luminance of 1600 cd/mm². The source has a spectral distribution close to that of a graybody at 6000 K. A wide range of spectral outputs is possible by using different materials in the core of the carbon rod. When the arc is enclosed in an atmosphere of vapor at high pressure, the lamp is a *compact short-arc source* and the radiation is divided between line and continuous spectra. See Figure 2-11 for a sketch of this type of lamp and its housing. Three useful types of compact arc lamps, designed to operate from 50 W to 25 kW, are the high-pressure *mercury arc lamp*, with comparatively weak background radiation but strong spectral lines and a good source of ultraviolet; the *xenon arc lamp*, with practically continuous radiation from the near-ultraviolet through the visible and into the near-infrared; and the *mercury-xenon arc lamp*, providing essentially the mercury spectrum but with xenon's contribution to the continuum and its own strong spectral emission in the 0.8 to 1-μm range. As mentioned earlier, the color quality of the xenon lamp is similar to that of sunlight at 6000 K color temperature. The spectral emission curve for a xenon lamp is shown in Figure 2-12. *Hydrogen*

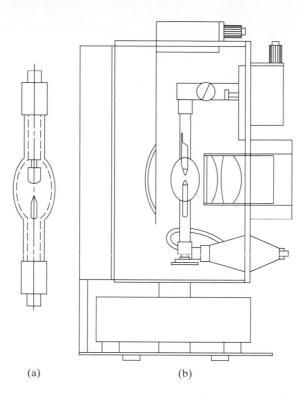

(a) (b)

Figure 2-11 High intensity, compact, short-arc light source. (a) Compact arc lamp. (b) Lamp installed in housing, showing back reflector and focusing system. (Ealing Electro-Optics. Used by permission.)

and *deuterium arc lamps* are ideal for ultraviolet spectroscopy because they produce a high radiance with a continuous background in the ultraviolet region.

Flash tubes represent a high output source of visible and near-infrared radiation, produced by a rapid discharge of stored electrical energy through a gas-filled tube. The gas is most often xenon. The *photoflash* tube, in contrast, provides high-intensity, short-duration illumination by the rapid combustion of metallic (aluminum or zirconium) foil or wire in a pure oxygen atmosphere.

When a high-intensity *point source* of radiation is desired in optical instrumentation, a useful lamp is the *concentrated zirconium arc lamp*, with wattages varying from 2 to 300 W. Zirconium vapor is formed by evaporation from an oxide-coated cathode in an argon atmosphere. Radiation originates both from the incandescence of a molten cathode surface and from the excited zirconium vapor and argon gas. It is viewed through a small hole, 0.13 to 2.75 mm in diameter, in a metallic anode and through an optically flat window sealed into the tube. The spectral distribution approximates that of a 3200-K graybody source.

The familiar *fluorescent lamps* use low-pressure, low-current electrical discharge in mercury vapor. The ultraviolet radiation from excited mercury atoms is converted to visible light by stimulating fluorescence in a phosphor coating on the inside of the glass-envelope surface. Spectral outputs depend on the particular phosphor used. "Daylight" lamps, for example, use a mixture of zinc beryllium silicate and magnesium tungstate.

Light-emitting diode. A very different type of light source is the low-intensity light-emitting diode (LED), a solid-state device employing a *p-n junction* in a semiconducting crystal. The device is hermetically sealed in an optically centered package. When a small bias voltage is applied in the forward direction, optical energy is produced by the recombination of electrons and holes in the

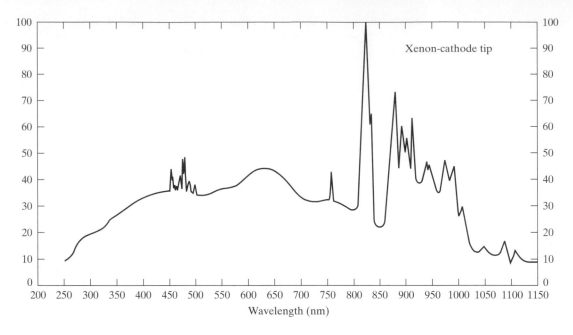

Figure 2-12 Spectral emission for xenon compact arc lamp. (Courtesy of Optical Radiation Corp. Used by permission.)

vicinity of the junction. Popular LEDs include the infrared gallium arsenide (GaAs) device, with a peak output wavelength near 900 nm, and the visible silicon carbide (SiC) device, with peak output at 580 nm. LEDs provide narrow spectral emission bands, as evident in Figure 2-13. Solid solutions of similar compound semiconductor materials produce outputs in a variety of spectral regions when the composition of the alloy is varied.

Coherent light sources—lasers. The *laser* is a very important modern source of coherent and extremely monochromatic radiation, capable of very high intensity. Lasers that emit radiation in the ultraviolet, visible, and infrared regions of the spectrum are readily available, there being currently over 50 dif-

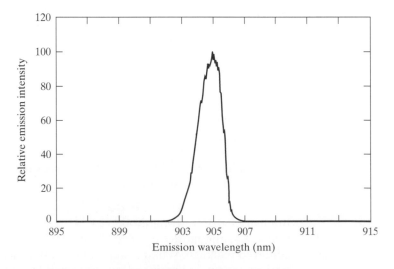

Figure 2-13 Spectral output from a GaAs light-emitting diode.

ferent varieties available on the market. Lasers emit radiation from the deep ultraviolet (around 200 nm) to the far infrared (near 10.6 μm). They operate with powers from milliwatts to kilowatts and produce nearly collimated beams of light, with beam divergences as small as 0.5 mrad. Because of the central role that lasers play in optical instrumentation, medicine, and technology, in general, they are treated at some depth in a separate chapter (Chapter 12).

2-7 DETECTORS OF RADIATION

Any device that produces a measurable physical response to incident radiant energy is a detector. The most common detector is, of course, the eye. Whereas the eye provides a qualitative and subjective response, the detectors discussed here provide a quantitative and objective response. In view of the unique role played by the eye in human vision, it is treated separately in Chapter 10, as well as in Chapters 5, 6, 7, and 12, where it plays a central role in the related discussion.

The most common detectors may be classified as follows:

A. Thermal detectors
 1. Thermocouples and thermopiles
 2. Bolometers and thermistors
 3. Pyroelectric
 4. Pneumatic or Golay
B. Quantum detectors
 1. Photoemissive-phototubes and photomultipliers
 2. Photoconductive
 3. Photovoltaic
 4. Photographic

Thermal detectors. When the primary measurable response of a detector to incident radiation is a rise in temperature, the device is called a *thermal detector*. The receptor is usually a blackened metal strip or flake that absorbs efficiently at all wavelengths. Such a device, in which an increase in temperature at a junction of two dissimilar metals or semiconductors generates a voltage, is called a *thermocouple* (Figure 2-14a). When the effect is enhanced by using an array of such junctions in series, the device is called a *thermopile* (Figure 2-14b). Thermal detectors also include bulk devices that respond to a rise in temperature with a significant change in resistance. Such an instrument may employ as

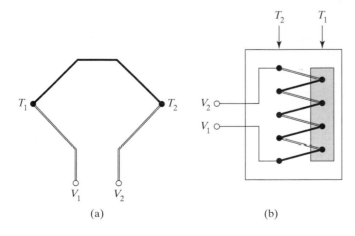

(a) (b)

Figure 2-14 (a) Thermocouple made of dissimilar materials (dark and light lines) joined at points T_1 and T_2, where a difference in temperature produces a voltage difference between terminals V_1 and V_2. (b) Thermocouple made of couples in series. Radiation is absorbed at the junctions T_1 in thermal contact with a black absorber and thermally insulated from the junctions T_2.

its sensitive element either a metal (*bolometer*) or, more commonly, a semiconductor (*thermistor*). Typically, two blackened sensitive elements are used in adjacent arms of a bridge circuit, one of which is exposed to the incident radiation. The imbalance in the circuit, due to the change in resistance, is indicated by a galvanometer deflection or current. In the *pyroelectric detector*, the temperature change causes a change in the surface charge of certain materials, like lithium tantalate or triglycerine sulfate (TGS) that exhibit the pyroelectric effect. The detector behaves like a capacitor whose charge is a function of the temperature. The *Golay cell*, instead, measures the thermal expansion of a gas. Heat absorbed by a blackened membrane is transmitted to the gas in an airtight chamber. The pressure rise in the gas is usually detected optically by the deflection of a mirror. Thermal detectors are generally characterized by a slow response to changes in the incident radiation. If the detector is expected to follow a changing input signal, such as that caused by a laser pulse, thermal detectors are not as desirable as the faster-responding quantum detectors to be discussed next. The speed of response is described by a *time constant*, a measure of the time required to regain equilibrium in output after a change in input. Thus quantum detectors, with smaller time constants, are better suited to high-frequency operation.

Quantum detectors.   Detectors that respond to the rate of incident photons rather than to thermal energy are called *quantum detectors*. Photons interact directly with the electrons in the detector material. When the measurable effect is the release of electrons from an illuminated surface, the device is called a *photoemissive detector*. A photosensitive surface, typically containing alkali metals, absorbs incident photons that transfer enough energy to enable some electrons to overcome the work function and escape from the surface. If the photoemitted electrons are simply collected by a positive-biased anode in an evacuated tube, enabling a current to be drawn into an external circuit, the detector is called a *diode phototube*. When the signal is internally amplified by secondary electron emission, the detector is a *photomultiplier*. See Figure 2-15. In this case the primary photoelectons are accelerated, undergoing a sequence of collisions. Each collision multiplies the current by the addition of secondary electrons, so that an avalanche of electrons becomes available at the output corresponding to each primary photoelectron.

Another means of amplification, used in the *gas-filled photocell*, allows the generation of additional electrons by ionization of the residual gas. In the case of energetic photons ($\lambda < 550$ nm), the sensitivity of photoemissive detectors is sufficient to allow the counting of individual photons. Such detectors possess superior sensitivity in the visible and ultraviolet spectral ranges. For wavelengths in the infrared over 1 μm, photoemitters are not available, and *photoconducting detectors* are used. In these detectors, photons absorbed into thin films or bulk material produce additional free charges in the form of electron-hole pairs. Both the negative (electrons) and positive (holes) charges increase the electrical conductivity of the sample. Without illumination, a bias voltage across such a material with high intrinsic resistivity produces a small or *dark* current. The presence of

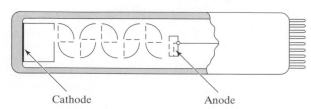

Cathode   Anode

Figure 2-15 One type of photomultiplier tube structure. Electrons photoemitted from the cathode are accelerated along zigzag paths down the tube so as to strike each of the curved *dynode* surfaces, each time producing additional secondary electrons. The multiplied current is collected at the anode.

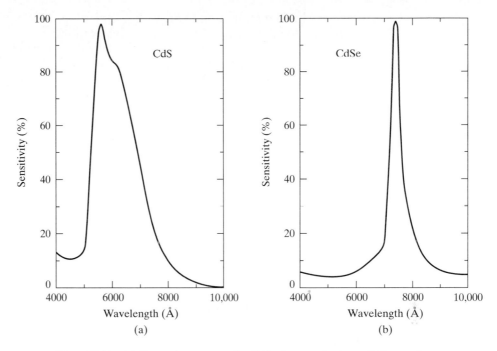

Figure 2-16 (a) Spectral response of a CdS photoconducting cell. The peak response at 5500 Å closely matches the response of the human eye. The cell is useful with incandescent, fluorescent, or neon lamps. (b) Spectral response of a CdSe photoconducting cell with peak at 7350 Å. The cell is sensitive to the near-infrared and is useful with incandescent or neon lamps.

illumination and the extra free-charge carriers so produced effectively lower the resistance of the material, and a larger photocurrent results. Semiconducting compounds such as cadmium sulfide (CdS) and cadmium selenide (CdSe) are often used in the visible and near-infrared regions. See Figure 2-16. Farther out in the near-infrared region, the compounds lead sulfide (PbS) and lead selenide (PbSe) are useful in the spectral ranges of 0.8–3 μm and 1–5 μm, respectively.

The most common *photovoltaic detector* is a *p-n* junction, the *semiconductor photodiode*. This device consists of a junction between doped *p*-type (rich in positive charge carriers) and doped *n*-type (rich in negative charge carriers) materials, most often silicon. Doping involves adding small amounts of an impurity to the semiconductor to provide either an excess (*n*-type) or deficiency (*p*-type) of conduction electrons. In a narrow region (junction) between these materials, a built-in electric field occurs as a consequence of current equilibrium. When photons are absorbed in the vicinity of the junction, the electron-hole pairs are separated by the field, causing a change in voltage, the *photovoltaic effect*. The solar cell and the photographic exposure meter are perhaps the best-known applications of this detector. A variation of the photovoltaic cell, the *avalanche diode*, provides an internal mechanism of amplification that results in enhanced sensitivity out to around 1.5 μm. In the region of 1 to 8 μm, the semiconductor compounds PbS, PbSe, and lead telluride (PbTe) possess a large photovoltaic effect and greater sensitivity than the thermocouple or the ordinary bolometer. As with other detectors that are designed to operate at longer wavelengths, photovoltaic detectors are often cooled to enable operation at greater sensitivity.

Two-dimensional arrays or panels of photodiodes allow the detection of images. Each photodiode, or *MOS* (metal-oxide-semiconductor) device, responds to the incident radiation to provide one *pixel* (picture element) of output. On exposure to light, each of the discrete devices fabricated on a silicon chip, say, stores photoinduced charge in a potential well created by an applied gate voltage. The stored charge contributed by each pixel, a measure of the local irradiance, is scanned to produce an electronic record of the image. Photodiode arrays are used in television cameras and have been used to replace photographic plates in astronomical telescopes and spectrographs.

Finally, a detector made popular by the widespread use of the camera is the *photographic film* or *plate*. Such photographic emulsions are available with spectral sensitivity that extends from the X-ray region into the near-infrared at around 1.2 μm. The sensitive material is an emulsion of silver halide crystals or grains. An incident photon imparts energy to the valence electron of a halide ion, which can then combine with the silver ion, producing a neutral silver atom. Even before developing, the emulsion contains a latent image, a distribution of reduced silver atoms determined by the variations in radiant energy received. The latent image is then "amplified," so to speak, by the action of the developer. The resulting chemical action provides further free electrons to continue the reduction process, with the latent image acting as a catalytic agent to further action. The density of the silver atoms, thus the opacity of the film, is a measure of both the irradiance and the time of exposure, so that photographic film, unlike other detectors, has the advantage of light-signal integration. Even weak radiation can be detected by the cumulative effect of a long exposure.

Detector response. In addition to a knowledge of the spectral range over which a particular detector is effective, it is important to know the actual sensitivity or, more precisely, the *responsivity S* of the detector, defined as the ratio of output to input:

$$S = \frac{\text{output}}{\text{input}}$$

Input may be radiant flux or irradiance. Output is almost always a current or voltage. For the responsivity to be a useful specification of a detector, it should be constant over the useful range of the instrument. In other words, the detector, together with its associated amplifier and circuits, should provide a linear response, with output proportional to input. In general, however, responsivity is not independent of wavelength. Curves of responsivity versus wavelength are provided with commercial detectors. When the responsivity is a function of λ, the detector is said to be *selective*. A *nonselective* detector is one that depends only on the radiant flux, not on the wavelength. Thermal detectors using a blackened strip as a receptor may be nonselective; however, entrance windows to such devices may well make them selective.

The *detectivity D* of a detector is the reciprocal of the minimum detectable power, called the *noise equivalent power*, Φ_N, of the detector:

$$D = \frac{1}{\Phi_N}$$

The minimum detectable power is limited by the *noise* inherent in the operation of the detector. The noise is that part of the signal or output not related to the desired input. Many sources of noise exist, including the statistical fluctuations of photons (or *radiation noise*) and the thermal agitation of current carriers, or *Johnson noise*, inherent in all detectors; the generation and recombination noise due to statistical fluctuations of current carriers in photoconductors; the *shot noise* due to random emission of electrons in photoemissive detectors; and the noise due to temperature fluctuations in thermal detectors. Mere amplification of a signal is not useful when it does not distinguish between signal and noise and results in the same signal-to-noise ratio, just as the mere magnification of a blurred optical image is not useful in clarifying its details.

PROBLEMS

2-1. Calculate the frequencies of electromagnetic radiation capable of producing a visual sensation in the normal eye.

2-2. A small, monochromatic light source, radiating at 500 nm, is rated at 500 W.
   (a) If only 2% of its total power is perceived by the eye as luminous power, what is its luminous flux output?
   (b) If the source radiates uniformly in all directions, determine its radiant and luminous intensities.
   (c) If the surface area of the source is 50 cm², determine the radiant and luminous exitances.
   (d) What are the irradiance and illuminance on a screen situated 2 m from the source, with its surface normal to the radiant flux?
   (e) If the screen contains a hole with a 5-cm diameter, how much radiant and luminous flux get through?

2-3. **(a)** A 50-mW He-Cd laser emits at 441.6 nm. A 4-mW He-Ne laser emits at 632.8 nm. With the help of Figure 2-5, compare the relative brightness of the two laser beams of equal diameter when projected side by side on a white piece of paper. Assume photopic vision.
   (b) What power argon laser emitting at 488 nm is required to match the brightness of a 0.5-mW He-Ne green laser at 543.5 nm, under the conditions of **(a)**?

2-4. A lamp 3 m directly above a point P on the floor of a room produces at P an illuminance of 100 lm/m².
   (a) What is the luminous intensity of the lamp?
   (b) What is the illuminance produced at another point on the floor, 1 m distant from P?

2-5. A lot is illuminated at night by identical lamps at the top of two poles 30-ft high and 40-ft apart. Assuming the lamps radiate equally in all directions, compare the illuminance at ground level for points directly under one lamp and midway between them.

2-6. A small source of 100 cd is situated at the focal point of a spherical mirror of 50-cm focal length and 10-cm diameter. What is the average illuminance of the parallel beam reflected from the mirror, assuming an overall reflectance of about 80%?

   Note: A portion of a spherical surface of radius r and area A subtends a solid angle of $\Delta\omega = A/r^2$ steradians at the center of the sphere. An equivalent (and more convenient) expression is given by

$$\Delta\omega = 2\pi(1 - \cos\theta)$$

where θ is the plane angle subtended at the center by the outer edge of the spherical surface. These expressions are also approximately valid when the surface area A is circular and flat, as long as the diameter of the circle is small compared with the distance r.

2-7. **(a)** The sun subtends an angle of 0.5° at the earth's surface, where the illuminance is about 10⁵ lx at normal incidence. Determine the luminance of the sun.
   (b) Determine the illuminance of a horizontal surface under a hemispherical sky with uniform luminance L. (See note, Problem 2-6.)

2-8. A photometer indicates that the intensity of an unknown light source matches that of a standard light source when the double-faced comparison screen is located 40 cm from the unknown and 60 cm from the standard. A second, unknown light source produces a match with the same standard source when the screen is located 60 cm from the unknown.

(a) What is the ratio of intensities of the second unknown to the first?

(b) What is the intensity of each unknown source if the intensity of the standard source is known to be 80 cd?

2-9. A point source of luminous power 1,000 lm radiates energy equally in all directions.

(a) How many lumens pass through the pupil of an eye if the pupil is 6 mm in diameter and the eye is 1 m from the light source?

(b) If the light passing through the pupil is brought to a focus at the retina, on a spot of diameter 0.01 mm, what is the illuminance of this spot?

2-10. For good vision, an illuminance of 7 ft-cd is needed on the surface of a work desk at a location directly beneath the lamp. If the lamp hangs 4 ft above the desk, what should be its illuminance if 30% of the required illuminance on the desk comes indirectly from window light? (See Table 2-2 for assistance with units.)

2-11. A small, flat, circular light source of area 6 mm^2 and luminance 10 cd/mm^2 is located 30 cm in front of a positive lens 5 cm in diameter and 20 D in power. The light intercepted by the lens comes to a focus 6 cm from the lens, forming a circular spot of area 0.24 mm^2 on the front corneal surface of an eye centered on the optical axis. The pupil of the eye receiving the light is open to a diameter of 2 mm. Note: 20 D power → 5 cm focal length.

(a) What fraction of the light incident on the cornea passes through the pupil and on to the retina?

(b) Assuming that the flat, circular light source approximates a point source, determine the total amount of flux that passes through the pupil.

2-12. The peak of the solar spectrum falls at about 500 nm. Determine the sun's surface temperature, assuming it radiates as a blackbody.

2-13. (a) At what wavelength does a blackbody at 6000 K radiate the most per unit wavelength?

(b) If the blackbody is a 1-mm diameter hole in a cavity radiator at this temperature, find the power radiated through the hole in the narrow wavelength region 5500–5510 Å.

2-14. At a given temperature, λ_{max} = 550 nm for a blackbody cavity. The cavity temperature is then increased until its total radiant exitance is doubled. What is the new temperature and the new λ_{max}?

2-15. What must be the temperature of a graybody with emissivity of 0.45 for it to have the same total radiant exitance as a blackbody at 5000 K?

REFERENCES

BAUER, GEORG. 1965. *Measurement of Optical Radiations*. New York: Focal Press.

BECHERER, RICHARD, and FRANC GRUM. 1982. *Optical Radiation Measurements*. New York: Academic Press, Inc.

BOYD, ROBERT W. 1983. *Radiometry and the Detection of Optical Radiation*. New York: John Wiley & Sons, Inc.

BUDDE, W. 1983. *Physical Detectors of Optical Radiation*. Optical Radiation Measurements Series, vol. 4. New York: Academic Press.

BYER, ALVIN. 1986. *Theoretical Optics for Clinicians.* Philadelphia: Pennsylvania College of Optometry.

CLARK JONES, R. 1968. "How Images are Detected." In *Lasers and Light. Readings from Scientific American*, pp. 81–88. San Francisco: W. H. Freeman and Company Publishers.

DRISCOLL, WALTER G., and WILLIAM VAUGHAN, eds. 1978. *Handbook of Optics*. Sponsored by the Optical Society of America. New York: McGraw-Hill Book Company. Stephen F. Jacobs, "Nonimaging Detectors." Jay F. Snell, "Radiometry and Photometry." George J. Zissis and Anthony J. Larocca, "Optical Radiators and Sources."

FITCH, JAMES MARSTON. 1968. "The Control of the Luminous Environment." In *Lasers and Light. Readings from Scientific American*, pp. 131–39. San Francisco: W. H. Freeman and Company Publishers.

IES Lighting Handbook. 1984. New York: Illuminating Engineering Society.

JAMES, T. H. 1952. "Photographic Development." *Scientific American* (Nov.): 30.

KINGSTON, R. H. 1980. *Detection of Optical and Infrared Radiation*. New York: Springer-Verlag.

MALACARA, ZACARIAS H., and MORALES R. ARQUIMEDES. 1988. "Light Sources." In *Geometrical and Instrumental Optics*, edited by Daniel Malacara. Boston: Academic Press.

MOREHEAD, FRED F., JR. 1967. "Light-Emitting Semiconductors." *Scientific American* (May): 108.

RESNICK, ROBERT. 1991. *Basic Concepts in Relativity and Early Quantum Theory*, 2d ed. Englewood Cliffs, N.J.: Prentice Hall.

STIMSON, A. 1974. *Photometry and Radiometry for Engineers*. New York: Wiley-Interscience.

3

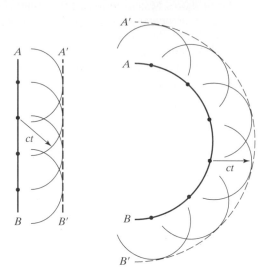

Geometrical Optics

INTRODUCTION

The treatment of light as wave motion allows for a region of approximation in which the wavelength is considered to be negligible compared with the dimensions of the relevant components of the optical system. This region of approximation is called *geometrical optics*. When the wave character of the light may not be so ignored, the field is known as *physical optics*. Thus geometrical optics forms a special case of physical optics in a way that may be summarized as follows:

$$\underset{\lambda \to 0}{\text{Limit}} \, \{\text{physical optics}\} = \{\text{geometrical optics}\}$$

Since the wavelength of light is very small compared with ordinary objects, early unrefined observations of the behavior of a light beam passing through apertures or around obstacles in its path could be handled by geometrical optics. Recall that the appearance of distinct shadows influenced Newton to assert that the apparent rectilinear propagation of light was due to a stream of light corpuscles rather than a wave motion. Wave motion characterized by longer wavelengths, such as those in water waves and sound waves, was known to give distinct bending around obstacles. Newton's model of light propagation, therefore, seemed not to allow for the existence of a wave motion with very small wavelengths. There was, in fact, already evidence of some degree of bending, even for light waves, in the time of Isaac Newton. The Jesuit, Francesco Grimaldi, had noticed the fine structure in the edge of a shadow, a structure not explainable in terms of the rectilinear propagation of light. This bending of light around the edges of an obstruction came to be called *diffraction*.

Within the approximation represented by geometrical optics, light is understood to travel out from its source along straight lines or rays. The ray is then simply the path along which light energy is transmitted from one point to another in an optical system. The ray is a useful construct, although abstract in the sense that a light beam, in practice, cannot be narrowed down indefinitely to approach a straight line. A pencil-like laser beam is perhaps the best actual approximation to a ray of light.[1] When a light ray traverses an optical system consisting of several homogeneous media in sequence, the optical path is a sequence of straight-line segments. Discontinuities in the line segments occur each time the light is reflected or refracted at an interface. The laws of geometrical optics that describe the subsequent direction of the rays are stated below and illustrated in Figure 3-1.

Law of Reflection. When a ray of light is reflected at an interface dividing two uniform media, the reflected ray remains within the *plane of incidence*, and the angle of reflection equals the angle of incidence. The plane of incidence includes the incident ray and the normal to the point of incidence.

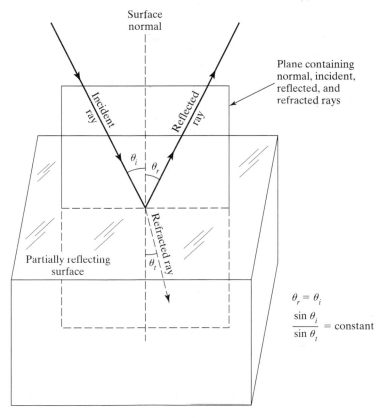

Figure 3-1 Illustration of the laws of reflection and refraction. The reflected and refracted rays remain in the plane containing the normal and incident ray.

[1]When an aperture through which the beam is passed is made small enough, however, even a laser beam begins to spread out in a characteristic diffraction pattern.

Law of Refraction (Snell's Law).   When a ray of light is refracted at an interface dividing two uniform media, the transmitted ray remains within the plane of incidence and the sine of the angle of refraction is directly proportional to the sine of the angle of incidence.

Both laws are summarized in Figure 3-1, in which an incident ray is partially reflected and partially transmitted at a plane interface separating two transparent media.

3-1   HUYGENS' PRINCIPLE

The Dutch physicist, Christian Huygens, envisioned light as a series of pulses emitted from each point of a luminous body and propagated in relay fashion by the particles of the ether, an elastic medium filling all space. Consistent with his conception, Huygens imagined each point of a propagating disturbance as capable of originating new pulses that contributed to the disturbance an instant later. His seminal principle showing how his model of light propagation implies the laws of geometrical optics can be stated as follows: Each point on the leading surface of a wave disturbance—the *wave front*—may be regarded as a secondary source of spherical waves (or *wavelets*), which themselves progress at the speed of light in the medium and whose envelope at a later time constitutes the new wave front. Simple applications of the principle are shown in Figure 3-2 for a plane and spherical wave. In each case, AB forms the initial wave disturbance or wave front, and $A'B'$ is the new wave front at a time t later. The radius of each wavelet is accordingly ct, where c is the speed of light in the medium. Notice that the new wave front is tangent to each wavelet at a single point. Huygens' approach is not without difficulties, since it does not allow for an explanation of diffraction (considered later). His principle also ignores the wave front formed by the back half of the wavelets, which would lead to a light disturbance traveling in the opposite direction. Despite weaknesses in this model, to be remedied later by Fresnel and others, Huygens was able to apply his principle to prove the laws of both reflection and refraction. We shall apply his principle to derive Snell's law, or the law of refraction, and leave the law of reflection to an alternate approach that follows later.

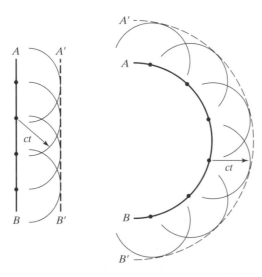

Figure 3-2   Illustration of Huygens' principle for plane and spherical waves.

Figure 3-3 illustrates Huygens' construction for the law of refraction. A change in the speed of light in the upper and lower media is taken into account. If the speed of light in vacuum is c, the speed of light in the upper medium is the ratio c/n_i, where n_i is a constant referred to as the *refractive index* that characterizes the medium of the incident ray. Similarly, the speed of light in the lower transmitting medium is c/n_t. Plane wave fronts such as AC and DF encounter a plane interface XY at an angle. The angle of incidence of the corresponding rays, AD, BE, and CF, relative to the perpendicular PD, is θ_i. The points D, E, and F on the incident wave front arrive at points D, J, and I of the interface XY at different times. In the absence of the refracting medium, the wave front GI is formed at the instant CF reaches I. During the progress of ray CF from F to I in time t, however, the ray AD has entered the lower medium, where its speed is, let us say, slower. Thus if the distance DG is $v_i t$, a wavelet of radius $v_t t$ is constructed with center at D. The radius DM can also be expressed as

$$DM = v_t t = v_t\left(\frac{DG}{v_i}\right) = \left(\frac{n_i}{n_t}\right)DG$$

Similarly, a wavelet of radius (n_i/n_t) JH is drawn centered at J. The new wave front KI includes point I on the interface and is tangent to the two wavelets at points M and N, as shown. The geometric relationship between the angles θ_i and θ_t formed by the representative incident ray AD and refracted ray DL, is Snell's law, as outlined in Figure 3-3. Snell's law of refraction may be expressed as

$$n_i \sin \theta_i = n_t \sin \theta_t \tag{3-1}$$

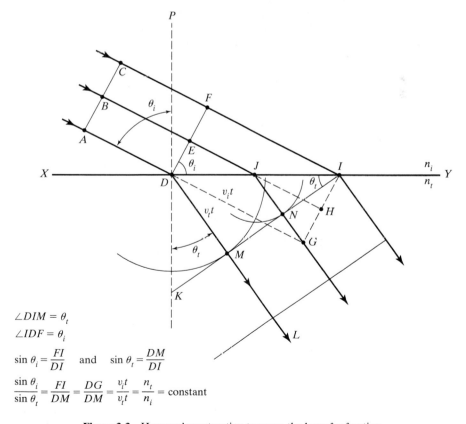

$\angle DIM = \theta_t$

$\angle IDF = \theta_i$

$\sin \theta_i = \dfrac{FI}{DI}$ and $\sin \theta_t = \dfrac{DM}{DI}$

$\dfrac{\sin \theta_i}{\sin \theta_t} = \dfrac{FI}{DM} = \dfrac{DG}{DM} = \dfrac{v_i t}{v_t t} = \dfrac{n_t}{n_i} = \text{constant}$

Figure 3-3 Huygens' construction to prove the law of refraction.

The laws of geometrical optics can also be derived, perhaps more elegantly, from a different fundamental hypothesis. The root idea was introduced by Hero of Alexandria, who lived in the second century B.C. According to Hero, when light is propagated between two points, it takes the shortest path. For propagation between two points in the same uniform medium, the path is clearly the straight line joining the two points. When light from the first point A (Figure 3-4) reaches the second point B after reflection from a plane surface, however, the same principle predicts the law of reflection, as follows. Figure 3-4 shows three possible paths from A to B, including the correct one, ADB. Consider, however, the arbitrary path ACB. If point A' is constructed on the perpendicular AO such that $AO = OA'$, the right triangles AOC and $A'OC$ are equal. Thus $AC = A'C$ and the shortest distance traveled by the ray of light from A' to B is obviously the straight line $A'DB$, so the path ADB is the correct choice taken by the actual ray. Elementary geometry shows that for this path, $\theta_i = \theta_r$. Note also that to maintain $A'DB$ as a single straight line, the reflected ray must remain within the plane of incidence, that is, the plane of the page.

The French mathematician, Pierre de Fermat, generalized Hero's principle to prove the law of refraction. In Figure 3-5, if the terminal point B lies below the surface of a second medium, the correct path is definitely not the shortest path or the straight line AB, for that would make the angle of refraction equal to the angle of incidence in violation of the empirically established law of refraction. Appealing to the "economy of nature," Fermat supposed instead that the ray of light traveled the path of least *time* from A to B, a generalization that included Hero's principle as a special case. If light travels more slowly in the second medium, as assumed in Figure 3-5, light bends at the interface so as to take a path that favors a shorter time in the second medium, thereby minimizing the overall transit time from A to B. With the help of a little calculus, the condition of minimal time leads again to Snell's law, as derived from Huygens' principle.[2]

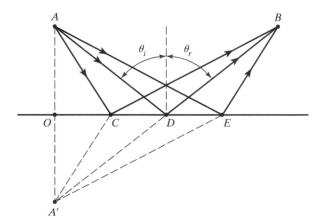

Figure 3-4   Construction to prove the law of reflection from Hero's principle.

[2]Extension of Huygens' principle to prove the law of reflection and of Fermat's principle to prove the law of refraction can be found in many optics texts, including *Introduction to Optics* by the authors of this book, Prentice-Hall, 1993.

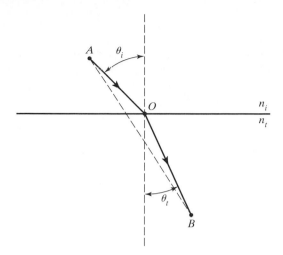

Figure 3-5 Construction illustrating Fermat's principle applied to refraction.

3-3 PRINCIPLE OF REVERSIBILITY

Refer again to the cases of reflection and refraction pictured in Figures 3-4 and 3-5. If the roles of points A and B are interchanged, so that B is the source of light rays, Fermat's principle of least time would demand that the light rays take the same path as before. In general then, any actual ray of light in an optical system, if reversed in direction, will retrace the same path backward. This principle will prove very useful in various applications to be dealt with later.

3-4 REFLECTION IN PLANE MIRRORS

Before discussing the formation of images in a general way, we treat the simplest—and experientially the most accessible—case of images formed by plane mirrors. In this context, it is important to distinguish between *specular reflection* from a perfectly smooth surface, and *diffuse reflection* from a granular or rough surface. In the former case, each of the parallel rays of a beam incident on the surface obeys the law of reflection and therefore reflects as a beam of parallel rays. In the latter case, though the law of reflection is obeyed at each point of the surface, the microscopically granular surface results in rays reflected in various directions and thus diffuse scattering of the originally parallel rays of light. Every plane surface produces some such scattering, since a perfectly smooth surface is not attainable in practice, but highly specular or mirror surfaces are common. The treatment that follows assumes the ideal case of specular reflection.

Consider the reflection of a single light ray OP from the xy-plane in Figure 3-6a. By the law of reflection, the reflected ray PQ remains within the plane of incidence, making equal angles with the normal at P. If the path OPQ is resolved into its x-, y-, and z-components, it is clear that the direction of ray OP is altered by the reflection only along the z-direction, and then in such a way that its z-component is simply reversed. If the rays before and after reflection are given by the equal length vectors, $\mathbf{r}_1$ and $\mathbf{r}_2$, we represent this reflection by writing

$$\mathbf{r}_1(a, b, c) \Rightarrow \mathbf{r}_2\,(a, b, -c)$$

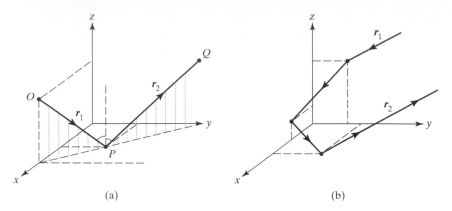

(a)   (b)

Figure 3-6   Geometry of a ray reflected from a plane. (a) The reflected ray remains within its plane of incidence and merely reverses its z-component. (b) Corner reflector. Three successive reflections from orthogonal planes send the ray back along its original direction.

It follows that if a ray is incident from such a direction as to reflect sequentially from all three rectangular coordinate planes, as in the "corner reflector" of Figure 3-6b,

$$\mathbf{r}_1(a, b, c) \Rightarrow \mathbf{r}_2(-a, -b, -c)$$

and the ray is simply reversed in direction, returning precisely parallel to the line of its original approach. A network of such corner reflectors ensures the exact return of a beam of light—a headlight beam from highway reflectors, for example, or a laser beam from a reflector on the moon.

Images are formed by many reflected rays that are not parallel to one another. Image formation in a plane mirror is illustrated in Figure 3-7a. A point ob-

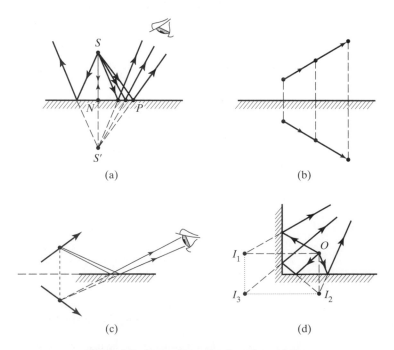

(a)   (b)

(c)   (d)

Figure 3-7   Image formation in a plane mirror.

ject S sends rays toward a plane mirror, which reflect as shown. The law of reflection ensures that pairs of triangles like SNP and $S'NP$ are equal, so that all reflected rays appear to originate at the *image point S'*, which lies along the normal line SN, and at such depth that the *image distance $S'N$* equals the *object distance SN*. The eye sees a point image at S' in exactly the same way it would see a real point object placed there. Since none of the actual rays of light lies below the mirror surface, the image is said to be a *virtual image*. The image S' cannot be projected on a screen as in the case of a *real image*. An extended object, such as the arrow in Figure 3-7b, is imaged point by point by a plane mirror surface in similar fashion. Each object point has its image point along its normal to the mirror surface and as far below the reflecting surface as the object point lies above the surface. Note that the image position does not depend on the position of the eye.

The construction in Figure 3-7b also makes clear that the image size is identical to the object size, giving a *magnification* of unity. In addition, the transverse orientation of object and image are the same. A right-handed object, however, appears left-handed in its image. In Figure 3-7c, where the mirror does not lie directly below the object, the mirror plane may be extended to determine the position of the image as seen by an eye positioned to receive reflected rays originating at the object. Figure 3-7d illustrates multiple images of a point object O formed by two perpendicular mirrors. Images I_1 and I_2 each results from a single reflection in one of the two mirrors, but a third image I_3 is also present, formed by sequential reflections from both mirrors. All parts of Figure 3-7 and the related discussion above should be understood clearly because they are fundamental to the optics of images.

3-5 REFRACTION THROUGH PLANE SURFACES

Consider light ray 1 in Figure 3-8a, incident at angle θ_1 at a plane interface separating two transparent media characterized, in order, by refractive indices n_1 and n_2. Let the angle of refraction be the angle θ_2. Snell's law, which takes the form,

$$n_1 \sin \theta_1 = n_2 \sin \theta_2 \qquad (3-2)$$

requires an angle of refraction θ_2 such that the refracted ray bends away from the normal when $n_2 < n_1$, as shown in Figure 3-8a for rays 1 and 2. For $n_2 > n_1$, on the other hand, the refracted ray bends *toward* the normal. The law also requires that ray 3, incident normal to the surface ($\theta_1 = 0$), be transmitted without change of direction ($\theta_2 = 0$), regardless of the ratio of refractive indices. In Figure 3-8a the three rays shown originate at a source point S below an interface and emerge into an upper medium of lower refractive index, as in the case of light emerging from water ($n_1 = 1.33$) into air ($n_2 = 1$). A unique image point is not determined by the three rays because they have no common intersection or virtual image point below the surface from which all three appear to originate after refraction, as shown by backward extensions of the refracted rays. For rays making a small angle with the normal to the surface, however, a reasonably good image can be located. In this approximation, where we allow only such *paraxial rays*[3] to form the image, the angles of incidence and refraction are both small and the approximation,

$$\sin \theta \cong \tan \theta \cong \theta \ (\text{in radians})$$

[3]In general, a paraxial ray is one that remains near the central axis of the image-forming optical system, thus making small angles with the optical axis.

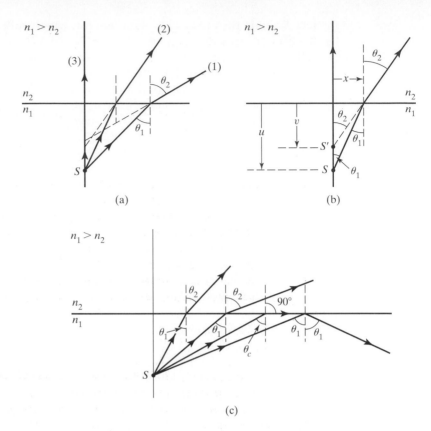

Figure 3-8 Geometry of rays refracted by a plane interface.

is valid. From Eq. (3-2), Snell's law can be approximated by

$$n_1 \tan \theta_1 \cong n_2 \tan \theta_2 \tag{3-3}$$

and taking the appropriate tangents from Figure 3-8b, we have

$$n_1\left(\frac{x}{u}\right) = n_2\left(\frac{x}{v}\right)$$

The image point occurs at the vertical distance v below the surface given by

$$v = \left(\frac{n_2}{n_1}\right)u \tag{3-4}$$

where u is the corresponding depth of the object. Thus objects underwater, viewed from directly overhead, appear to be nearer the surface than they actually are, since in this case $v = (1/1.33)u = 3/4\,u$. Even when the viewing angle θ_2 is not small, a reasonably good retinal image of an underwater object is formed because the aperture or pupil of the eye admits only a small bundle of rays while forming the image. Since these rays differ very little in direction, they appear to originate from approximately the same image point. However, the depth of this image is not $\frac{3}{4}$ the object depth, as for paraxial rays, and in general varies with the angle of viewing.

Rays from the object that make increasingly larger angles of incidence with the interface must, by Snell's law, refract at increasingly larger angles, as shown in Figure 3-8c. A *critical angle* of incidence θ_c is reached when the angle of refraction reaches $90°$. Thus, from Snell's law,

$$\sin \theta_c = \left(\frac{n_2}{n_1}\right)\sin 90 = \frac{n_2}{n_1}$$

or

$$\theta_c = \sin^{-1}\left(\frac{n_2}{n_1}\right) \tag{3-5}$$

For angles of incidence $\theta_1 > \theta_c$, the incident ray experiences *total internal reflection*, as shown. This phenomenon is essential in the transmission of light along glass fibers by a series of total internal reflections, as discussed in Chapter 17. Note that the phenomenon does not occur unless $n_1 > n_2$, so that θ_c can be determined from Eq. (3-5).

We will return to the nature of images formed by refraction at a plane surface when we treat it as a special case of refraction from a spherical surface.

3-6 IMAGING BY AN OPTICAL SYSTEM

We address now the subject of image quality, in general, and indicate the practical and theoretical factors that render an image less than perfect. In Figure 3-9, let the region labeled "optical system" include any number of reflecting and/or refracting surfaces, of any curvature, that may alter the direction of rays leaving an *object point O*. This region may include any number of intervening media, but we shall assume that each individual medium is homogeneous and isotropic,[4] and so characterized by its own refractive index. Thus rays spread out radially in all directions from object point O, as shown, in real *object space*, which precedes the first reflecting or refracting surface of the optical system. The family of spherical surfaces normal to the rays are the *wave fronts*, the locus of points such that each ray contacting a wavefront represents the same transit time of light from the source. In real object space the rays are diverging and the spherical wave fronts are expanding. Suppose now that the optical system redirects these rays in such a way that on leaving the optical system and entering real *image space* the wave fronts are contracting and the rays are converging to a common point that we define to be the *image point, I*. In the spirit of Fermat's principle, we can say that since every such ray starts at O and ends at I, every such ray requires the same

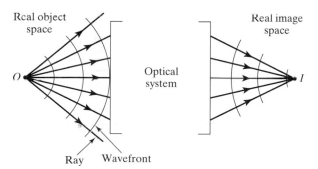

Figure 3-9 Image formation by an optical system.

[4]Homogeneity requires the medium to be optically uniform throughout, whereas isotropy requires its optical properties to be the same in any direction.

transit time. These rays are said to be *isochronous*. Further, by the principle of reversibility, if I is the object point, each ray reverses its direction but maintains its path through the optical system, and O is the corresponding image point. The points O and I are said to be *conjugate* points for the optical system. In an ideal optical system, *every* ray from O intercepted by the system—and only these rays—also pass through I. To image an actual object, this requirement must hold for every object point and its conjugate image point.

Nonideal images are formed in practice because of (1) light scattering, (2) aberrations, and (3) diffraction. Some rays leaving O do not reach I due to reflection losses at refracting surfaces, diffuse reflections from reflecting surfaces, and scattering by inhomogeneities in transparent media. Loss of rays by such means merely diminishes the brightness of the image; however, some of these rays are scattered through the image point from nonconjugate object points, degrading the image. When the optical system itself cannot produce the one-to-one relationship between object and image rays required for perfect imaging of all object points, we speak of system *aberrations*. Such aberrations are treated later. Finally, since every optical system intercepts only a portion of the wave front emerging from the object, the image cannot be perfectly sharp. Even if the image were otherwise perfect, the effect of processing a limited portion of the wave front leads to *diffraction* and a blurred image, which is said to be *diffraction limited*. This source of imperfect images, discussed further in the sections under diffraction, represents a fundamental limit to the sharpness of an image that cannot be entirely overcome. This difficulty arises from the wave nature of light. Only in the unattainable limit of geometric optics, where $\lambda \rightarrow 0$, would diffraction effects disappear entirely.

Reflecting or refracting surfaces that form perfect images are called *Cartesian surfaces*. In the case of reflection, such surfaces are the conic sections, as shown in Figure 3-10. In each of these figures, the roles of object and image points

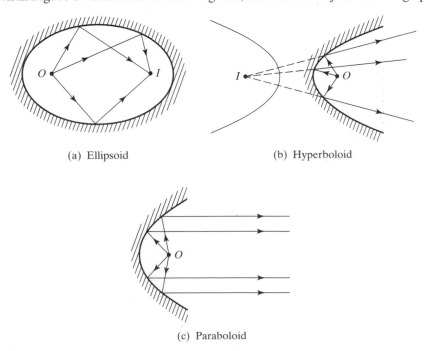

(a) Ellipsoid   (b) Hyperboloid

(c) Paraboloid

Figure 3-10 Cartesian reflecting surfaces showing conjugate object and image points.

may be reversed by the principle of reversibility. Notice that in Figure 3-10b, the image is virtual. In Figure 3-10c, the parallel reflected rays are said to form an image "at infinity." In each case, one can show that Fermat's principle, requiring isochronous rays between object and image points, leads to a condition that is equivalent to the geometric definition of the corresponding conic section.

Cartesian surfaces that produce perfect imaging by refraction are called *Cartesian ovoids of revolution*, shown in Figure 3-11a. In most cases, however, the image is desired in the same optical medium as the object. This goal is achieved by a lens that refracts light rays twice, once at each surface, producing a real image outside the lens. Thus it is of particular interest to determine the Cartesian surfaces that render every object ray parallel after the first refraction. Such rays incident on the second surface can then be refracted again to form an image. The solutions to this problem are illustrated in Figures 3-11b and c. Depending on the relative magnitudes of the refractive indices, the appropriate refracting surface is either a hyperboloid ($n_i > n_o$) or an ellipsoid ($n_o > n_i$), as shown. The first of these corresponds to the usual case of an object in air. A double hyperbolic lens then functions as shown in Figure 3-12. Note, however, that the aberration-free imaging so achieved applies only to object point O at the correct distance from the lens and on axis. For points near to O, imaging is not perfect. The larger the actual object, the less precise is its image. Because images of actual objects are not free from aberrations and because hyperboloid surfaces are difficult to grind exactly,[5] most optical surfaces are spherical. The *spherical aberrations* so introduced are accepted as a compromise when weighed against the cost and difficulty of fabricating nonspherical surfaces. In the remainder of our treatment of geometrical optics, we concentrate on the case of spherical reflecting and refracting surfaces with a *radius of curvature r*. Of course, plane surfaces and cylindrical surfaces are also both important, but can be treated as special cases of spherical surfaces.

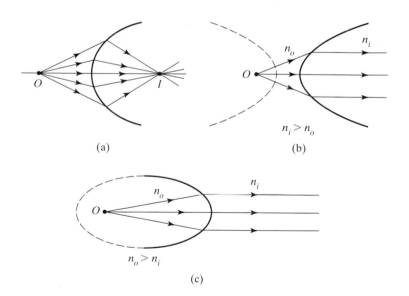

Figure 3-11 Cartesian refracting surfaces. (a) Cartesian ovoid images O at I by refraction. (b) Hyperbolic surface images object point O at infinity when O is at one focus and $n_i > n_o$. (c) Ellipsoidal surface images object point O at infinity when O is at one focus and $n_o > n_i$.

[5]Modern computer-driven machines have facilitated the process.

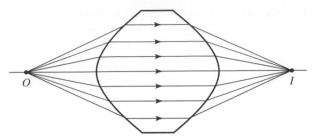

Figure 3-12 Aberration-free imaging of point object O by a double hyperbolic lens.

3-7 REFLECTION AT A SPHERICAL SURFACE

Spherical mirrors may be either concave or convex relative to an object point O, depending on whether the center of curvature C is on the same or opposite side of the surface. In Figure 3-13a, the mirror shown is convex, and two rays of light originating at O are drawn, one normal to the spherical surface at its vertex V

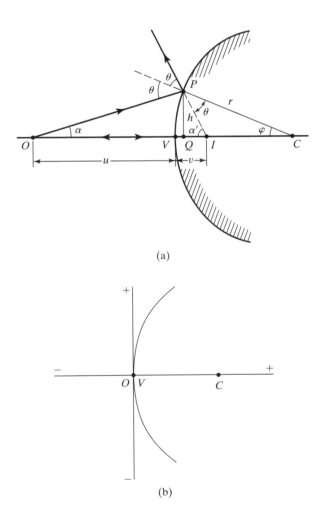

(a)

(b)

Figure 3-13 Reflection at a spherical surface. (a) Location of image point. (b) Sign convention adopted.

and the other an arbitrary ray incident at P. The first ray reflects back along itself; the second reflects at P as if incident on a plane tangent at P, according to the law or reflection. Relative to each other, the two reflected rays diverge as they leave the mirror. The intersection of the two rays (extended backward) determines the image point I conjugate to object point O. The image is virtual and located behind the mirror surface.

Object and image distances from the vertex are shown as u and v, respectively. A perpendicular of height h is drawn from P to the axis at Q. We seek a relationship between u and v that depends only on the radius of curvature r of the mirror. As we shall see, such a relation is possible only to a first-order approximation of the sines and cosines of angles such as α and φ made by the object and image rays at various points on the spherical surface. This means that, in place of expansions of $\sin \varphi$ and $\cos \varphi$ in series as shown here,

$$\sin \varphi = \varphi - \frac{\varphi^3}{3!} + \frac{\varphi^5}{5!} - \cdots$$

$$\cos \varphi = 1 - \frac{\varphi^2}{2!} + \frac{\varphi^4}{4!} - \cdots$$

we consider the first terms only and write

$$\sin \varphi \cong \varphi \quad \text{and} \quad \cos \varphi \cong 1 \quad \text{so that} \quad \tan \varphi = \frac{\sin \varphi}{\cos \varphi} \cong \varphi \quad (3\text{-}6)$$

These relations are accurate to 1% or less if the angle φ is 10° or smaller. This approximation leads to first-order (or Gaussian) optics, after Karl Friedrich Gauss, who in 1841 developed the foundations of this subject. Returning now to the problem at hand—that of relating u, v and r—notice that two angular relationships may be obtained from Figure 3-13a, because the exterior angle of a triangle equals the sum of its interior angles. Thus

$$\theta = \alpha + \varphi \quad \text{in } \Delta OPC \quad \text{and} \quad 2\theta = \alpha + \alpha' \quad \text{in } \Delta OPI$$

which combine to give

$$\alpha - \alpha' = -2\varphi \tag{3-7}$$

Using the small-angle approximation, the angles of Eq. (3-7) can be replaced by their tangents, yielding

$$\frac{h}{u} - \frac{h}{v} = -2\frac{h}{r}$$

Note that we have neglected the axial distance VQ, small when φ is small. Cancellation of h produces the desired relationship,

$$\frac{1}{u} - \frac{1}{v} = -\frac{2}{r} \tag{3-8}$$

If the spherical surface is chosen to be concave instead, the center of curvature would be to the left. For certain positions of the object point O, it is then possible to find a real image point, also to the left of the mirror. In these cases, the resulting geometric relationship analogous to Eq. (3-8) consists of the same terms, but with different algebraic signs, depending on the *sign convention* employed. One such choice of sign conventions leads to the single equation

$$\frac{1}{u} + \frac{1}{v} = \frac{2}{r} \tag{3-9}$$

to represent all cases. The sign convention to be used in conjunction with Eq. (3-9) is summarized by Figure 3-13b. If Cartesian axes are drawn at the reflecting (or refracting) surface such that the origin O of the axes coincides with the vertex V of the surface, then

- object and image distances u and v are negative to the left of the vertex and positive to the right.
- The radius of curvature r is positive when the center of curvature C is to the right of the vertex (convex surface case) and negative when C is to the left (concave surface case).
- Vertical dimensions are positive above the horizontal axis and negative below.

In the application of these rules, light is assumed to be directed from left to right.[6] Applying these sign conventions to Figure 3-13a, the object distance u is negative, the image distance v is positive, and the radius of curvature r is positive. Making u negative in Eq. (3-8) results in the mirror equation given by Eq. (3-9).

According to this sign convention, notice that for the case of reflection, negative object and image distances correspond to real objects and images, and positive object and image distances correspond to virtual objects and images. Virtual objects occur only with a sequence of two or more reflecting or refracting elements and are considered later.

The spherical mirror described by Eq. (3-9) yields, for a plane mirror $(r \to \infty)$, $v = -u$, in agreement with the previous discussion. The negative sign implies a virtual image (v positive) for a real object (u negative). Notice also in Eq. (3-9) that object and image distances appear symmetrically, implying their interchangeability as conjugate points. For an object at infinity ($u \to \infty$), incident rays are parallel (wave fronts are plane) and $v = r/2$, as illustrated in Figure 3-14a and b for both concave ($r < 0$) and convex ($r > 0$) mirrors. The image distance in each case is defined as the *focal length* of the mirrors. Thus the focal length is half the radius of curvature in magnitude and has the same sign.

$$f = \frac{r}{2} \begin{cases} > 0 \text{ for convex mirror} \\ < 0 \text{ for concave mirror} \end{cases} \tag{3-10}$$

With this in mind, the mirror equation can then be written more compactly as

$$\frac{1}{u} + \frac{1}{v} = \frac{1}{f} \quad \text{where} \quad f = \frac{r}{2} \tag{3-11}$$

In Figure 3-14c, a construction is shown that helps us determine the transverse magnification. The object is an extended arrow of transverse dimension h_o.

---

[6]In the case of reflection, the light reverses direction, a point that must be kept in mind. Thus, for example, a distance to the right of the surface would be positive for light from left to right, but negative for light from right to left.

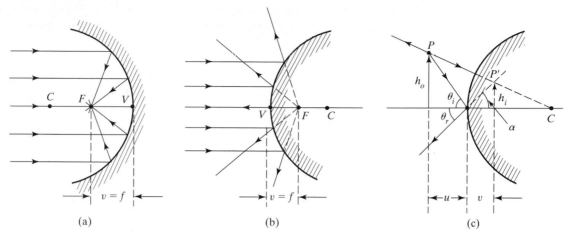

Figure 3-14 (a) and (b) Location of focal points. (c) Construction to determine magnification of a spherical mirror.

The image of object point P at the top of the arrow is located by two rays whose behavior on reflection is known. The ray incident at the vertex must reflect to make equal angles with the axis. The other ray is directed toward the center of curvature along a normal to the surface and so must reflect back along itself. The intersection of the two reflected rays occurs *behind* the mirror at P' and locates a virtual image there. Because of the equality of the three angles shown—θ_i, θ_r, and α—it follows, from similar right triangles, that

$$\frac{h_o}{u} = \frac{h_i}{v}$$

The lateral magnification m is defined as the ratio of lateral image size to corresponding lateral object size, giving

$$m = \frac{h_i}{h_o} = \frac{v}{u} \tag{3-12}$$

Extending the sign convention to include magnification, we assign a $(+)$ magnification when the image has the same orientation as the object, as in Figure 3-14c, and a $(-)$ magnification when the image is inverted relative to the object. To produce a $(+)$ magnification in the use of Eq. (3-12), where u is negative and v is positive, we introduce a negative sign in the equation and write as the general form,

$$m = -\frac{v}{u} \tag{3-13}$$

Once the points C and F are located, image formation by a spherical mirror may also be determined (approximately) by graphical methods. Figure 3-15 illustrates several examples that should be examined carefully. The validity of each ray reflection has been established by the discussion above. In each case the image point P' of the top point P of the arrow is located by the intersection of three reflected rays. In Figure 3-15a, the reflected rays (converging) intersect directly at P'. In Figures 3-15b and 3-15c, the reflected rays (diverging) must be extended backward (see dashed lines) to locate the intersection points P'.

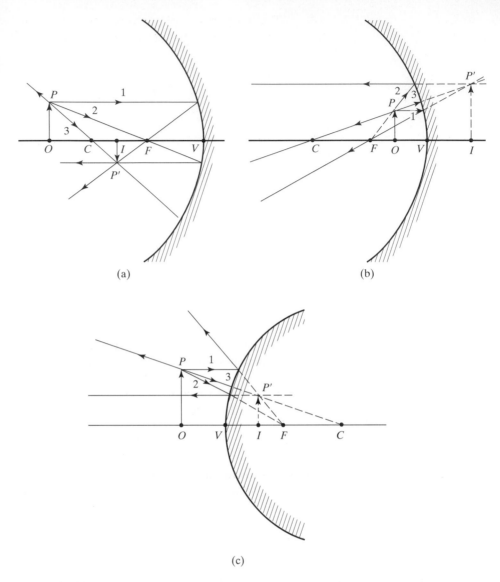

Figure 3-15 Ray diagrams for spherical mirrors. (a) Real image, concave mirror. (b) Virtual image, concave mirror. (c)Virtual image, convex mirror.

❏ Example

An object 3 cm high is placed 20 cm to the left of a (a) convex, and (b) concave spherical mirror, each of 10-cm focal length. Determine the position and nature of the image in each case.

Solution

(a) Convex mirror: $f = +10$ cm and $u = -20$ cm.

$$\frac{1}{u} + \frac{1}{v} = \frac{1}{f} \quad \text{or} \quad v = \frac{fu}{u - f} = \frac{(10)(-20)}{-20 - 10} = +6.67 \text{ cm}$$

$$m = -\frac{v}{u} = -\frac{6.67}{-20} = +0.333 = +\frac{1}{3}$$

The image is *virtual* (because v is positive), the image is located 6.67 cm behind the mirror, is *erect* (because m is positive) and one-third the size of the object, or 1 cm high.

(b) Concave mirror: $f = -10$ cm and $u = -20$ cm

$$v = \frac{fu}{u-f} = \frac{(-10)(-20)}{(-20)-(-10)} = -20 \text{ cm}$$

$$m = \frac{v}{u} = -\frac{-20}{-20} = -1$$

The image is *real* (because v is negative), the image is located 20 cm in front of the mirror, is *inverted* (because m is negative), and is the same size as the object, or 3 cm high. Image and object both happen to be at $2f = -20$ cm, the center of curvature of the mirror. ■

3-8 CURVATURES AND VERGENCES

Another useful way of looking at object-image formation is in terms of the change in curvature of the wave fronts as they diverge from an object point and converge toward an image point. Wave fronts from a point source at O consist of spherical envelopes that grow less curved as they arrive at more distant points. Figure 3-16 shows a series of such wave fronts, indicating their radial distance in centimeters from the source and also their *curvature*. The curvature is usually expressed in reciprocal meters, given the name of *diopters*. Thus a wave front of 10 cm radius has a curvature of 1/0.1 or 10 diopters, written 10 D. If this wave front is diverging, as from an object point, its curvature is -10 D, whereas if it is converging, as toward an image point, its curvature is $+10$ D. This choice of signs is consistent with the convention already adopted for object and image distances. The curvature in diopters is also indicated for each of the diverging wave fronts in Figure 3-16. Note in the figure that, as the distance r grows larger, the curvature $1/r$ of the wave front decreases in magnitude, approaching the value zero as r approaches infinity and the distant wave fronts approach a flat or plane shape.

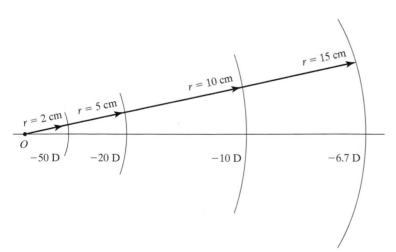

Figure 3-16 Change in curvature of wave fronts *diverging* from a point source at O. If the light direction is reversed, the wave fronts *converge* to an image point at O.

Now let us write Eq. (3-9) for mirrors in the language of curvatures rather than the language of object-image distances. Beginning with the original form,

$$\frac{1}{u} + \frac{1}{v} = \frac{2}{r}$$

Notice that the first term $1/u$ is just the curvature of the wavefront at object distance u from the source point and the second term $1/v$ is the curvature of the wave front at distance v from the image point. We define these special curvatures as *vergences*, with $V \equiv 1/v$ *and* $U \equiv 1/u$. Because the light direction is reversed on reflection, we reverse the sign of the curvature associated with the image point, writing $V = -1/v$ for the case of reflection. The mirror equation can now be written as

$$U - V = \frac{2}{r} \quad \text{or} \quad U - \frac{2}{r} = V \tag{3-14}$$

In the second form, the equation is more intuitive. It states that the initial vergence U of the wave front at the mirror is modified by the mirror to produce the final vergence V at the mirror. This modification is due to the action of the mirror, and is represented by the term $-2/r$, which we define to be the mirror *Power P*. Thus, we can write, simply

$$U + P = V \quad \text{where} \quad P = -\frac{2}{r} \tag{3-15}$$

This equation, now expressed in terms of wave front curvatures or vergences and the shaping power of the mirror, is equivalent to Eq. 3-9. We can also rewrite the magnification equation in this language. From Eq. (3-13),

$$m = -\frac{v}{u} = -\frac{1/u}{1/v} = -\frac{U}{-V} = \frac{U}{V}$$

and so the equation,

$$m = \frac{U}{V} \tag{3-16}$$

is equivalent to that of Eq. 3-13.

To show how a mirror problem can be solved from this new point of view, we repeat the mirror example following Eq. (3-13).

❏ **Example**

(a) Convex mirror: $r = 2f = +20$ cm

$U = 1/u = 1/(-20 \text{ cm}) = 1/(-0.2 \text{ m}) = -5 \text{ D}$, and

$P = -2/(20 \text{ cm}) = -2/(0.2 \text{ m}) = -1/(0.1 \text{ m}) = -10 \text{ D}$

Then $V = U + P = -5 \text{ D} - 10 \text{ D} = -15 \text{ D}$

and $v = -1/V = -1/(-15 \text{ D}) = 0.0667 \text{ m} = 6.67 \text{ cm}$

Furthermore, $m = U/V = (-5\text{D})/(-15 \text{ D}) = +1/3$

(b) Concave mirror: $r = 2f = -20$ cm
$$U = -5 \text{ D, as in (a),}$$
$$P = +10 \text{ D, as in (a) except for a sign reversal in } r$$
Then $V = U + P = -5 \text{ D} + 10 \text{ D} = +5 \text{ D}$
and $v = -1/V = -1/5 \text{ D} = -(0.2 \text{ m}) = -20$ cm
Furthermore, $m = U/V = (-5 \text{ D})/(+5 \text{ D}) = -1$

These are the same results obtained by using Eqs. (3-9) and (3-13) directly. ■

3-9   REFRACTION THROUGH SPHERICAL SURFACES

We turn now to a similar treatment of refraction at a spherical surface, choosing in this case the concave surface of Figure 3-17. Two rays are shown emanating from object point O. One is an axial ray normal to the surface at its vertex and so it refracts there without change in direction. The other ray is an arbitrary ray incident at P and refracts there according to Snell's law,

$$n_1 \sin \theta_1 = n_2 \sin \theta_2 \qquad (3\text{-}17)$$

The two refracted rays, diverging as they leave the surface, appear to emerge from their common intersection, the image point I to the *left* of the surface. In triangle CPO, the exterior angle $\alpha = \theta_1 + \varphi$. In triangle CPI, the exterior angle $\alpha' = \theta_2 + \varphi$. Using again the small angle approximation for paraxial rays and substituting for θ_1 and θ_2 in Eq. (3-17), we have

$$n_1(\alpha - \varphi) = n_2 (\alpha' - \varphi) \qquad (3\text{-}18)$$

Next, writing the tangents for the angles by inspection of Figure 3-17 and again neglecting the distance QV because of the small angle approximation, we obtain

$$n_1\left(\frac{h}{u} - \frac{h}{r}\right) = n_2\left(\frac{h}{v} - \frac{h}{r}\right)$$

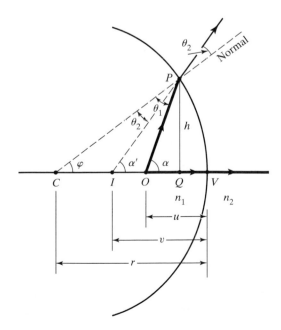

Figure 3-17   Refraction at a concave spherical interface for which $n_2 > n_1$.

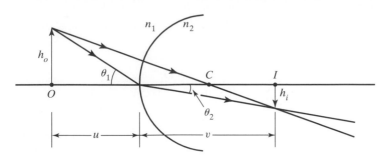

Figure 3-18 Construction to determine lateral magnification at a convex spherical refracting surface.

By canceling h throughout and rearranging, we get

$$\frac{n_1}{u} - \frac{n_2}{v} = \frac{n_1 - n_2}{r} \tag{3-19}$$

Employing the same sign convention as introduced for mirrors, all three quantities u, v, and r in Figure 3-18 are negative. Equation (3-19) above then becomes

$$\frac{n_2}{v} - \frac{n_1}{u} = \frac{n_2 - n_1}{r} \tag{3-20}$$

which holds equally well for convex surfaces. When $r \to \infty$, the spherical surface becomes a plane refracting surface, and

$$v = \left(\frac{n_2}{n_1}\right)u \tag{3-21}$$

where v is the *apparent depth*, as found previously. For a real object ($u < 0$), it follows from the equation that $v < 0$ also, and the image is to the left of the refracting surface. Since this is a case of refraction, not reflection, the image is virtual. The lateral magnification of an extended object is simply determined by inspection of Figure 3-18. For the ray incident at the vertex at angle θ_1, Snell's law requires that

$$n_1 \sin \theta_1 = n_2 \sin \theta_2$$

which in the small-angle approximation, using tangents for sines, becomes

$$n_1 \left(\frac{h_o}{u}\right) = n_2 \left(\frac{h_i}{v}\right)$$

The lateral magnification is then

$$m = \frac{h_i}{h_o} = \frac{n_1 v}{n_2 u} \tag{3-22}$$

This result must be adjusted, if necessary, to indicate the image inversion of Figure 3-18, which is associated with a negative magnification. Since v is positive and u is negative, m has the negative value required as the equation stands so the equation serves in general as the magnification for a spherical refracting surface. For the case of a plane refracting surface, Eq. (3-21) may be incorporated into Eq. (3-22), giving $m = +1$. Thus images formed by plane refracting surfaces have the same lateral dimensions and orientations as the object.

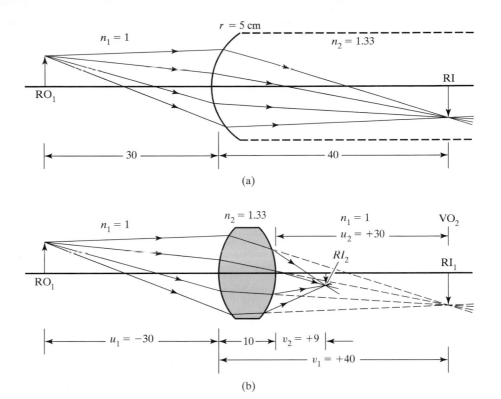

Figure 3-19 Example of refraction by spherical surfaces. (a) Refraction by a single spherical surface. (b) Refraction by a thick lens. Subscripts 1 and 2 refer to refractions at the first and second surfaces, respectively.

❏ **Example**

As an extended example of refraction by spherical surfaces, refer to Figure 3-19. In (a), a real object RO_1 is positioned in air, 30 cm from a convex spherical surface of radius 5 cm. To the right of the interface, the refractive index is that of water. Before constructing representative rays, we first find the image distance and lateral magnification of the image, using Eqs. (3-20) and (3-22). Equation (3-20) becomes

$$\frac{1.33}{v} - \frac{1}{-30} = \frac{1.33 - 1}{5}$$

giving $v = +40$ cm. The positive sign indicates that the image is real, being located beyond the surface where real rays of light are refracted. Considering next the magnification, Eq. (3-22) becomes

$$m = \frac{(1)(+40)}{(1.33)(-30)} = -1$$

indicating an inverted image, equal in size to that of the object. Figure 3-19a shows the image as well as several rays, which are now determined. In this example we have assumed that the medium to the right of the spherical surface extends far enough so that the image is formed inside it without further refraction. Let us suppose now (Figure 3-19b) that the second medium is only

10 cm thick, forming a *thick lens*, with a second, concave spherical surface also of radius 5 cm. The refraction by the first surface is, of course, unaffected by this change. Inside the lens, just beyond the first surface, the rays are directed as before, on their way to form an image 40 cm from the first surface. However, these rays are intercepted by the second surface and are again refracted to produce a different image, as shown. Since the convergence of the rays striking the second surface is determined by the tentative position of the first image, its location (never actually formed) nevertheless serves as the appropriate object distance to be used for the second refraction. We call the real image *RI*, tentatively formed by surface 1, a *virtual object VO$_2$* for surface 2. Then, by the sign convention established previously, we make the virtual object distance, relative to the second surface, a positive quantity when using Eqs. (3-20) and (3-22). For the second refraction then, Eq. (3-20) becomes

$$\frac{1}{v_2} - \frac{1.33}{+30} = \frac{1 - 1.33}{-5}$$

or $v_2 = +9$ cm, indicating a final real image to the right of the second surface. Notice that the roles of n_1 and n_2 are now reversed, since light moves—relative to the second surface—from water ($n = 1.33$) into air ($n = 1$). The magnification, according to Eq. (3-22), is

$$m = \frac{(1.33)(+9)}{(1)(+30)} = +\frac{2}{5}$$

The final image RI_2 is then $\frac{2}{5}$ the lateral size of its (virtual) object RI_1 and appears with the same orientation. Relative to the *original* object RO_1, however, the *overall* magnification is $m = m_1 m_2 = (-1)(+\frac{2}{5}) = -\frac{2}{5}$, so that the final image RI_2 is $\frac{2}{5}$ as large as RO_1 and inverted. ∎

In general, whenever a train of reflecting or refracting surfaces is involved in the processing of a final image, the individual reflections and/or refractions are considered in the order in which light is actually incident upon them. The object distance for the *n*th step is determined by the image distance for the previous or $(n - 1)$ step. If the image of the $(n - 1)$ step is not actually formed, as in the example above, it serves as a *virtual object* for the *n*th step.

3-10   VERGENCES IN REFRACTION

The problem of refraction at spherical surfaces can also be described by the method of wave front curvatures introduced above. Recall that the refraction equation in terms of object and image distance is given by

$$\frac{n_2}{v} - \frac{n_1}{u} = \frac{n_2 - n_1}{r} \tag{3-23}$$

Here light progresses left to right from a medium of refractive index n_1 into a medium of index n_2, separated by a spherical interface whose radius is r. Now let us define $U \equiv n_1/u$ as the vergence of wavefronts originating at the source and arriving at the spherical surface. Similarly, we set $V \equiv n_2/v$ to represent the vergence of wave fronts at the spherical surface associated with the formation of the image.[7] These definitions are slightly more general than those introduced for mirrors, where $n_1 = n_2 = 1$ because we were then dealing with light in a single medium. Eq. 3-23 can be rewritten as

$$V - U = P \quad \text{or} \quad U + P = V \qquad (3\text{-}24)$$

This is the form encountered previously and has the same intuitive interpretation: The curvature U of the object wave front is modified by the power P of the spherical surface to produce the curvature V of the image wave front. In other words,

$$\text{Object Vergence} + \text{Surface Power} = \text{Image Vergence}$$

In this case the power of the surface can be read from Eq. 3-23:

$$P = \frac{n_2 - n_1}{r} \qquad (3\text{-}25)$$

The magnification can also be recast in terms of vergence as follows:

$$m = \frac{n_1 v}{n_2 u} = \frac{(n_1/u)}{(n_2/v)} = \frac{U}{V}$$

Thus, lateral magnification is simply expressed as the ratio of object vergence to image vergence:

$$m = \frac{U}{V} \qquad (3\text{-}26)$$

❏ **Example**

We repeat the example of refraction at two successive spherical surfaces using the vergence approach. Of course, we should get the same answers!

Solution   For the first spherical surface, we had $u_1 = 30$ cm, $r = +5$ cm, $n_1 = 1$, and $n_2 = 1.33$. Then:

$$U = n_1/u = 1/(-30 \text{ cm}) = 1/(-0.3 \text{ m}) = -3.33 \text{ D}$$
$$P = (n_2 - n_1)/r = (1.33 - 1)/(0.05 \text{ m}) = 1/(0.15 \text{ m}) = +6.67 \text{ D}$$
$$V = U + P = -3.33 \text{ D} + 6.67 \text{ D} = +3.33 \text{ D, and}$$
$$v = n_2/V = 1.33/(3.33 \text{ D}) = 1.33(0.3 \text{ m}) = 0.4 \text{ m} = +40 \text{ cm}$$
$$\text{Also, } m = U/V = (-3.33 \text{ D})/(+3.33 \text{ D}) = -1$$

For the second spherical surface, $n_1 = 1.33$, $n_2 = 1$, $u = +30$ cm, $r_2 = -5$ cm $= -0.05$ m

$$U = n_2/u = 1.33/(+0.3 \text{ m}) = 1/(0.225 \text{ m}) = +4.43 \text{ D}$$
$$P = (1 - 1.33)/(-0.05 \text{ m}) = 1/(0.1515 \text{ m}) = +6.60 \text{ D}$$
$$V = U + P = 4.43 \text{ D} + 6.60 \text{ D} \cong 11 \text{ D, and}$$

---

[7]Notice that since light never reverses direction in the case of refraction, the image vergence is defined without the negative sign that appeared previously in reflection.

$$v = n_2/V = 1/(11 \text{ D}) \cong 0.09 \text{ m} = 9 \text{ cm}$$
$$\text{Finally, } m = U/V = (4.43 \text{ D})/(11 \text{ D}) \cong 0.4 \text{ or } 2/5 \qquad\blacksquare$$

3-11 THIN LENSES

We now apply the preceding method to the case of the thin lens. As in the example of Figure 3-19, two refractions at spherical surfaces are involved. The simplification we make is to neglect the thickness of the lens in comparison with the object and image distances, an approximation that is justified in most practical situations. At the first refracting surface, of radius r_1,

$$\frac{n_2}{v_1} - \frac{n_1}{u_1} = \frac{n_2 - n_1}{r_1} \qquad (3\text{-}27)$$

and at the second surface, of radius r_2,

$$\frac{n_1}{v_2} - \frac{n_2}{u_2} = \frac{n_1 - n_2}{r_2} \qquad (3\text{-}28)$$

We have assumed that the lens of index n_2 faces the same medium of refractive index n_1 on both sides. Now the second object distance, in general, is given by

$$u_2 = v_1 - t \qquad (3\text{-}29)$$

where t is the thickness of the lens. Notice that this relation is perfectly general, producing the correct sign of u_2, as in Figure (3-19), and also in cases where the intermediate image falls inside or to the left of the lens. In the thin-lens approximation, neglecting t,

$$u_2 = v_1$$

When this value of u_2 is substituted into Eq. (3-28) and the equation is added to Eq. (3-27), the terms in n_2/v_1 cancel and the result is

$$\frac{n_1}{v_2} - \frac{n_1}{u_1} = (n_2 - n_1)\left(\frac{1}{r_1} - \frac{1}{r_2}\right)$$

Now u_1 is the original object distance and v_2 is the final image distance, so we may drop their subscripts and simply rewrite the equation for the thin lens as a whole,

$$\frac{1}{v} - \frac{1}{u} = \frac{n_2 - n_1}{n_1}\left(\frac{1}{r_1} - \frac{1}{r_2}\right) \qquad (3\text{-}30)$$

The *focal length f* of the thin lens is defined as the image distance v for an object at infinity. Accordingly, for $v = f$ and $1/u = 0$, we have

$$\frac{1}{f} = \frac{n_2 - n_1}{n_1}\left(\frac{1}{r_1} - \frac{1}{r_2}\right) \qquad (3\text{-}31)$$

This is shown for a converging lens in Figure 3-20a, where F is the image point, called the *focal point*. We know that if the light rays shown there are reversed in direction, the rays would follow the same paths. Then F would act as the object point for an image formed at infinity. Thus *focal length* can also be defined as the

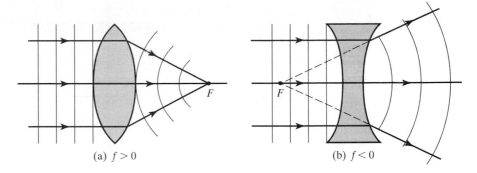

(a) $f > 0$ (b) $f < 0$

Figure 3-20 Action of a converging lens (a) and diverging lens (b) on plane wave fronts of light.

object distance for an image at infinity. Of course, in regarding ray reversal of Figure 3-20a in this way, we are looking at a case where light is moving from right to left. According to our sign convention, f would have to be negative. This is exactly what Eq. (3-30) gives for the case $1/v = 0$ and $u = f$. If we call the first focal length f_1 and the second f_2, then $f_2 = f_1$.

Equation (3-31) is called the *lensmaker's equation* because it predicts the magnitude of the focal length of a lens fabricated with a given refractive index and radii of curvature and used in a medium of refractive index n_1. In most cases, the ambient medium is air and $n_1 = 1$. The thin lens equation, in terms of the focal length, is then

$$\frac{1}{v} - \frac{1}{u} = \frac{1}{f} \tag{3-32}$$

Wave front analysis for plane wavefronts, as shown in Figure 3-20, indicates that a lens thicker in the middle causes incident parallel rays to converge, while one thinner in the middle causes them to diverge. The portion of the wave front that passes through the thicker region is delayed relative to other parts of the wavefront. Converging lenses are characterized by positive focal lengths and diverging lenses by negative focal lengths (as is evident from the figure) where the images are real and virtual, respectively.

Sample ray diagrams for converging (or *convex*) and diverging (or *concave*) lenses are shown in Figure 3-21. The thin lenses are best represented for purposes of ray construction by a vertical line with edges suggesting the general shape of the lens. Ray 1 from the top of the object is incident parallel to the axis and converges, in Figure 3-21a, through the focal point or diverges, in Figure 3-21b, as if originating at the focal point. A second ray 2 is simply the inverse of the first. Although two rays are sufficient to locate the image, a third ray 3 may also be drawn through the center of the lens without bending. The midsection of the lens behaves as a parallel plate, which does not alter the direction of the incident ray, and because it is thin, displaces the ray by a negligible amount. In constructing ray diagrams, observe that, except for the central ray, every ray refracted by a convex lens bends *toward* the axis, and every ray refracted by a concave lens bends *away* from the axis. From either diagram, the angles subtended by object

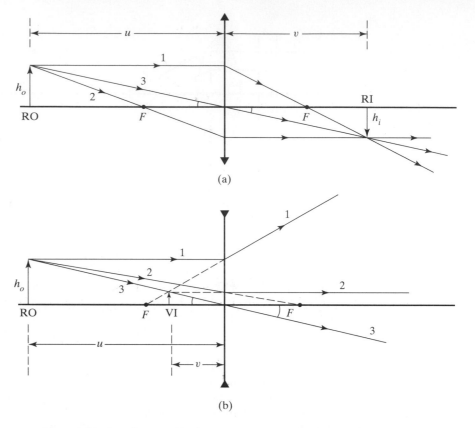

Figure 3-21 Ray diagrams for image formation by a convex lens (a) and a concave lens (b).

and image at the center of the lens are seen to be equal. Thus for either the real image in (a) or the virtual image in (b), it follows that

$$\frac{h_o}{u} = \frac{h_i}{v}$$

and lateral magnification

$$m = \frac{h_i}{h_o} = \frac{v}{u}$$

This equation should be tested to see whether it conforms to our sign convention, or whether a negative sign needs to be added. In case (a), $u < 0$, $v > 0$, and $m < 0$. The negative result is desired since the image is inverted. In case (b), $u < 0$, $v < 0$, and $m > 0$. The positive result is desired since the image is erect relative to its object. The magnification equation is correct as it stands, therefore, and we can write for all cases,

$$m = \frac{v}{u} \tag{3-33}$$

Further examples of ray diagrams for a train of two lenses are illustrated in Figure 3-22. Table 3-1 and Figure 3-24 at the end of the chapter provide a con-

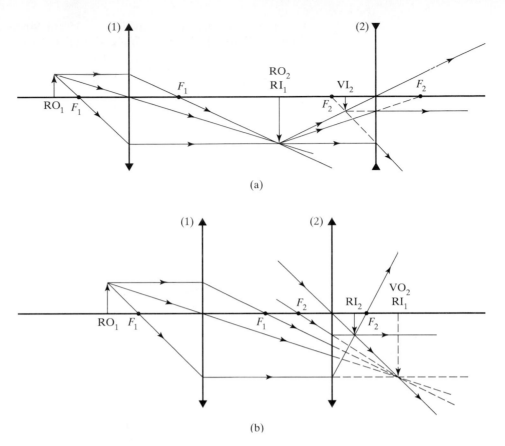

(a)

(b)

Figure 3-22 (a) Formation of a virtual image by a two-element train of a convex lens 1 and concave lens 2. The intermediate image RI_1 serves as a real object RO_2 for the second lens. (b) Formation of a real image RI_2 by a train of two convex lenses. The intermediate image RI_1 serves as a virtual object VO_2 for the second lens.

venient summary of image formation by reflection and refraction with spherical surfaces.

❏ Example

Find and describe the intermediate and final images produced by a two-lens system such as the one sketched in Figure 3-22a. Both are 15-cm focal length lenses, and their separation is 60 cm. Let the object RO_1 be 25 cm from the first lens.

Solution   The first lens is convex: $f_1 = +15$ cm, $u_1 = -25$ cm.

$$\frac{1}{v_1} - \frac{1}{u_1} = \frac{1}{f} \quad \text{or} \quad v_1 = \frac{fu_1}{f + u_1} = \frac{(15)(-25)}{(15 - 25)} = +37.5 \text{ cm}$$

$$m_1 = \frac{v_1}{u_1} = \frac{+37.5}{-25} = -1.5$$

Thus the first image is real (because v_1 is positive), 37.5 cm to the right of the first lens, inverted (because m is negative), and 1.5 times the size of the object.

The second lens is concave: $f_2 = -15$ cm. Since real rays of light diverge from the first real image, it serves as a real object for the second lens, with $u_2 = -(60 - 37.5) = -22.5$ cm or 22.5 cm to the left of lens (2). Then

$$v_2 = \frac{fu_2}{f + u_2} = \frac{(-15)(-22.5)}{-15 - 22.5} = -9 \text{ cm}$$

$$m_2 = \frac{v_2}{u_2} = \frac{-9}{-22.5} = +0.4$$

Thus the final image is *virtual* (because v_2 is negative), 9 cm to the *left* of the second lens, erect *relative to its own object* (because m is positive), and 0.4 times its size. The *overall* magnification m is given by the product $m_1 m_2 = (-1.5)(0.4) = -0.6$. The final image is therefore inverted relative to the *original* object and $\frac{6}{10}$ its lateral size. All these features are exhibited qualitatively in the ray diagram of Figure 3-22a. ∎

3-12 VERGENCES FOR THE THIN LENS

To write a vergence equation for the thin lens, we can make use of the fact that the power of the lens as a whole is just the sum of the powers of the two surfaces, which we already know how to write. Since the lens is thin, we are making the assumption that the wave front curvature does not change appreciably as the wave moves through the thickness of the lens. Thus we can write

$$\left\{ \begin{array}{c} \text{Object vergence} \\ \text{at first surface} \end{array} \right\} + \left\{ \begin{array}{c} \text{Power of} \\ \text{first surface} \end{array} \right\} + \left\{ \begin{array}{c} \text{Power of} \\ \text{second surface} \end{array} \right\} = \left\{ \begin{array}{c} \text{Image vergence} \\ \text{at second surface} \end{array} \right\}$$
$$U \quad + \quad P_1 \quad + \quad P_2 \quad = \quad V$$

Let us assume the most general case, in which object space has a refractive index n_1, the lens has an index n_2, and image space has an index n_3. Then

$$\frac{n_1}{u} + \frac{n_2 - n_1}{r_1} + \frac{n_3 - n_2}{r_2} = \frac{n_3}{v}$$

or $U + P = V$, where $U = n_1/u$, $V = n_3/v$, and

$$P = P_1 + P_2 = \frac{n_2 - n_1}{r_1} + \frac{n_3 - n_2}{r_2} \tag{3-34}$$

Commonly, the lens is situated with the same medium on either side, that is, $n_1 = n_3$. Then

$$P = (n_2 - n_1)\left(\frac{1}{r_1} - \frac{1}{r_2} \right) \tag{3-35}$$

A further simplification occurs when the lens is situated in air, with $n_1 = 1$.

The magnitude of the focal length can also be expressed in terms of the lens power, using the lensmaker's equation:

$$\frac{1}{f} = \frac{n_2 - n_1}{n_1}\left(\frac{1}{r_1} - \frac{1}{r_2} \right) = \frac{P}{n_1} \tag{3-36}$$

or

$$f = \frac{n_1}{P} \qquad (3\text{-}37)$$

Finally, the lateral magnification is expressible by

$$m = \frac{v}{u} = \frac{1/u}{1/v} = \frac{U/n_1}{V/n_3} = \frac{n_3 U}{n_1 V} \qquad (3\text{-}38)$$

When the first and final media are identical, $m = U/V$.

❏ **Example**

Recalculate the preceding example (see Figure 3-22a) using the vergence technique.

Solution   We solve in a two-step process. Considering the first (convex) lens, $f = +15$ cm, $u = -25$ cm, the lens separation is 60 cm, and the lens is in air, so that

$n_1 = 1$. Then $U = n_1/u = 1/(-0.25 \text{ m}) = -4$ D
$P = n_1/f = 1/(0.15 \text{ m}) = +6.67$ D
$V = U + P = -4 \text{ D} + 6.67 \text{ D} = 2.67$ D
$v = n_3/V = 1/(2.67 \text{ D}) = 0.375 \text{ m} = +37.5$ cm
$m = U/V = (-4 \text{ D})/(+2.67 \text{ D}) = -1.5$

For the second, or concave lens, $f = -15$ cm and $u = 37.5 - 60 = -22.5$ cm

Then $U = n_1/u = 1/(-0.225 \text{ m}) = -4.44$ D
$P = n_1/f = 1/(-0.15 \text{ m}) = -6.67$ D
$V = U + P = -4.44 \text{ D} - 6.67 \text{ D} = -11.11$ D
$v = n_3/V = 1/(-11.11 \text{ D}) = -0.090 \text{ m} = -9.0$ cm
$m = U/V = (-4.44 \text{ D})/(-11.11 \text{ D}) = +0.40$

These results are in agreement with those calculated previously and are to be interpreted in the same way. ■

3-13   THIN LENSES IN CONTACT

The vergence approach we have been using provides an intuitive way of dealing with the effect of reflecting and refracting surfaces on the convergence or divergence of light rays. The approach is useful for another reason that is of particular interest to optometrists. When thin lenses are placed together, the focal length f of the combination, treated as a single lens, can be found in terms of the focal lengths $f_1, f_2, \ldots$ of the individual lenses. For example, with two such lenses back to back, we can write the individual lens equations as

$$\frac{1}{v_1} - \frac{1}{u_1} = \frac{1}{f_1} \quad \text{and} \quad \frac{1}{v_2} - \frac{1}{u_2} = \frac{1}{f_2}$$

Since the image distance for the first lens plays the role of the object distance for the second lens, we may also write

$$u_2 = v_1$$

and, adding the two equations with this substitution made,

$$\frac{1}{v_2} - \frac{1}{u_1} = \frac{1}{f_1} + \frac{1}{f_2} = \frac{1}{f_{eq}}$$

The reciprocals of the individual focal lengths, therefore, add to give the reciprocal of the overall or equivalent focal length f_{eq} of the pair. In general, for several thin lenses in contact,

$$\frac{1}{f_{eq}} = \frac{1}{f_1} + \frac{1}{f_2} + \frac{1}{f_3} + \cdots \tag{3-39}$$

Expressed in diopters, the refractive powers simply add:

$$P = P_1 + P_2 + P_3 + \ldots \tag{3-40}$$

In a nearsighted person, the refractive (converging) power of the eye is too great, so that a real image is formed in front of the retina. By reducing the convergence with a number of diverging test lenses placed together in front of the eye until an object is clearly focused on the retina, an optometrist can determine the net diopter specification of the single corrective lens needed by simply adding the diopters of these test lenses. In a farsighted person, the natural converging power of the eye is not strong enough, and additional converging power must be added in the form of spectacles with a converging lens. These matters will be discussed at much greater length in chapters to come.

3-14 NEWTONIAN EQUATION FOR THE THIN LENS

When object and image distances are measured relative to the focal points of a lens, as by the distances x and x' in Figure 3-23, an alternative form of the thin lens equation results, called the *Newtonian* form. In the figure, the two rays shown determine two right angles, joined by the focal point, on each side of the lens.

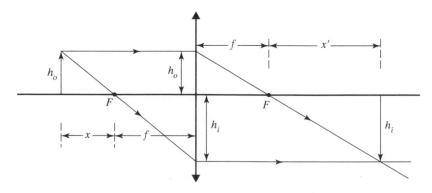

Figure 3-23 Construction used to derive Newton's equations for the thin lens.

Since each pair constitutes similar triangles, we may set up proportions between sides that represent the lateral magnification:

$$\frac{h_i}{h_o} = \frac{f}{x} \quad \text{and} \quad \frac{h_i}{h_o} = \frac{x'}{f}$$

Introducing a negative sign for the magnification, due to the inverted image, we can write

$$m = -\frac{f}{x} = -\frac{x'}{f} \tag{3-41}$$

Equating the last two parts of Eq. (3-41) and rearranging, we obtain the Newtonian form of the thin lens equation,

$$xx' = f^2 \tag{3-42}$$

This equation is somewhat simpler than Eq. (3-32) and is found to be more convenient in certain applications.

TABLE 3-1 SUMMARY OF PARAXIAL OPTICS EQUATIONS

	Spherical Surface		Plane Surface
	(Object-Image)	(Vergence)	
REFLECTION	$\dfrac{1}{u} + \dfrac{1}{v} = \dfrac{1}{f}$ $f = \dfrac{r}{2}$ $m = -\dfrac{v}{u}$ Concave: $r < 0, f < 0$ Convex: $r > 0, f > 0$	$V = U + P$ $P = -\dfrac{2}{r}$ $m = \dfrac{U}{V}$	$v = -u$ $m = +1$
REFRACTION (SINGLE SURFACE)	$\dfrac{n_2}{v} - \dfrac{n_1}{u} = \dfrac{n_2 - n_1}{r}$ $m = \dfrac{n_1 v}{n_2 u}$ Concave: $r < 0$ Convex: $r > 0$	$V = U + P$ $P = \dfrac{n_2 - n_1}{r}$ $m = \dfrac{U}{V}$	$v = \dfrac{n_2}{n_1} u$ $m = +1$
REFRACTION (THIN LENS)	$\dfrac{1}{v} - \dfrac{1}{u} = \dfrac{1}{f}$ $\dfrac{1}{f} = \dfrac{n_2 - n_1}{n_1}\left(\dfrac{1}{r_1} - \dfrac{1}{r_2}\right)$ $m = \dfrac{v}{u}$ Converging: f positive Diverging: f negative	$V = U + P$ $P = \dfrac{n_1}{f}$ $m = \dfrac{U}{V}$	

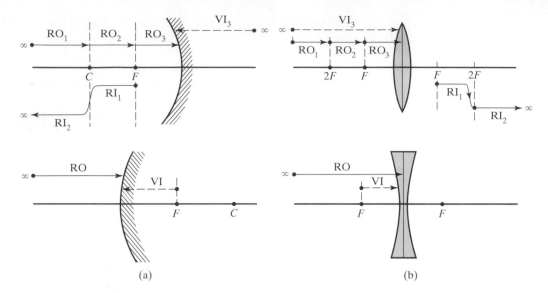

Figure 3-24 Summary of image formation by (a) spherical mirrors and (b) thin lenses. The location, nature, magnification, and orientation of the image are indicated or suggested. For example, as a real object RO_2 moves from the center of curvature to the focal point of a concave mirror, its conjugate image RI_2, real, inverted, and magnified, moves from the center of curvature to infinity. As a real object RO moves from infinity to a concave lens, its conjugate image VI, virtual, erect, and demagnified, moves from the focal point to the lens.

PROBLEMS

3-1. Derive an expression for the transit time of a ray of light that travels a distance x_1 through a medium of index n_1, a distance x_2 through a medium of index $n_2, \ldots,$ and a distance x_m through a medium of index n_m. Use a summation to express your result.

3-2. A double convex lens has a diameter of 5 cm and zero thickness at its edges. A point object on an axis through the center of the lens produces a real image on the opposite side. Both object and image distances are 30 cm, measured from a plane bisecting the lens. The lens has a refractive index of 1.52. Using the equivalence of optical paths through the center and edge of the lens, determine the thickness of the lens at its center.

3-3. Determine the minimum height of a wall mirror that will permit a 6-ft person to view his or her entire height. Sketch rays from the top and bottom of the person and determine the proper placement of the mirror such that the full image is seen, regardless of the person's distance from the mirror.

3-4. A ray of light makes an angle of incidence of 45° at the center of the top surface of a transparent cube of index 1.414. Trace the ray through the cube.

3-5. To determine the refractive index of a transparent plate of glass, a microscope is first focused on a tiny scratch in the upper surface, and the barrel position is recorded. Upon further lowering the microscope barrel by 1.87 mm, a focused image of the scratch is seen again. The plate thickness is 1.50 mm. What is the reason for the second image, and what is the refractive index of the glass?

3-6. A small source of light at the bottom face of a rectangular glass slab 2.25 cm thick is viewed from above. Rays of light totally reflected at the top surface outline a circle of 7.60 cm in diameter on the bottom surface. Determine the refractive index of the glass.

3-7. Show that the lateral displacement s of a ray of light penetrating a rectangular plate of thickness t is given by

$$s = \frac{t \sin (\theta_1 - \theta_2)}{\cos \theta_2}$$

where θ_1 and θ_2 are the angles of incidence and refraction, respectively. Find the displacement when $t = 3$ cm, $n = 1.50$, and $\theta_1 = 50°$.

3-8. A meter stick lies along the optical axis of a convex mirror of focal length 40 cm, with its nearer end 60 cm from the mirror face. How long is the image of the meter stick?

3-9. A glass hemisphere is silvered over its curved surface. A small bubble in the glass is located on the central axis through the hemisphere 5 cm from the plane surface. The radius of curvature of the spherical surface is 7.5 cm, and the glass has an index of 1.50. Looking along the axis into the plane surface, one sees two images of the bubble. How do they arise and where do they appear?

3-10. A concave mirror forms an image on a screen twice as large as the object. Both object and screen are then moved to produce an image on the screen that is three times the size of the object. If the screen is moved 75 cm in the process, how far is the object moved? What is the focal length of the mirror?

3-11. A sphere 5 cm in diameter has a small scratch on its surface. When the scratch is viewed through the glass from a position directly opposite, where does the scratch appear and what is its magnification? Assume $n = 1.50$ for the glass. Solve by both image location and vergence models.

3-12. **(a)** At what position in front of a spherical refracting surface must an object be placed so that the refraction produces parallel rays of light? In other words, what is the focal length of a single spherical refracting surface? Solve by both image location and vergence methods.

(b) Since real object distances are positive, what does your result imply for the cases $n_2 > n_1$ and $n_2 < n_1$?

3-13. A small goldfish is viewed through a spherical glass fishbowl 30 cm in diameter. Determine the apparent position and magnification of the fish's eye when its actual position is **(a)** at the center of the bowl, and **(b)** nearer to the eye, halfway from center to glass, along the line of sight. Assume that the glass is thin enough so that its effect on the refraction may be neglected.

3-14. A small object faces the convex spherical glass window of a small water tank. The radius of curvature of the window is 5 cm. The inner back side of the tank is a plane mirror, 25 cm from the window. If the object is 30 cm outside the window, determine the nature of its final image, neglecting any refraction due to the thin glass window itself.

3-15. A plano-convex lens having a focal length of 25.0 cm is to be made with glass of refractive index 1.520. Calculate the radius of curvature of the grinding and polishing tools to be used in making this lens.

3-16. Calculate the focal length of a thin, meniscus lens whose spherical surfaces have radii of curvature of magnitudes 5 and 10 cm. The glass is of index 1.50. Sketch both positive and negative versions of the lens.

3-17. One side of a fish tank is built using a large-aperture thin lens made of glass ($n = 1.50$). The lens is equiconvex, with radii of curvature 30 cm. A small fish in the tank is 20 cm from the lens. Where does the fish appear when viewed through the lens? What is its magnification?

3-18. Two thin lenses have focal lengths of -5 and $+20$ cm. Determine their equivalent focal lengths when (a) cemented together and (b) separated by 10 cm.

3-19. Two identical, thin, plano-convex lenses with radii of curvature of 15 cm are situated with their curved surfaces in contact at their centers. The intervening space is filled with oil of refractive index 1.65. The index of the glass is 1.50. Determine the focal length of the combination. (*Hint:* Think of the oil layer as an intermediate thin lens.)

3-20. An eyepiece is made of two thin lenses, each of $+20$-mm focal length, separated by a distance of 16 mm.

(a) Where must a small object be positioned so that light from the object is rendered parallel by the combination?

(b) Does the eye see an image erect relative to the object? Is it magnified? Use a ray diagram to answer these questions by inspection.

3-21. A diverging thin lens and a concave mirror have focal lengths of equal magnitude. An object is placed $3f/2$ from the diverging lens, and the mirror is placed a distance $3f$ on the other side of the lens. Using Gaussian optics, determine the *final* image of the system after two refractions **(a)** by a three-ray diagram, and **(b)** by calculation.

3-22. A small object is placed 20 cm from the first of a train of three lenses with focal lengths, in order, of 10, 15, and 20 cm. The first two lenses are separated by 30 cm and the last two by 20 cm. Calculate the final image position relative to the last lens and its linear magnification relative to the original object when

    (a) all three lenses are positive.

    (b) the middle lens is negative.

    (c) the first and last lenses are negative.

    Provide diagrams for each case.

3-23. A convex thin lens with refractive index of 1.50 has a focal length of 30 cm in air. When immersed in a certain transparent liquid, it becomes a negative lens with a focal length of 188 cm. Determine the refractive index of the liquid.

3-24. It is desired to project onto a screen an image that is four times the size of a brightly illuminated object. A plano-convex lens with $n = 1.50$ and $R = 60$ cm is to be used. Employing the Newtonian form of the lens equations, determine the appropriate distance of the object and screen from the lens. Is the image erect or inverted? Check your results using the ordinary lens equations.

3-25. Three lenses of focal lengths 10 cm, 20 cm, and -40 cm are placed in contact to form a single, compound lens.

    (a) Determine the powers of the individual lenses and that of the unit, in diopters.

    (b) Determine the vergence of an object point 12 cm from the unit and that of the resulting image. Convert the result to an image distance in centimeters.

3-26. **(a)** A point source is in air. What is the vergence of the light 6 cm from the point source?

    (b) How does this change if the point source is under water ($n = 1.33$)?

    (c) How far is a diverging wavefront of -8.00 D from the point object in glass ($n = 1.50$)?

3-27. A concave mirror has a radius of 12 cm.

    (a) What is the power of the mirror?

    (b) A small object is situated on axis 5 cm from the mirror. What is the vergence of rays originating at the object after reflection from the mirror?

    (c) Describe the final image.

3-28. A thin lens of power $+5.0$ D is placed in the path of a converging pencil of rays of $+3.00$ D vergence. Find the image position.

3-29. A lens is moved along the optical axis between a fixed object and a fixed image screen. The object and image positions are separated by a distance L that is more than four times the focal length of the lens. Two positions of the lens are found for which an image is in focus on the screen, magnified in one case and reduced in the other. If the two lens positions differ by distance D, show that the focal length of the lens is given by

$$f = \frac{(L^2 - D^2)}{4L}$$

This is *Bessel's method* for finding the focal length of a lens.

3-30. An image of an object is formed on a screen by a lens. Leaving the lens fixed, the object is moved to a new position and the image screen moved until it again receives a focused image. If the two object positions are u_1 and u_2 and if the transverse magnifications of the image are m_1 and m_2, respectively, show that the focal length of the lens is given by

$$f = \frac{(u_1 - u_2)}{(1/m_1 - 1/m_2)}$$

This is *Abbe's method* for finding the focal length of a lens.

3-31. Determine the ratio of focal lengths for two identical, thin plano-convex lenses when one is silvered on its flat side and the other on its curved side. Light is incident on the unsilvered side.

3-32. A ray of light traverses successively a series of plane interfaces, all parallel to one another and separating regions of differing thickness and refractive index.

   (a) Show that Snell's law holds between the first and last regions, as if the intervening regions did not exist.

   (b) Calculate the net lateral displacement of the ray from point of incidence to point of emergence.

3-33. A parallel beam of light is incident on a plano-convex lens that is 4 cm thick. The radius of curvature of the spherical side is also 4 cm. The lens has a refractive index of 1.50 and is used in air. Determine where the light is focused for light incident on each side.

3-34. A spherical interface, with radius of curvature 10 cm, separates media of refractive index 1 and 4/3. The center of curvature is located on the side of the higher index. Find the focal lengths for light incident from each side. How do the results differ when the two refractive indices are interchanged?

3-35. An airplane is used in aerial surveying to make a map of ground detail. If the scale of the map is to be 1:50,000 and the camera used has a focal length of 6 in., determine the proper altitude for the photograph.

REFERENCES

BENNETT, ARTHUR G. 1984. *Clinical Visual Optics.* Boston: Butterworth-Heinemann.

BYER, ALVIN. 1986. *Theoretical Optics for Clinicians.* Philadelphia: Pennsylvania College of Optometry.

FEYNMAN, RICHARD P., ROBERT B. LEIGHTON, and MATTHEW SANDS. 1975. *The Feynman Lectures on Physics*, Vol. 1. Reading, Mass.: Addison-Wesley Publishing Co. Ch. 26, 27.

HEAVENS, O. S., and R. W. DITCHBURN. *Insight into Optics.* New York: John Wiley & Sons.

KEATING, MICHAEL P. 1988. *Geometric, Physical and Visual Optics.* Boston: Butterworth-Heinemann.

LONGHURST, R. S. 1967. *Geometrical and Physical Optics*, 2d ed. New York: John Wiley and Sons. Ch. 1, 2.

ROSSI, BRUNO. 1957. *Optics.* Reading, Mass.: Addison-Wesley Publishing Company. Ch. 1, 2.

SMITH, F. Dow. 1968. "How Images are Formed." In *Lasers and Light. Readings from Scientific American*, pp. 59–70. San Francisco: W. H. Freeman and Company Publishers.

TUNNACLIFFE, ALAN. 1989. *Introduction to Visual Optics.* New York: State Mutual Book & Periodical Service, Limited.

WALDMAN, GARY. 1983. *Introduction to Light.* Englewood Cliffs, N.J.: Prentice Hall.

WELFORD, W. T. 1981. *Optics*, 2d ed. Oxford University Press.

WINSTON, ROLAND. 1991. "Nonimaging Optics." *Scientific American* (March): 76.

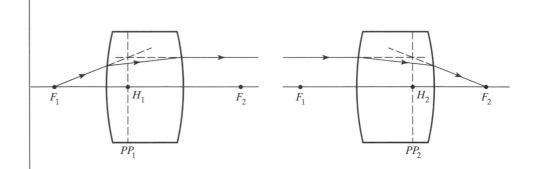

4

The Thick Lens

INTRODUCTION

This chapter describes methods of analyzing complex optical systems involving thick lenses and/or a number of reflecting and refracting elements. Of course, it is possible to use the equations developed in the preceding chapters for refraction and reflection in a stepwise fashion. As pointed out previously, each image becomes the object for the next step and the new image is calculated. It is most useful, in dealing with such systems, to characterize the thick lens or optical system in terms of its *cardinal points*. When applied to a single, thick lens, cardinal points provide a way of interpreting the effects of refraction in which the thickness of the lens is not negligible and so enters into the computations. The definition and application of cardinal points for a thick lens is discussed in this chapter. The use of cardinal points to analyze the behavior of more complex systems, however, is most easily accomplished using matrix methods and is not treated here.[1] It should be remembered that all these methods involve the paraxial approximation. When exact results are needed, it is possible to trace a ray of light through an optical system precisely, step by step. One such method, well adapted to computer programming, is presented in this chapter.

---

[1]A treatment using 2×2 matrices can be found in Chapter 4 of *Introduction to Optics*, 2nd edition, Frank L. & Leno S. Pedrotti, Prentice Hall, 1993.

4-1 THE THICK LENS

Consider a spherical *thick lens*, that is, a lens whose thickness along its optical axis cannot be ignored without leading to serious errors in analysis. Just when a lens moves from the category of *thin* to *thick* clearly depends on the accuracy required. The thick lens can be treated by the methods of Chapter 3. The glass medium is bounded by two spherical refracting surfaces. The image of a given object, formed by refraction at the first surface, becomes the object for refraction at the second surface. The object distance for the second surface takes into account the thickness of the lens. The image formed by the second surface is then the final image due to the action of the composite thick lens.

The thick lens can also be described in a way that allows graphical determination of images corresponding to arbitrary objects, much like the ray rules for a thin lens. This description, in terms of the *cardinal points* of the lens, is useful also because it can be applied to more complex optical systems. An optical system consisting of several reflectors and/or refractors is characterized by one set of cardinal points. Thus, even though we are at present interested in a single thick lens, the following description is applicable—once the positions of the cardinal points are known—to an arbitrary optical system that we can imagine to be contained within the outlines of the thick lens.

There are six cardinal points $(F_1, F_2, H_1, H_2, N_1, N_2)$ on the axis of a thick lens from which its imaging properties can be deduced. See Figures 4-1 and 4-2. Planes[2] normal to the axis at these points are called the *cardinal planes*. The six cardinal points consist of the first and second *system focal points* (F_1 and F_2), which are already familiar; the first and second *principal points* (H_1 and H_2); and the first and second *nodal points* (N_1 and N_2).

A ray entering the thick lens from the first focal point F_1 emerges parallel to the axis (Figure 4-1a), and a ray parallel to the axis as it enters the lens is refracted by the lens through the second focal point F_2 (Figure 4-1b). The extensions of the incident and emerging rays in each case intersect, by definition, in the *principal planes*, and these cross the axis at the principal points, H_1 and H_2. If the thick lens were a single thin lens, the two principal planes would coincide at the vertical line that is usually drawn to represent the lens. Principal planes, in general, do not coincide and may even be located outside the optical system itself. Once the locations of the principal planes are known, accurate ray diagrams

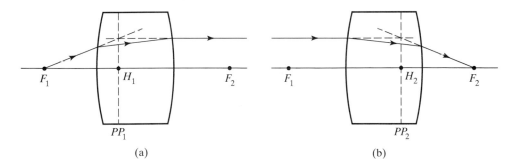

(a)   (b)

Figure 4-1 Illustration showing focal points F_1, F_2, principal points H_1, H_2, and principal planes PP_1, PP_2, for an optical system.

[2]These "planes" are actually slightly curved surfaces that can be considered plane in the paraxial approximation.

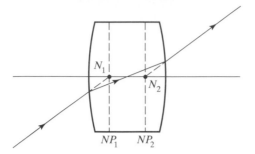

Figure 4-2 Illustration of the nodal points N_1 and N_2 and the nodal planes NP_1 and NP_2 for an optical system.

can be drawn. The two rays associated with the focal points, as used in the thin-lens ray diagrams, are represented here as bending at their intersections with corresponding principal planes (dashed lines in Figures 4-1a and 4-1b). Though the dashed lines are not the actual path of the ray *inside* the lens (solid line), this construction produces the correct exit ray. The third ray usually drawn for thin-lens diagrams is one through the lens center, undeviated and negligibly displaced. The nodal points of a thick lens, or of any optical system, permit the correction to this ray, as shown in Figure 4-2. Any ray directed toward the first nodal point N_1 emerges from the optical system parallel to the incident ray, but is displaced so that it appears to come from the second nodal point on the axis, N_2.

The positions of all six cardinal points are indicated in Figure 4-3. Distances are *directed*, positive or negative, by our usual sign convention that makes distances directed to the left negative and directed to the right, positive. Notice that for the thick lens, the distances a and b determine the positions of the principal points H_1 and H_2 relative to the vertices V_1 and V_2, while f_1 and f_2 determine the focal point positions F_1 and F_2 relative to the principal points H_1 and H_2. Note carefully that these focal lengths are *not* measured from the vertices of the lens. Figure 4-4 represents an arbitrary optical system of any number of reflectors and refractors and indicates the system's cardinal points. Such an optical system might be the complete human eye, for example, or the multielement lens of a quality camera. In any case, the optical system included between the input plane and the output plane can be characterized by one set of cardinal points. Shown are the three rays that can always be drawn. If the three rays originate at a given object point, the final outcome of these rays, in direction and position, establishes the conjugate image point.

We summarize the basic equations for the thick lens without proof. These derivations involve combining the refraction equations for the two surfaces of

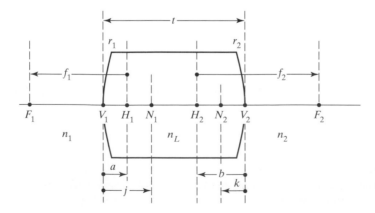

Figure 4-3 Symbols used to signify the cardinal points and locations for a thick lens. Axial points include focal points (F), vertices (V), principal points (H), and nodal points (N). Directed distances separating their corresponding planes are defined in the drawing.

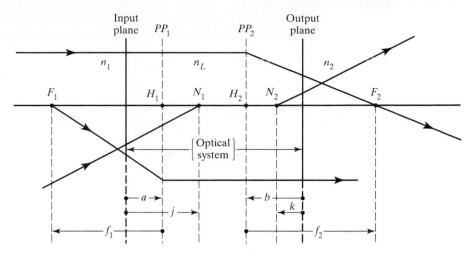

Figure 4-4 Location designations for the six cardinal points of an optical system. Rays associated with the focal points, nodal points, and principal planes are also shown.

the lens and require algebraic manipulations that are straightforward, yet rather tedious. Utilizing the symbols defined in Figure 4-3 or Figure 4-4, the focal length f_1 is given by

$$\frac{1}{f_1} = \frac{n_L - n_2}{n_1 r_2} - \frac{n_L - n_1}{n_1 r_1} - \frac{(n_L - n_1)(n_L - n_2)}{n_1 n_L} \frac{t}{r_1 r_2} \tag{4-1}$$

and the focal length f_2 is conveniently expressed in terms of f_1 by

$$f_2 = -\frac{n_2}{n_1} f_1 \tag{4-2}$$

Notice that the two focal lengths have the same magnitude if the lens is surrounded by a single refractive medium, so that $n_1 = n_2$. The principal planes can be located next, using

$$a = \frac{n_L - n_2}{n_L r_2} f_1 t \quad \text{and} \quad b = -\frac{n_L - n_1}{n_L r_1} f_2 t \tag{4-3}$$

The positions of the nodal points are given by

$$j = \left(1 - \frac{n_2}{n_1} + \frac{n_L - n_2}{n_L r_2} t\right) f_1 \quad \text{and} \quad k = \left(1 - \frac{n_1}{n_2} - \frac{n_L - n_1}{n_L r_1} t\right) f_2 \tag{4-4}$$

Image and object distances u and v and lateral magnification m are related by

$$-\frac{f_1}{u} + \frac{f_2}{v} = 1 \quad \text{and} \quad m = -\frac{n_1 v}{n_2 u} \tag{4-5}$$

as long as the distances u and v, as well as the focal lengths f_1 and f_2, are measured respectively from the corresponding principal points H_1 and H_2. In the ordinary case of a lens in air, with $n_1 = n_2 = 1$, notice that $a = j$ and $b = k$: First and second principal points are superimposed over corresponding nodal points. Also, first and second focal lengths are equal in magnitude, and the equations

$$-\frac{1}{u} + \frac{1}{v} = \frac{1}{f} \quad \text{and} \quad m = \frac{v}{u} \tag{4-6}$$

are valid, provided symbols are properly reinterpreted. These are identical with the form of the thin-lens equations and similarly require positive f for a convex lens and negative f for a concave lens.

❏ Example

Determine the focal lengths f_1 and f_2 and the location of the principal points H_1 and H_2 for a 4-cm thick, biconvex lens with refractive index of 1.52 and radii of curvature of 25 cm, when the lens caps the end of a long cylinder filled with water ($n = 1.33$). See Figure 4-5.

Solution   Insert numerical values into Eqs. (4-1) to (4-4) for the thick lens in the order given:

$$\frac{1}{f_1} = \frac{1.52 - 1.33}{1(-25)} - \frac{1.52 - 1}{1(+25)} - \frac{(1.52 - 1)(1.52 - 1.33)}{1(1.52)} \frac{4}{(-25)(+25)}$$

or $f_1 = -35.74$ cm to the *left* of the first principal plane. Then

$$f_2 = -\left(\frac{1.33}{1}\right)(-35.74) = 47.53 \text{ cm}$$

to the *right* of the second principal plane, and

$$a = \frac{1.52 - 1.33}{(1.52)(-25)}(-35.74)(4) = 0.715 \text{ cm}$$

$$b = -\frac{1.52 - 1}{(1.52)(+25)}(47.53)(4) = -2.60 \text{ cm}$$

Thus the principal point H_1 is situated 0.715 cm to the *right* of the left vertex of the lens, and H_2 is situated 2.60 cm to the *left* of the right vertex.

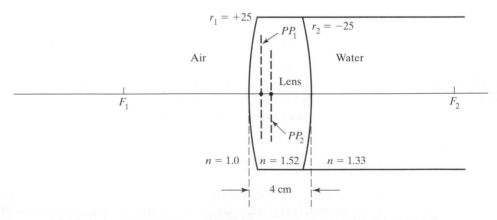

Figure 4-5   Thick biconvex lens surrounded by air on the left and water on the right.

The nodal points are calculated from

$$j = \left[1 - \frac{1.33}{1} + \frac{1.52 - 1.33}{1.52(-25)}4\right](-35.74) = 12.51 \text{ cm}$$

$$k = \left[1 - \frac{1}{1.33} - \frac{1.52 - 1}{1.52(25)}4\right](47.53) = 9.19 \text{ cm}$$

Thus the first nodal point N_1 occurs at a distance of 12.51 cm to the *right* of the left vertex of the lens, and the second nodal point N_2 is situated at a distance of 9.19 cm to the *right* of the right vertex. ∎

In the following examples, we make use of these equations to find the cardinal points for several thick lenses. We also use ray diagrams to find the image location and size of a given object.

___

❏ **Example**

(a) Find the cardinal points and sketch a ray diagram for the hemispherical glass lens shown in Figure 4-6. The radii of curvature are $r_1 = 3$ cm and $r_2 \to \infty$, and the lens in air has a refractive index of 1.50.

(b) Sketch a ray diagram to locate the image of a 1-cm tall object located 3 cm to the left of the first principal point for this lens.

Solution

(a) Calculate the primary and secondary focal points, using $r_1 = 3$ cm, $r_2 \to \infty$, $n_L = 1.50, n_1 = n_2 = 1$. Notice that the thickness of the hemispherical lens is identical to its radius, or $t = 3$ cm. Treating the infinity symbol ∞ as an arbitrarily large number, we substitute into Eqs. (4-1) through (4-4), as follows:

$$\frac{1}{f_1} = \frac{1.50 - 1}{(1)(\infty)} - \frac{1.50 - 1}{(1)(3)} - \frac{(1.50 - 1)(1.50 - 1)}{(1)(1.50)}\frac{3}{3(\infty)}$$

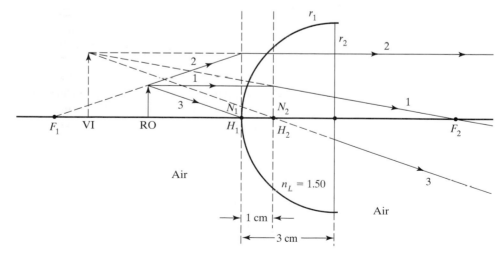

Figure 4-6 Ray construction for a hemispherical lens, using cardinal points.

The two terms with ∞ in the denominator vanish, and

$$\frac{1}{f_1} = -\frac{1}{6} \quad \text{or} \quad f_1 = -6 \text{ cm}$$

Then

$$f_2 = -\frac{1}{1}(-6) = +6 \text{ cm}$$

The principal points are at distances a and b from the vertices, given by

$$a = \frac{1.50 - 1}{1.50(\infty)}(-6)(3) = 0 \quad \text{and} \quad b = -\frac{1.50 - 1}{1.50(3)}(6)(3) = -2 \text{ cm}$$

Since the lens is in air, the nodal points coincide with the principal points, that is, $j = a = 0$ and $k = b = -2$ cm. The lens and the calculated cardinal points are drawn to scale in Figure 4-6.

(b) The ray diagrams for locating the image of object RO, using the principal planes and nodal points, are drawn as follows: Ray 1 from the top of the object is directed parallel to the optical axis and bends at the second principal plane so as to pass through the second focal point F_2. Ray 2 from the top of the object is in line with F_1, bends at the first principal plane, and emerges parallel to the axis. Ray 3 is directed toward the first nodal point N_1, and after refraction maintains its original direction, but is displaced as if originating at the second nodal point, N_2. In this case, the emerging rays 1, 2, and 3 diverge from the lens and locate the image to the left of the lens as shown: virtual, erect, and slightly magnified.

Now, let us also use the Eqs. (4-6), which have the appearance of the thin-lens equations, to solve for the location of the image just determined by the ray diagrams. For the thick lens, Eqs. (4-6) require that object, image, and focal points be located relative to the corresponding principal points rather than to the front and back surfaces of the lens. Accordingly, the 1-cm object shown in Figure 4-6 is located 3 cm to the left of the first principal plane at H_1. Note that the first principal plane is tangent to the front surface of the lens at its vertex. Then, according to our sign convention, $u = -3$ cm and, substituting into Eqs. (4-6), we have

$$\frac{1}{v} - \frac{1}{(-3)} = \frac{1}{6} \quad \text{or} \quad v = -6 \text{ cm}$$

Thus, the virtual image is located 6 cm to the left of the second principal plane, in agreement with the results of the ray diagrams. The magnification is given by

$$m = \frac{v}{u} = \frac{-6}{-3} = +2.0$$

The image is erect (because m is positive) and twice the size of the object, again in agreement with the ray diagram in Figure 4-6. In this example we have seen that ray diagrams using the principal planes and nodal points can determine the position and nature of the image VI. Further, if the thick lens is in air, we have seen also that the thin-lens-like Eqs. 4-6 can be used to calculate the image size and location in perfect agreement with the ray diagrams. ∎

❏ **Example**

Find the cardinal points and sketch a ray diagram for the thick lens shown in Figure 4-7. The radii of curvature are $r_1 = -10$ cm and $r_2 = 25$ cm and its thickness on axis is 5 cm. The lens is situated in air and has a refractive index of 1.60.

Solution   One way to find the focal length of this lens, which locates cardinal point F_2, would be to use the two-step object-image approach, finding the final image point for an initial object at infinity. By definition, this locates the focal point F_2, measured from the second (final) surface. As usual in this approach, the image position for refraction at the first surface becomes the object position for refraction at the second surface.

Another way to find the focal length is to use the method of vergences to determine the conjugate image of some specified object. The object vergence is first modified by the power of the first refracting surface ($V = U + P$). The image vergence so produced is not immediately modified by the power of the second refracting surface, however, because the thickness of the lens is not negligible. In passing from the first to the second surface, the vergence of the wavefront changes. Such change in vergence, due to pure translation, is analyzed in Chapter 5, where the appropriate formula is derived. Once the new vergence is found, the effect of the second surface can be calculated, as usual, to complete the analysis.

At this point, however, we choose to solve the problem using Eqs. (4-1) through (4-4), which also have the advantage of yielding all the cardinal points for the thick lens. Equation (4-1) for the primary focal length can be simplified somewhat since for this case, $n_1 = n_2 = 1$. The result is

$$\frac{1}{f_1} = (n_L - 1)\left(\frac{1}{r_2} - \frac{1}{r_1}\right) - \frac{(n_L - 1)^2}{n_L}\frac{t}{r_1 r_2}$$

Notice that for a thin lens, $t \cong 0$; this causes the second term on the right to vanish, leaving the first term as the standard thin-lens formula for the primary focal length. The correction term—the second term—is proportional to the lens

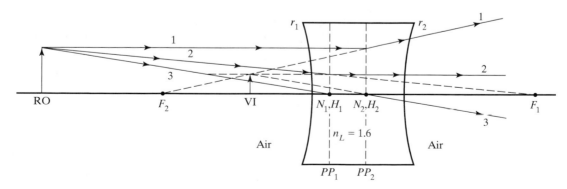

Figure 4-7   Cardinal points and ray construction for a thick lens.

thickness and inversely proportional to the product of the radii of curvature of the two lens surfaces. Substituting numerical values, we have

$$\frac{1}{f_1} = (1.60 - 1)\left(\frac{1}{25} - \frac{1}{-10}\right) - \frac{(1.60 - 1)^2}{1.60} \frac{5}{(-10)(25)}$$

from which we calculate $f_1 = 11.30$ cm. Then $f_2 = -f_1 = -11.30$ cm. The principal and nodal points are found from Eqs. (4-3):

$$a = j = \frac{1.60 - 1}{1.6(25)}(11.30)(5) = +0.848 \text{ cm}$$

and

$$b = k = -\frac{1.60 - 1}{1.6(-10)}(-11.30)(5) = -2.12 \text{ cm}$$

The cardinal points are shown, approximately to scale, in Figure 4-7, together with ray diagrams for an arbitrary object. The conjugate image is located as shown—virtual, erect, and reduced in size. ∎

4-2 RAY TRACING

The assumption of paraxial rays greatly simplifies the description of the progress of light rays through an optical system, because trigonometric terms do not appear in the equations. For many purposes, this treatment is sufficient. In practice, rays of light contributing to an image in an optical system are, in fact, usually rays in the near neighborhood of the optical axis. If the quality of the image is to be improved, however, ways must be found to reduce the ever-present *aberrations* that arise from the presence of rays deviating, more or less, from this ideal assumption. To determine the actual path of individual rays of light through an optical system, each ray must be traced independently, using only the laws of reflection and refraction, together with geometry. This technique is called *ray tracing* because it was formerly done by hand, graphically, with ruler and compass, in a step-by-step process through an accurately scaled sketch of the optical system. Today, with the help of computers, the necessary calculations yielding the progressive changes in a ray's altitude and angle is done more easily and quickly. Graphics techniques are used to actually draw the optical system and to trace the ray's progress through the optical system on the monitor.[3]

Ray-tracing procedures, such as the one to be described here, are often limited to *meridional rays*, that is, rays that pass through the optical axis of the system. Since the law of refraction requires that refracted rays remain in the plane of incidence, a meridional ray remains within the same meridional plane throughout its trajectory. Thus the treatment in terms of meridional rays is a two-dimensional treatment,[4] greatly simplifying the geometrical relationships required. Rays contributing to the image that do not pass through the optical axis are called *skew rays* and require three-dimensional geometry in their calculations. The added complexity is no problem for the computer, once the ray-trac-

---

[3]An example of commercial software available, well-adapted for academic purposes, and inexpensive, is BEAM2, a product of Stellar Software, Berkeley, California. It is IBM-PC, XT, AT, and PS/2 compatible and supports CGA, EGA, VGA, and Hercules graphics adapters.

[4]The two dimensions are those of the page on which we have been drawing our ray diagrams. Without emphasizing this, we have been using meridional rays in all diagrams.

ing program is written. Analysis of various aberrations, such as spherical aberration, astigmatism, and coma (described in the following chapter), require knowledge of the progress of selected nonparaxial rays and skew rays. The design of a complex lens system, such as a photographic lens with four or five elements, is a combination of science and skill. By alternating ray tracing with small changes in the positions, focal lengths, and curvatures of the surfaces involved and in refractive indices of the elements, the design of the lens system is gradually optimized.

For our present purposes, it will be sufficient to show how the appropriate equations for meridional ray tracing can be developed and how they can be repeated in stepwise fashion to follow a ray through any number of spherical refracting surfaces that constitute an optical system. The technique is well adapted to reiterative loops handled by computer programs.

Figure 4-8 shows a single, representative step in the ray-tracing analysis. By incorporating a sign convention, the equations developed from this diagram can be made to apply to any ray and to any spherical surface. The ray selected originates (or passes through) point A, making an angle α with the optical axis. The ray passes through the optical axis at O and then intersects the refracting surface at P, where it is refracted into a medium of index n', cutting the axis again at I. The angles of incidence θ and refraction θ' are related by Snell's law. Points O and I are conjugate points with distances u and v from the surface vertex at V. The radius of curvature of the surface is also shown, passing through the center of curvature C. Other points and lines are added to help in developing the necessary geometrical relationships.

The sign convention is the same as that used previously in this chapter. Distances to the left of the vertex V are negative and to the right, positive. If we use light rays progressing from left to right, their angles of inclination have the same sign as their slopes. Distances measured above the axis are positive and below,

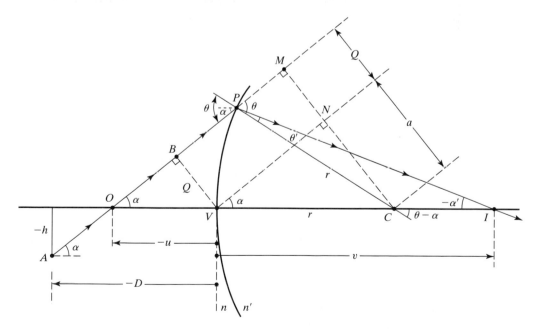

Figure 4-8 Single refraction at a spherical surface. The figure defines the symbols and shows the geometrical relationships that lead to ray-tracing equations for a meridional ray.

negative. An important quantity in the calculations, also subject to this sign convention, is the parameter Q, the perpendicular distance from the vertex to the ray, as shown.

The input parameters for the ray are assumed to be its elevation h, angle α, and distance D. Figure 4-8 shows that the magnitude of the object distance, u, is related to D by

$$u = D - \frac{h}{\tan \alpha} \tag{4-7}$$

Also, in ΔOBV:

$$\sin \alpha = \frac{Q}{-u} \tag{4-8}$$

In ΔPMC:

$$\sin \theta = \frac{a + Q}{r}$$

In ΔVNC:

$$\sin \alpha = \frac{a}{r}$$

Eliminating the length a from the last two equations, we get

$$\sin \theta = \frac{Q}{r} + \sin \alpha \tag{4-9}$$

Snell's law at P gives:

$$n \sin \theta = n' \sin \theta' \tag{4-10}$$

In ΔCPI:

$$\theta - \alpha = \theta' - \alpha' \tag{4-11}$$

The Q parameter for the refracted ray, designated Q', is shown in Figure 4-9a. Analogous to the relations just found, we see that in ΔCMV:

$$\sin (-\alpha') = \frac{a'}{r}$$

In ΔPLC:

$$\sin \theta' = \frac{Q' - a'}{r}$$

As before, when a' is eliminated, there results

$$Q' = r(\sin \theta' - \sin \alpha') \tag{4-12}$$

In ΔITV:

$$\sin (-\alpha') = \frac{Q'}{v} \qquad \text{or} \qquad v = \frac{-Q'}{\sin \alpha'} \tag{4-13}$$

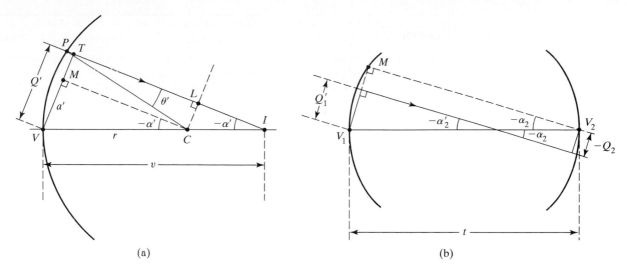

(a)   (b)

Figure 4-9 (a) Geometrical relationship of refracted-ray parameters with the distance Q. (b) Geometrical relationships illustrating the transfer between Q and α after one refraction and before the next.

The relevant equations describing the first refraction are included in Table 4-1 under the first column for the general case. The calculations lead to new values of α, Q, and u, which prepare for the next refraction in the sequence. The geometrical *transfer* to the next surface, at distance t from the first, is shown in Figure 4-9b, where, in $\Delta V_2 M V_1$,

$$\sin(-\alpha_2) = \frac{V_1 M}{t} = \frac{Q_1' - Q_2}{t}$$

or

$$Q_2 = Q_1' + t \sin \alpha_2 \tag{4-14}$$

TABLE 4-1   MERIDIONAL RAY-TRACING EQUATIONS (INPUT: n, n', r, α, h, D)

General case	Ray ∥ axis: $\alpha = 0$	Plane surface: $r \Rightarrow \infty$
$u = D - \dfrac{h}{\tan\theta}$	—	$u = D - \dfrac{h}{\tan\theta}$
$Q = -u\sin\alpha$	$Q = h$	$Q = -u\sin\alpha$
$\theta = \sin^{-1}\left(\dfrac{Q}{r} + \sin\alpha\right)$	$\theta = \sin^{-1}\left(\dfrac{Q}{r} + \sin\alpha\right)$	—
$\theta' = \sin^{-1}\left(\dfrac{n\sin\theta}{n'}\right)$	$\theta' = \sin^{-1}\left(\dfrac{n\sin\theta}{n'}\right)$	—
$\alpha' = \theta' - \theta + \alpha$	$\alpha' = \theta' - \theta + \alpha$	$\alpha' = \sin^{-1}\left(\dfrac{n}{n'\sin\alpha}\right)$
$Q' = r(\sin\theta' - \sin\alpha')$	$Q' = r(\sin\theta' - \sin\alpha')$	$Q' = Q\dfrac{\cos\alpha'}{\cos\alpha}$
$v = \dfrac{-Q'}{\sin\alpha'}$	$v = \dfrac{-Q'}{\sin\alpha'}$	$v = \dfrac{-Q'}{\sin\alpha'}$

Transfer: Input: t $n = n'$
$Q = Q' + t\sin\alpha'$ *Input:* new n', r
$\alpha = \alpha'$ *Return:* to calculate θ

❏ **Example**

Do a ray trace for two rays through a *Rapid landscape* photographic lens of three elements. The rays enter the lens from a distant object at altitudes of 1 and 5 mm above the optical axis. The lens specifications (all dimensions are in mm) are as follows.

$$R_1 = -120.8$$

$$t_1 = 6 \qquad n_1 = 1.521$$

$$R_2 = -34.6$$

$$t_2 = 2 \qquad n_2 = 1.581$$

$$R_3 = -96.2$$

$$t_3 = 3 \qquad n_3 = 1.514$$

$$R_4 = -51.2$$

Solution Since the rays are parallel to the axis, the second column of Table 4-1 is used to calculate the progress of the ray. These calculations can be tabulated as follows.

Input	Results: Ray at $h = 1$	Results: Ray at $h = 5$
First surface:		
$n = 1, n_1 = 1.521$	$Q = 1$	$Q = 5$
$\alpha = 0$	$\alpha' = 0.1625°$	$\alpha' = 0.8128°$
$h = 1$ or 5	$v = -352.66$	$v = -352.53$
$R = -120.8$	$Q = 1.0000$	$Q' = 5.0010$
Second surface:		
$t = 6$	$Q = 1.0170$	$Q = 5.0861$
$n_2 = 1.581$	$\alpha' = 0.2202°$	$\alpha' = 1.1041°$
$R = -34.6$	$v = -264.59$	$v = -264.03$
	$Q' = 1.0170$	$Q' = 5.0876$
Third surface:		
$t = 2$	$Q = 1.0247$	$Q = 5.1261$
$n_3 = 1.514$	$\alpha' = 0.2030°$	$\alpha' = 1.0178°$
$R = -96.2$	$v = -289.26$	$v = -288.58$
	$Q' = 1.0247$	$Q' = 5.1260$
Final surface:		
$t = 3$	$Q = 1.0353$	$Q = 5.1793$
$n = 1$	$\alpha' = -0.2883°$	$\alpha' = -1.4520°$
$R = -51.2$	$v = 205.72$	$v = 203.91$
	$Q' = 1.0353$	$Q' = 5.1672$

Thus the two parallel rays intersect the optical axis at 205.72 and 203.91 mm beyond the final surface, missing a common focus by 1.8 mm. ∎

PROBLEMS

4-1. A biconvex lens of 5 cm thickness and index 1.60 has surfaces of radius 40 cm. If this lens is used for objects in water, with air on its opposite side, determine its effective focal length and principal points.

4-2. A double concave lens of glass with $n = 1.53$ has surfaces of 5 D (diopters) and 8 D, respectively. The lens is used in air and has an axial thickness of 3 cm.

   (a) Determine the position of its focal and principal planes.

   (b) Also find the position of the image, relative to the lens center, corresponding to an object at 30 cm in front of the first lens vertex.

(c) Calculate the paraxial image distance assuming the thin-lens approximation. What is the percent error involved?

4-3. A biconcave lens has radii of curvature of 20 cm and 10 cm. Its refractive index is 1.50 and its central thickness is 5 cm. Describe the image of a 1-in tall object, situated 8 cm from the first vertex.

4-4. An equiconvex lens having spherical surfaces of radius 10 cm, a central thickness of 2 cm, and a refractive index of 1.61 is situated between air and water ($n = 1.33$). An object 5 cm high is placed 60 cm in front of the lens surface. Find the cardinal points for the lens and the position and size of the image formed.

4-5. A hollow, glass sphere of radius 10 cm is filled with water. Refraction due to the thin glass walls is negligible for paraxial rays.

   (a) Determine its cardinal points and make a sketch to scale.

   (b) Calculate the position and magnification of a small object 20 cm from the sphere.

   (c) Verify your analytical results by drawing appropriate rays on your sketch.

4-6. A lens has the following specifications. $r_1 = +15$ cm $= r_2$, $n = 1.00$, $n_L = 1.60$, $n' = 1.30$, t (thickness) $= 2.0$ cm. Find the principal points. Include a sketch, roughly to scale, and do a ray diagram for a finite object of your choice.

4-7. A glass lens 3 cm thick along the axis has one convex face of radius 5 cm and the other, also convex, of radius 2 cm. The former face is on the left in contact with air and the other in contact with a liquid of index 1.40. The refractive index of the glass is 1.50. Find the positions of the foci, principal planes, and nodal points of the system.

4-8. A gypsy's crystal ball has a refractive index of 1.50 and a diameter of 8 in.

   (a) Determine the location of its cardinal points.

   (b) Where will sunlight be focused by the crystal ball?

4-9. A thick lens in air presents two concave surfaces, each of radius 5 cm, to incident light. The lens is 1 cm thick and has a refractive index of 1.50.

   (a) Find its cardinal points.

   (b) Do a ray diagram for some object.

REFERENCES

ATNEOSEN, RICHARD, and RICHARD FEINBERG. 1991. "Learning Optics with Optical Design Software." *American Journal of Physics* 59 (March): 242–47.

FINCHAM, W. H. A., and M. H. FREEMAN. 1980. *Optics*, 9th ed. Boston: Butterworth Publishers. Ch. 8, 9, 19.

KINGSLAKE, RUDOLF. 1978. *Lens Design Fundamentals*. New York: Academic Press. Ch. 2, 3, 7.

NUSSBAUM, ALLEN. 1968. *Geometric Optics: An Introduction.* Reading, Mass.: Addison-Wesley Publishing Company. Ch. 2–4.

NUSSBAUM, ALLEN, and RICHARD A. PHILLIPS. 1976. *Contemporary Optics for Scientists and Engineers*. Englewood Cliffs, N.J.: Prentice-Hall. Ch. 1.

SMITH, WARREN J. 1978. "Image Formation: Geometrical and Physical Optics." In *Handbook of Optics*, edited by Walter G. Driscoll and William Vaughan. New York: McGraw-Hill Book Company.

TUNNACLIFFE, ALAN. 1989. *Introduction to Visual Optics*. New York State Mutual Book and Periodical Service, Limited. Ch. 1.

5

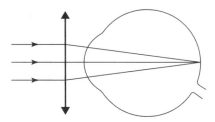

Vergence and Vision

INTRODUCTION

In this chapter we apply the vergence method to a few applications that deal with the eye. We consider the eye in more detail in Chapter 10. Here, we assume the eye is spherically symmetric so that astigmatic aberrations do not arise. These are treated in Chapters 6 and 7. This chapter may be skipped without loss in continuity by those who wish to give priority to other topics in this book.

5-1 FOCUSING BY THE EYE

To deal with the applications discussed in this chapter, a simple schematic of the eye, as shown in Figure 5-1, and a general explanation of the optics of vision will suffice. Light from external objects is focused on the retina. Refraction of light rays entering the eye is accomplished primarily by a single-surface spherical refraction at the cornea and also, to some extent, by the crystalline lens of the eye. The reason the lens itself is not more effective is that it is surrounded on both sides by a fluid whose refractive index does not differ greatly from that of the crystalline lens, so that minimal bending of light rays occurs at each interface. The regions of fluids, or *humors*, as they are called, do not concern us at this time. The fine adjustments made by the eye when focusing on objects at various distances is accomplished by muscular control of the shape of the crystalline lens. The lens

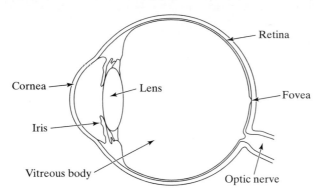

Figure 5-1 Schematic representation of the main features of the eye.

can be made more or less convex in shape, thereby varying its focusing power. This adjustment, which occurs automatically in the act of looking at things, is called *accommodation*. The eye is most relaxed when viewing distant objects and under increasing muscular control when viewing objects nearer to the eye. Accommodation taken into account, the point farthest from the eye at which an object is well-focused on the retina is called the eye's *far point*. Similarly, the point nearest to the eye at which an object is well-focused on the retina is called the *near point*. The eye's ability to accommodate—to see clearly objects both far and near—is a property that can vary both with the individual and with age.

For example, for a normal, relaxed, adult eye, the length of the eye—the distance from cornea to retina—is about 23 mm, but is found to vary somewhat among individuals. The optical condition in which sharp focus occurs in front of the retina (eye too long or convergence too strong) is called *nearsightedness* or *myopia*, resulting in defective vision especially for distant objects. On the other hand, the condition in which a sharp focus would only take place behind the retina (eye too short or convergence too weak) is called *farsightedness* or *hyperopia*, resulting in defective vision especially for near objects. In addition, the elasticity of the lens of the eye generally lessens with age, impairing the accommodation that might once have been possible. Both myopia and hyperopia are forms of *ametropia*, the name given to the abnormal condition in which images fail to focus precisely on the retina. This term contrasts with *emmetropia*, when such an abnormal condition does not exist. The defects in vision described above are easily corrected with the help of an accessory or external lens, either a contact lens placed on the cornea or a spectacle lens positioned a short distance in front of the eye.

5-2 FOCUSING IN EMMETROPIC AND AMETROPIC EYES

Figure 5-2 shows the simplest possible model of the eye, consisting of a single convex thin lens and a screen. The thin lens represents the optical action of the cornea and lens together, while the screen represents the retina, where the images that are "seen" are formed. The refractive media on both sides of the lens is air.

❏ **Example**

Suppose the power of the single lens is +60.0 D. The (+) sign simply indicates that the lens tends to cause rays of light to converge and form a retinal image. The relaxed eye without accommodation is focused on a distant object. Rays

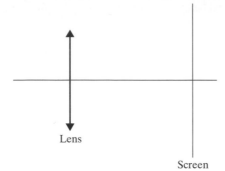

Lens

Screen

Figure 5-2 Simple model of the eye, consisting of a single convex lens (representing cornea and crystalline lens) and a screen (retina) on which images are formed.

from a point on the distant object enter the lens as parallel rays, without convergence or divergence. The following discussion and calculations are illustrated in Figure 5-3.

(a) What is the lens-retinal separation for the unaccommodated, *emmetropic* eye, that is, for the eye that focuses distant objects perfectly?

(b) Where does the image of a distant object occur if the lens has, instead, a power of +50.0 D? Of +70.0 D?

(c) How much dioptric power does each require in the form of an additional contact lens to correct the problem?

Solution

(a) Note that, in other words, we are simply asking for the focal length of the lens. Since the rays are parallel, the object vergence $U = 0$, making the image vergence $V = U + P = +60.0$ D. The image distance is then

$$v = \frac{1}{V} = \frac{1}{60} = 0.01667 \text{ m} = 16.67 \text{ mm}$$

The diagram for this condition is shown in Figure 5-3(a).

(b) The image is no longer focused on the retina, and we are dealing now with an *ametropic* eye. If $P = +50.0$ D, $V = +50.0$ D and

$$v = \frac{1}{V} = \frac{1}{50} = 0.02 \text{ m} = 20 \text{ mm}$$

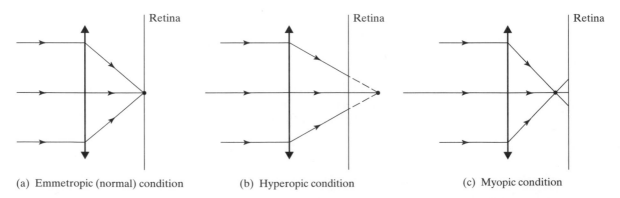

(a) Emmetropic (normal) condition   (b) Hyperopic condition   (c) Myopic condition

Figure 5-3 Dependence of image vergence on lens power. Parallel rays of light enter the eye from an object "at infinity." Behavior is that of (a) an emmetrope, (b) a hyperope, and (c) a myope.

Since the eye in question is only 16.67 mm long[1] (cornea to retina), the image would form beyond the retina. The rays are not yet focused when they reach the retina [see Figure 5-3(b)], so the person sees a blur. This is the condition of a farsighted or *hyperopic* person, whose dioptric power is too weak for sharp vision. If, on the other hand, $V = P = +70.0$ D,

$$v = \frac{1}{V} = \frac{1}{70} = 0.01429 \text{ m} = 14.29 \text{ mm}$$

This distance is shorter than the length of the eye, so the image point falls in front of the retina. After converging to a focus, however, the rays diverge and again form a blur on the retina, the condition of a nearsighted or *myopic* person. See Figure 5-3(c).

(c) The farsighted person needs $+60.0$ D, but only has $+50.0$ D of converging power. He needs an additional lens of $60 - 50$ or $+10.0$ D to see a distant object clearly. The myopic person also needs $+60.0$ D, but has $+70.0$ D, more converging power than needed. He requires an additional lens of $60 - 70 = -10.0$ D to counteract the action of his unaided eye. Thus the hyperopic eye requires 10.0 D more of *convergence*, while the myopic eye requires 10.0 D more of *divergence*. These refractive errors are fully compensated by a positive (convex) lens of $+10.0$ D power to aid in the first case, and a negative (concave) lens of -10.0 D to aid in the second. ■

❏ **Example**

In the simple thin lens-retina model of the unaided eye, let us suppose the distance between lens and retina is 18 mm for a female patient. What dioptric power is required for an emmetrope to focus clearly on a distant object?

Solution For a distant object (at infinity), $U = 0$ and $P = V$. Since $v = 18$ mm $= 0.18$ m, $V = 1 / 0.18 = +55.55$ D, the required power. ■

❏ **Example**

If, instead, the unaided eye has a more strongly convergent power, say, $P = +62.0$ D, the light focuses before it reaches the retina, and the 18-mm eye is myopic. What then is its far point?

Solution In this case, $V = +55.55$ D and $P = +62.0$ D, so that

$$U = V - P = 55.55 - 62.0 = -6.45 \text{ D}$$

Then the object distance

$$u = \frac{1}{U} = \frac{1}{(6.45)} = -0.155 \text{ m} = -15.5 \text{ cm}$$

---

[1]This focal length is appropriate for the simple model of the eye considered here. For a normal, relaxed, adult eye, the length of the eye from cornea to retina is about 23 mm, but varies somewhat among individuals. More accurate models of the eye are considered in Chapter 10, *Optics of the Eye.*

that is, 15.5 cm in front of the eye. Because this person is myopic, her eye is too strongly convergent. She cannot focus on a distant object, with parallel rays entering the eye. To see distant objects clearly—in focus on the retina—she needs light that is diverging with a minimum vergence of -6.45 D on entering the eye. This puts her farthest manageable object distance (far point) at 15.5 cm from the eye! Thus, the far point of a myopic eye is always at a finite distance. This is, of course, why the person is said to be *nearsighted*. By contrast, an emmetrope has a far point at infinity and can focus on distant objects without accommodation, that is, with a relaxed eye. Like an emmetrope, a hyperope can also see distant objects clearly, but only by increasing the converging power of the eye, that is, only with accommodation via muscular control. These situations are discussed in more detail in what follows.

❏ **Example**

What correction is needed for the myope described above?

Solution   We have seen that our myope needs -6.45 D of divergence in order to focus on distant objects without accommodation. Thus the required divergence of -6.45 D must be supplied by an external or correcting lens. If the far point of a myope is determined by testing, the required power of the correcting lens is automatically known. More formally, if we require the correcting lens of power P_c to substitute for the object vergence U, then

$$P_c = U = V - P = 55.55 \text{ D} - 62.00 \text{ D} = -6.45 \text{ D} \qquad ■$$

5-3 EFFECTIVE POWER

Suppose a single thin lens in a medium of index n intercepts a parallel beam of light from a distinct object ($U = 0$). The vergence of the light leaving the lens is given by the power of the lens ($V = P$). What is the vergence of the wave front at some distance x beyond the lens? In order to derive this result, we consider a negative lens of focal length f, as shown in Figure 5-4. The spherical wave fronts emerging on the right side of the lens appear to be centered at the focal point F of the lens, at the distance f, as shown. The vergence upon exiting the lens (point O) is $V_o = n/f$, and the vergence at point x is $V_x = n/(x + f)$. The negative value of the focal length gives the vergence V_o the appropriate negative value for a diverging wave front. To get the correct value of V_x, we must attend to our sign convention. V_x should also be negative and smaller than V_0. Therefore the magnitude

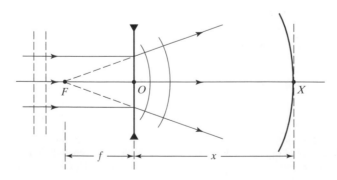

Figure 5-4   Schematic to find the wave front vergence at distance x beyond the negative lens.

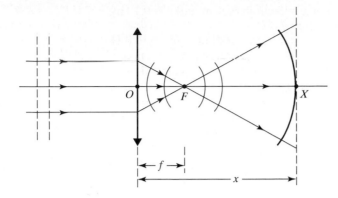

Figure 5-5 Schematic to illustrate change in vergence at a distance x beyond the positive lens.

of $(x + f)$ must be greater than the magnitude of f. Since x is positive according to our usual sign convention, we can satisfy the requirement on V_x by introducing a negative sign in front of x, writing $V_x = n/(f - x)$. Next we can rewrite the formula as

$$V_x = \frac{n}{f - x} = \frac{1}{\dfrac{f}{n} - \dfrac{x}{n}} = \frac{1}{\dfrac{1}{P} - \dfrac{x}{n}} = \frac{nP}{n - xP} \tag{5-1}$$

where P is the power of the lens. When the lens is in air, $n = 1$, and

$$V_x = \frac{P}{1 - xP} \tag{5-2}$$

This formula is also correct for a positive lens, with P a positive number (see Figure 5-5). V_x is often called the *effective power* or *effectivity* of the thin lens of power P at the distance x beyond the lens.

❑ **Example**

A negative thin lens in air, of focal length 10.0 cm, is subject to incident plane waves. What is the curvature or vergence of the wave front 25 cm beyond the lens?

Solution The lens power in diopters for $f = -0.1$ m is $1/f = -10.0$ D, and $x = +25$ cm $= 0.25$ m. Then

$$V_x = \frac{P}{1 - xP} = \frac{-10}{1 - (0.25)(-10)} = \frac{-10}{3.5} = -2.86 \text{ D}$$

The wave front is thus gradually losing curvature as it proceeds beyond the lens,[2] decreasing from a curvature of 10 diopters at the lens to 2.86 diopters 25 cm beyond the lens. ∎

❑ **Example**

A positive thin lens in air, of focal length 10 cm, is subject to incident plane waves. Find the curvature or vergence of the wave front both at 5 cm and at 20 cm beyond the lens.

[2]Notice that we must express x in meters in order to work with the reciprocal meters inherent in the unit, Diopters.

Solution For $x = 5$ cm $= 0.05$ m and $f = 0.10$ m,

$$V_x = \frac{+10.0\ D}{1 - (0.05)(10)} = \frac{+10.0\ D}{0.5} = +20.0\ D$$

For $x = 20$ cm $= 0.20$ m,

$$V_x = \frac{+10.0\ D}{1 - (0.20)(10)} = \frac{+10.0\ D}{-1.0} = -10.0\ D$$

The reason for the change in curvature from positive (converging) to negative (diverging) should be clear upon examination of Figure 5-5. ∎

Lens Corrections. The prescription or correction for a given eye depends on whether the correcting lens is a contact lens, or a spectacle lens positioned in frames in front of the eye. The reason is that the contact lens is positioned at the cornea, while the spectacle lens is several millimeters away. Between the two positions, the vergence of the wave front of light is changing. Of course, from the cornea on, the optics of the two cases is identical.

We shall see that the formula, Eq. (5-2), just derived, permits us to compare the two corrections. Suppose we know the contact lens correction P_c and wish to find the corresponding lens correction, P_s. Since at the cornea the vergence is P_c, the vergence P_s a distance x in front of the cornea is given by

$$P_s = \frac{P_c}{1 - xP_c} \tag{5-3}$$

which is just Eq. (5-2) applied to this situation. Here x is measured backward from the cornea and so is *negative*. If, on the other hand, we know the spectacle lens correction P_s and wish to find the contact lens correction P_c, again using Eq. (5-2), we get

$$P_c = \frac{P_s}{1 - xP_s} \tag{5-4}$$

where now x is measured as a *positive* distance from the spectacle lens position to the cornea.

❏ **Example**

Suppose the prescription power of a contact lens for hyperopia is $+10$ D. If the person is to be fitted instead with spectacle lenses, worn at a distance of 14 mm from the cornea, what is the required prescription? See Figure 5-6.

Solution Here, $P_c = +10$ D, $x = -14$ mm $= -0.014$ m. Then

$$P_s = \frac{P_c}{1 - xP_c} = \frac{10}{1 - (-0.014)(10)} = \frac{10}{1 + 0.14} = +8.77\ D$$

Notice that the spectacle lens correction is of lesser power because some of the required convergence occurs naturally as light progresses from the positive lens to the cornea. ∎

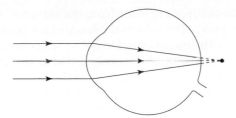

(a) Unaided hyperopic eye

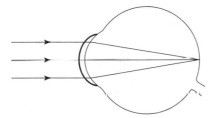

(b) With contact lens

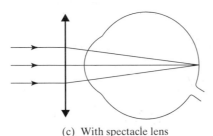

(c) With spectacle lens

Figure 5-6 Equivalent correction of (a) a hyperopic eye by using either (b) a contact lens at the cornea or (c) a spectacle lens a short distance from the cornea. Prescription power differs slightly due to the difference in position of the two lenses.

❏ **Example**

Suppose instead that a myope uses spectacle lenses of -10 D and wishes to change to contact lenses. His cornea is at a distance of 13 mm from the spectacle lens. What should be the contact lens power to provide equivalent correction?

Solution   Now, $x = 13$ mm $= 0.013$ m, and $P_s = -10$ D. Then

$$P_c = \frac{P_s}{1 - xP_s} = \frac{-10}{1 - (0.013)(-10)} = -8.85 \text{ D}$$

The contact lens must be of lesser negative power than the spectacle lens due to the decrease in vergence of the light as it spreads from the spectacle lens to the cornea. ■

5-4   ACCOMMODATION

As explained earlier, accommodation is the adjustment a normal eye must make, under muscular control, to focus clearly on an object when the object is not at the far point and the eye is brought, accordingly, under some tension. This adjustment is a small increase in the converging power of the crystalline lens. The increase in vergence cancels exactly the decrease in vergence between the zero vergence for the plane wave front from the far point to a slightly negative value for the diverging wave front from the less distant point. There is, however, a limit to the degree of adjustment possible for a particular eye. As the object moves in from the far point, a position is reached where the eye is under maximum tension to focus clearly on the object. This is the near point of the eye. If the object is moved closer than the near point, the unaided eye is not able to focus clearly. For a young emmetrope, the near point is typically around 25 cm. Thus the range of clear vision would be from 25 cm to infinity.

By definition, the accommodation A_0 called for is the difference between the power P_a of the accommodated eye and the power P_u of the unaccommodated eye, or

$$A_o \equiv P_a - P_u \qquad (5\text{-}5)$$

Since the accommodation is matched exactly by the change in wave front vergence between the vergence U_{fp} from the far point and the vergence U_x from the new object position, it is also true that

$$A_o = U_{fp} - U_x \qquad (5\text{-}6)$$

Let us now use Eqs. (5-5) and (5-6) to calculate accommodation for several representative cases.

❑ **Example**

An emmetrope with $+60.0$ D of unaccommodated power P_u focuses on an object at his near point of 25.0 cm. How much accommodation P_a is needed to see the object at the near point clearly?

Solution We use Eq. (5-5) to calculate A_o, relying also on the basic relation $P = V - U$ for objects with vergence U at the eye. In this case, the eye is emmetropic so that we know the eye can focus light from an object at infinity $(U = 0)$ with unaccommodated power of $P_u = 60.0$ D. Thus $P_u = V = 60.0$ D. This emmetrope's eye, therefore, requires a vergence of 60.0 D for *any* object if the wave front leaving the crystalline lens is to converge to a clear image on the emmetrope's retina. At the nearpoint, $u = -25$ cm $= -0.25$ m, so that the object vergence $U_{np} = 1/-0.25 = -4.0$ D. To focus the nearpoint object, V must be $+60.0$ D as we have indicated. The power required in accommodation P_a is then

$$P_a = V - U_{np} = 60.0 - (-4.0) = +64.0 \text{ D}$$

and, using now Eq. (5-5),

$$A_o = P_a - P_u = +64.0 \text{ D} - 60.0 \text{ D} = +4.0 \text{ D}$$

Thus the eye must increase its converging power by 4.0 D in order to focus at its near point. It accomplishes this by applying increased muscular tension, causing a slight bulging of the crystalline lens. ∎

❏ **Example**

A young female myope with a far point at 50 cm views an object at 25 cm from the eye. What accommodation does she require?

Solution   Here it is more convenient to use Eq. (5-6): $A_o = U_{fp} - U_x$. Calculating the two vergences,

$U_{fp} = 1/u_{fp} = 1/(-50 \text{ cm}) = 1/(-0.5 \text{ m}) = -2.0 \text{ D}$
$U_x = 1/u_x = 1/(-0.25 \text{ m}) = -4.0 \text{ D}$
Finally, $A_0 = U_{fp} - U_x = -2.0 \text{ D} - (-4.0 \text{ D}) = +2.0 \text{ D}$

The myope must add two diopters of converging power if she is to see the object clearly at 25 cm from the eye.   ∎

When an object is at the far point (infinity), $U_{fp} = U_x$, and Eq. (5-6) indicates that $A_0 = 0$. This is, of course, consistent with our description of a normal eye in a perfectly relaxed state when viewing distant objects. The accommodation A_0, in fact, is always greater than zero, because in accommodating, the eye can only *add* converging power, that is, can only cause a bulging of the crystalline lens. What happens when the object position x is farther from the eye than the far point, as it can frequently be for a nearsighted person? In this case the magnitude of the wave front vergence $|U_{fp}|$ from the farpoint is greater than the wave front vergence $|U_x|$ from the more distant point x. Since both wave fronts are diverging and therefore negative, this means that $A_0 = U_{fp} - U_x$ is negative. A negative accommodation A_0 would seem to imply that the crystalline lens can continue to focus sharply while becoming more diverging than in its relaxed state. Of course, this cannot be! By Eq. (5-5), also, we see that a negative A_0 would require that $P_a < P_u$. Again, this would mean that the eye loses, rather than gains, convergent power during accommodation. This condition is not physically possible, since applied muscular control can only add tension and cause the lens to bulge. The apparent discrepancy between Eqs. (5-5) and (5-6) with the points discussed is resolved by admitting virtual objects at positive distances from the cornea, or positions behind the cornea. This situation occurs when light rays, in order to form a clear image on the retina, must begin by converging to a point behind the retina. In such cases, the image point on the retina has as its conjugate object a point that lies behind the retina. Instances of this kind occur for a hyperopic eye and are considered in the next section, where we discuss the "range of vision" experienced by emmetropes, myopes, and hyperopes.

Range of Vision.   Before describing hyperopic vision in more detail, it may be good to do a little summary here that will help to compare and contrast the hyperope with the myope and the emmetrope. According to the preceding discussion, an emmetrope can focus clearly on objects from a far point of "infinity" to a near point that varies with age. When viewing objects at infinity, the eye is relaxed and focuses plane or zero-curvature wave fronts sharply on the retina. For all object positions between the far point and near point, the eye accommodates by increasing its converging power. Figure 5-7 shows how the near point and the range of vision vary with age in a typical emmetrope. If the eye is myopic, on the other hand, its far point is not at infinity, but at some finite distance in front of the eye. In its relaxed state, the myopic eye is too convergent, so that light from the object must arrive with some divergence in order to offset the eye's

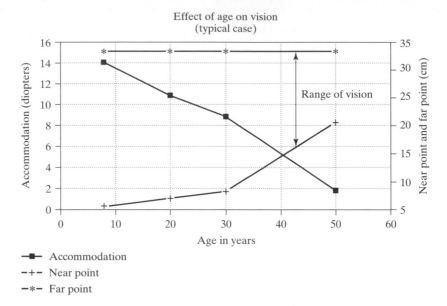

Figure 5-7 Loss of accommodation with age for a typical emmetrope is shown using the vertical axis on the left. On the right, the vertical axis shows typical variation in the near point of a myopic eye having a 3.0 D refractive error. The far point is assumed constant. The vertical separation of the near and far points is the range of clear vision, which also decreases with age.

excessive converging power. The myope's range of vision then extends from a finite far point to his near point. Again, because his near point varies with age, so will his *range of vision*. In general, with advancing age, accommodation diminishes and the near point moves toward the far point. The range of vision for the emmetrope and the myope is shown schematically in Figures 5-8(a) and 5-8(b).

Now consider the hyperopic eye, without accommodation, for which the analysis is a bit more complicated. Recall that the relaxed lens for the hyperope lacks sufficient converging power to focus sharply on an object at infinity. This means, in effect, that light at the retina is still converging toward an image point behind the retina. There is no point, then, in real object space, that can serve as the far point. For the hyperopic, unaccommodated eye, the far point would have to move further from the eye, beyond infinity, so to speak. But since plus infinity and minus infinity are equivalent, we can imagine this point as approaching the eye from the plus infinity side beyond the retina. The far point is then *behind* the retina and is said to be *virtual*, coinciding with the intersection point of parallel rays entering the eye, as shown previously, in Figure 5-6(a). It is a virtual object in the usual sense: If light rays were converging *toward* the virtual object, in our simple model of the eye, the lens would then be able to use this initial convergence, together with its own converging power, to form an image on the retina. Here, as with an emmetrope or a myope, the far point and the retinal image are optically conjugate points.

We have been considering the hyperopic eye without accommodation. With accommodation, the eye can itself provide the needed convergence, which means that it may manage to focus on distant objects up to its own near point. As its converging power gradually increases from the relaxed state, we can imagine the focused object point moving from its virtual position at the far point behind the retina toward plus infinity, equivalent to minus infinity and—if there is sufficient accommodation available—approaching the eye from real object space. When

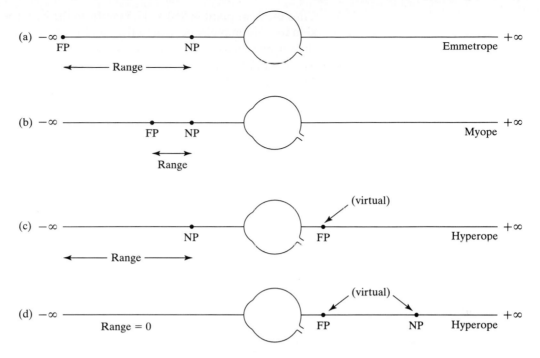

Figure 5-8 Location of near and far points for (a) an emmetropic eye, (b) a myopic eye, (c) and (d) hyperopic eyes with weak accommodation. Points to the right of the eye signify virtual distances. The range of clear vision for real object points is also indicated.

accommodation runs out, the object point is at the near point of the hyperopic eye. Thus, depending on the eye's ability to add convergence to the incoming wave front, several ranges of vision are possible, as indicated in Figure 5-8. Let us calculate ranges of vision for the four cases illustrated there.

❏ **Example**

Determine the near points and far points for the following eyes, given that each can provide a maximum accommodation of 8.0 D.

  (a) An emmetrope
  (b) A myope with a "refractive error" (need for a refractive correction) of −5.0 D
  (c) A hyperope with a refractive error of 5.0 D
  (d) A hyperope with a refractive error of 12.0 D

Solution

  (a) Emmetrope: The far point of an emmetrope is, by definition, at infinity, so its object vergence at the eye is $U_{fp} = 0$. Let us denote the wave front vergence from the near point as U_x. Then, by Eq. (5-6),

$$U_x = U_{np} = U_{fp} - A_0 = 0 - A_0$$
$$U_{np} = -A_0 = -8.0 \text{ D and } u_{np} = 1/(-8.0 \text{ D}) = -0.125 \text{ m} = -12.5 \text{ cm}$$

Thus the near point is real at 12.5 cm from the eye. The range of clear vision for this emmetrope is then from infinity to 12.5 cm at full accommodation. See Figure 5-8(a).

(b) Myope: Let the far point vergence be U_x. Then, at zero accommodation ($A_o = 0$),

$U_{fp} = U_x = -5.0$ D and
$u_{fp} = 1/(-5.0$ D$) = -0.2$ m $= -20$ cm

At full accommodation ($A_0 = 8.0$ D), the vergence from the nearpoint is

$U_{np} = U_x = U_{fp} - A_0 = -5$ D -8 D $= -13$ D and
$u_{np} = 1/(-13$ D$) = -0.077$ m $= -7.7$ cm

The range of clear vision for this myope is thus rather short, from the real far point at 20 cm from the eye to the real near point at 7.7 cm from the eye. See Figure 5-8(b).

(c) Hyperope: Let the far point vergence be U_x. Then, at zero accommodation ($A_0 = 0$),

$U_{fp} = U_x = +5.0$ D and
$u_{fp} = 1/(+5.0$ D$) = +0.2$ m $= +20$ cm. Since u_{fp} is positive, the far point is behind the retina, a *virtual object*.

At full accommodation ($A_0 = 8.0$ D), the vergence from the near point is

$U_{np} = U_x = U_{fp} - A = +5$ D $- 8$ D $= -3$ D and
$u_{np} = 1/(-3$ D$) = -0.333$ m $= -33.3$ cm

Since the far point of this hyperope is virtual, the range of clear vision is from infinity to the hyperope's real near point at 33.3 cm from the eye. See Figure 5-8(c).

(d) Hyperope: This hyperope has need for a stronger correction of 12.0 D. Following the same steps as in cases (b) and (c) above, we have

$U_{fp} = U_x = +12$ D and
$u_{fp} = 1/(+12$ D$) = +0.0833$ m $= +8.33$ cm (behind the retina).
$U_{np} = U_x = U_{fp} - A = +12$ D -8 D $= +4$ D and
$u_{np} = 1/(+4$ D$) = +0.25$ m $= +25.0$ cm (also behind the retina).

Both near point and far point for this hyperope are virtual. Thus this individual cannot clearly focus on an object at any distance without the help of external lenses. See Figure 5-8(d). ■

PROBLEMS

5-1. An emmetrope with an unaccommodated eye is represented by a +64.00 D thin lens and retina (screen).

(a) What is the lens-retinal separation for this unaccommodated eye?

(b) Where does the image of a distant object occur if the lens has instead a power of +60.00 D? Of +68.00 D?

(c) How much dioptric power does each require in the form of an additional contact lens to correct his vision problem?

5-2. In the simple, thin lens-retina model of the eye, let us suppose the distance between lens and retina is 16.5 mm for a particular eye.

(a) What dioptric power is required for an emmetrope to focus on a distant object?

(b) If, instead, the lens has a more strongly convergent power of $+65.00$ D, what is its far point?

(c) What correction is needed in an external lens?

5-3. A person suffers from myopia requiring a correction of $+4.50$ D. The maximum accommodation possible is $+6.00$ D. Without correction, can this myope read the fine print in a book held 40 cm from the eye?

5-4. Using a $+4.00$ D lens in spectacles, a person can see clearly objects from 20 cm to 100 cm from the lens. What effect on the range of vision results from replacing the spectacle lens with one of $+6.00$ D?

5-5. Consider the thin lens-screen model of the eye, with air on both sides of the lens. Let the unaccommodated power of the lens P_u be $+60.00$ D. If the maximum accommodation possible is A_0, determine—for each of the following cases— the near and far points of the uncorrected eye, as well as the distance vision correction needed at the cornea and at a spectacle lens situated a distance x in front of the cornea.

(a) $A_0 = 8.00$ D; lens-retina separation = 20.0 mm.

(b) $A_0 = 8.50$ D; lens-retina separation = 16.0 mm.

(c) $A_0 = 2.50$ D; lens-retina separation = 17.2 mm.

5-6. The vergence of a converging wave front is $+6.00$ D at a particular point along its path. What is its vergence "downstream" at distances of 10 cm and 20 cm?

5-7. Solve Problem 5-6 when the given wave front is diverging instead.

5-8. Solve Problem 5-6 when the distances of 10 and 20 cm are "upstream" instead.

5-9. Solve Problem 5-7 when the distances of 10 and 20 cm are "upstream" instead.

5-10. Incident plane waves are refracted by a positive thin lens of focal length 25 cm. Find the wave front curvature or vergence at 15 and 30 cm beyond the lens.

5-11. Solve Problem 5-10 when the thin lens is negative.

5-12. The prescription power of a contact lens for hyperopia is $+12.0$ D. If the person is switching to spectacle lenses, at a distance of 12 mm from the cornea, what is the required prescription?

5-13. The prescription power of a spectacle lens for a myope is -12.0 D. If the myope switches to contact lenses from the spectacle lenses at 14 mm from the cornea, what is the new prescription?

5-14. An emmetrope with $+58.0$ D unaccommodated power focuses on an object at a near point of 25.0 cm from the eye. How much accommodation is required?

5-15. A myope with a far point at 60.0 cm views an object at 25 cm from the eye. What accommodation is needed?

5-16. An emmetrope with an unaccommodated eye is represented by a $+62.00$ D thin lens and retinal screen. Determine the retinal size of the image of a distant object that subtends an angle of 20′ at the lens. (*Hint:* Draw a ray from the object, through the center of the lens, and onto the retina. Then use geometry to determine the image size).

REFERENCES

BENNETT, ARTHUR G. 1984. *Clinical Visual Optics*. Boston: Butterworth-Heinemann.

BYER, ALVIN. 1986. *Theoretical Optics for Clinicians*, 2nd ed. Philadelphia: Pennsylvania College of Optometry.

HORNE, D. F. 1978. *Spectacle Lens Technology*. Philadelphia: I. O. P. Publishing.

KEATING, MICHAEL P. 1988. *Geometric, Physical and Visual Optics*. Boston: Butterworth-Heinemann.

TUNNACLIFFE, ALAN. 1989. *Introduction to Visual Optics*. New York State Mutual Book & Periodical Service, Limited.

6

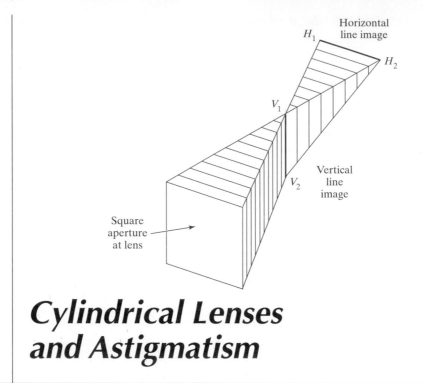

Cylindrical Lenses and Astigmatism

INTRODUCTION

The spherical lenses treated so far have been characterized by spherical symmetry. This means that if the lens is rotated by an arbitrary angle about the optical axis through its center, an *axis of symmetry*, it looks just as it did before. Because the orientation of surface curvatures has not changed, its optical behavior remains unchanged. On the other hand, if the radius of curvature of one or both surfaces differs along different cross sections of the lens, it loses its spherical symmetry and some of its simplicity of analysis. In particular, a *cylindrical lens* is shaped like half a beer can, sliced down the middle, with one plane surface (zero curvature or $r \rightarrow \infty$) and one surface of semicircular cross section. Such a lens has asymmetric focusing properties, as will be seen in greater detail. Whereas a spherical lens produces a point image of a point object, a cylindrical lens produces a line image of a point object. Because of this property, a spherical lens is said to be *stigmatic*, and the cylindrical lens, *astigmatic*. In this chapter we treat astigmatic lenses. As will become evident, such lenses are useful in correcting a common defect of vision called *astigmatism*.

6-1 CYLINDRICAL LENSES

Consider first a spherical lens, as shown in Figures 6-1(a) and (b). Orthogonal axes are shown as solid lines through the geometric lens center. Parallel rays of light through the vertical axis [see Figure 6-1(a)] and through the horizontal axis [see Figure 6-1(b)] are handled identically by the lens, converging them to a common focus at F. The lens can be rotated through any arbitrary angle about its optical axis with the same result. The lens is said to have spherical symmetry about an axis through its center. Another way of saying this is that the focusing properties of a spherical lens are invariant to rotation about its central (optical) axis.

Next, consider the convex and concave cylindrical lenses shown in Figure 6-2. One surface of the lens is cylindrical while the opposite surface is plane.[1] Thus the first surface has a definite, finite radius of curvature while the plane surface has an infinite radius of curvature. In Figure 6-3, two vertical slices or sections are shown perpendicular to the axis of a convex cylindrical lens. Through each section, three representative rays are drawn. The operation of this lens is clearly asymmetric. Focusing occurs for rays along a vertical section but not for rays along a horizontal section, where the lens presents no curvature. Thus rays 1, 2, and 3 focus to point A, and rays 4, 5, and 6 focus to point B, but there is no focusing of rays in a horizontal section, such as the pairs of rays 1 and 4, 2 and 5, or 3 and 6. Other vertical sections would produce other points along the focused line image AB in the same way. Notice that the line image AB so formed is always parallel to the cylinder axis. This important feature is also shown in Figure 6-4, where the line image is real for a convex lens and virtual for a concave lens. From these figures, it is evident that the length of the line image is equal to the axial length of the lens, assuming rays of light parallel to the optical axis enter along the entire length of the lens. If an aperture is placed in front of the lens to limit

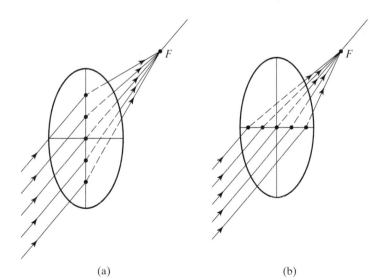

(a) (b)

Figure 6-1 Parallel rays of light focused by a spherical lens. Because of its symmetry relative to rotation about an axis through its center, the lens treats (a) vertical, and (b) horizontal fans of rays similarly, producing in each case a point image at the same location. Each ray refracts twice through the lens, once at each surface. For simplicity, only one refraction is shown.

[1]To be more precise, we are speaking of a *plano-convex* or *plano-concave* lens. When one surface is cylindrical and the other plane, the lens is plano-convex. Generally speaking, the first surface might be cylindrical while the second surface has some other curvature. In this case, the behavior of the lens as a whole, due to the addition of the powers of the two surfaces, may not reduce to that of a simple plano-convex lens as described here.

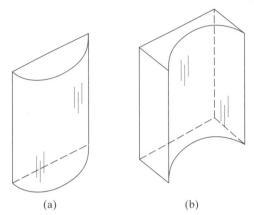

Figure 6-2 Cylindrical lenses shown as sections of a cylindrical rod. The lens in (a) is convex, and in (b), concave.

(a) (b)

the bundle of light rays through the lens, the length of the line image is just the aperture dimension along the cylinder axis, or the *effective length* of the lens.

In Figure 6-3, the line image formed is the result of an object point "at infinity." In Figure 6-5, the object point O is near the lens, producing diverging rays of light incident on the lens. Still, if the lens is thin, focusing occurs along the vertical sections, as shown. Rays 1 and 3, in the left vertical section, focus at A; rays 2 and 4 in the right vertical section focus at B. However no focusing occurs for rays 1, 5, and 2 along the horizontal section. Because of the divergence of the rays entering the lens, however, the length of the focused line image is no longer equal to the effective length of the lens. The divergence of the extreme rays at each end of the lens now determines an image line that is *longer* than the length of the lens. The length can be found from a simple geometrical argument that is apparent in Figure 6-6(a), a view of the central horizontal section in Figure 6-5 as seen from above. If the effective length of the cylindrical lens is CL, then by similar triangles, it follows that

$$\frac{AB}{CL} = \frac{u + v}{u} \qquad \text{or} \qquad AB = \left(\frac{u + v}{u}\right)CL$$

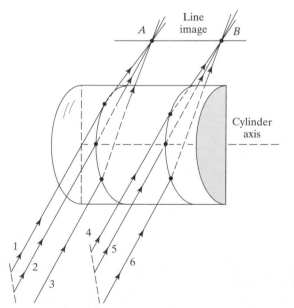

Figure 6-3 Focusing property of a convex cylindrical lens. Rays through a vertical section, such as rays 1, 2, and 3, come to a common focus, but rays through a horizontal section, such as rays 1 and 4, do not focus. Parallel rays form a line image that is parallel to the cylinder axis. For simplicity, refraction is shown only at the front surface.

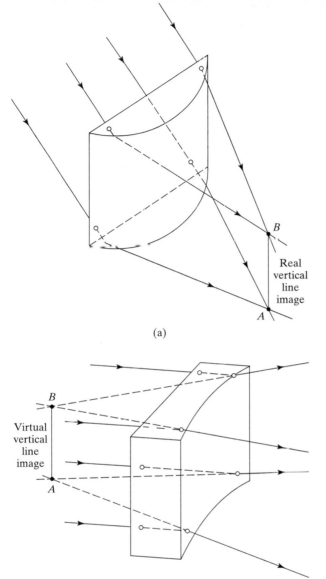

(a)

(b)

Figure 6-4 Formation of line images by cylindrical lenses for light incident from a distant object. In (a) the convex lens forms a real image. In (b), the concave lens forms a virtual image. In either case, the line image is parallel to the cylinder axis.

These relations are purely geometrical, without sign convention. If we modify this result to take into account our sign convention for u and v, we have a general form for the plano-cylindrical lens that handles all cases:

$$AB = \left| \frac{u - v}{u} \right| CL \tag{6-1}$$

where u and v are object and image distances, as usual, with appropriate signs. The vertical bars direct one to take the absolute value of the quotient to keep AB always positive. The image distance v is found from the vergence equality $U + P = V$, where P is the usual spherical power of the lens due to its vertical

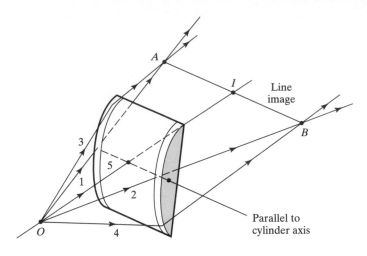

Figure 6-5 Formation of a line image by a convex cylindrical lens when the object is a point at a finite distance from the lens. In this case the line image is longer than the axial length of the lens.

focusing section, as shown in Figure 6-6(b). From Eq. (3-35), for a lens with two refracting surfaces, the power is given by

$$P = (n_2 - n_1)\left(\frac{1}{r_1} - \frac{1}{r_2}\right) \tag{6-2}$$

Since $r_2 \to \infty$ for the plane surface, this reduces to

$$P = \frac{n_2 - n_1}{r} \tag{6-3}$$

where n_2 is the index of refraction for the lens and n_1 for the surrounding medium.

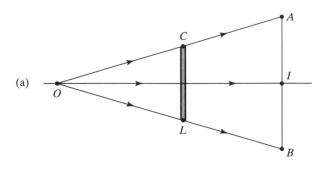

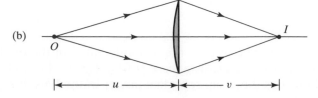

Figure 6-6 (a) Light rays in a top view of the horizontal (nonfocusing) section of the lens in Figure 6-5. (b) Light rays in a side view of the vertical (focusing) section of the lens in Figure 6-5.

❑ Example

A thin plano-cylindrical lens in air has a radius of curvature of 10 cm, a refractive index of 1.50, and an axial length of 5 cm. Light from a point object is incident on the convex, cylindrical surface from a distance of 25 cm to the left of the lens. Find the position and length of the line image formed by the lens.

Solution The object vergence $U = 1/u = 1/(-0.25) = -4.0$ D.
The power of the lens is $P = (1.50 - 1)/(+0.10) = +5.0$ D.
The image vergence is then $V = U + P = -4.0$ D $+ 5.0$ D $= +1.0$ D, so that the image distance is $v = 1/V = 1/(+1.0$ D$) = 1$ m.
Also, the line image length

$$AB = \left| \frac{u - v}{u} \right| CL = \left| \frac{-0.25 - 1.0}{-0.25} \right| (0.05) = 0.25 \text{ m} = 25 \text{ cm}$$

The line image is parallel to the cylindrical axis, enlarged to 25 cm, and located 1 meter from the lens. If the lens is rotated about its optical axis, the line image also rotates, remaining parallel to the cylindrical axis. ∎

Looking again at Figure 6-5, imagine a screen placed on the exit side of the lens so as to catch the light from the lens. We have argued that when the screen is at the distance v from the lens, one sees a focused line image AB on the screen, in this case with a horizontal orientation. As the screen is moved farther away from the lens, one sees an unfocused blur that has the general shape of the aperture—either of the rectangular cross section of the lens alone or of the lens with an aperture placed against it. It should be evident from Figure 6-5 that, as the screen moves from the lens toward the line image AB, the horizontal dimension (width) of the blur increases and its vertical dimension (height) decreases. As the screen moves beyond the line image, its width continues to increase, but its height now also increases due to the divergence of the rays after focusing. If the aperture is circular, these blur images are elliptical in shape, with changing major and minor axes formed by the width and height of the blur. If the aperture is square, the blurs are rectangular in shape. Widths and heights of the blur pattern can be found at any position of the screen using the geometry apparent in Figures 6-6(a) and (b), respectively. This behavior can be observed easily in the laboratory.

❑ Example

A thin, cylindrical lens of $+10.0$ D power and a *vertical* cylinder axis is located 25 cm from a point source of light. A square aperture 1 cm on a side is placed directly in front of the lens.

(a) Describe the image of the point source formed by the lens.
(b) Describe the light pattern on a screen positioned halfway between lens and line image. (Use Figures 6-6, but note the change in orientation of the cylinder axis from horizontal to vertical in this example.)

Solution

(a) The object vergence is $U = 1/u = 1/(-0.25 \text{ m}) = -4.0$ D.
The image vergence $V = U + P = -4.0$ D $+ 10.0$ D $= +6.0$ D.

The image distance is then $v = 1/V = 1/6.0 = 0.167$ m $= 16.7$ cm. The length of the line image is

$$AB = \left|\frac{u-v}{u}\right|CL = \left|\frac{-25.0 - 16.7}{-25.0}\right|(1\ \text{cm}) = 1.67\ \text{cm}$$

Thus the image is a vertical line, 1.67 cm in length, at a distance of 16.7 cm from the lens.

(b) At a distance x from the lens, the vertical dimension h of the rectangular light pattern is found by replacing v by x, and AB by h in Figure 6-6 and Eq. (6-1). For $x = 16.7/2 = 8.35$ cm, we have

$$h = \left|\frac{u-x}{u}\right|CL = \left|\frac{-25.0 - 8.35}{-25.0}\right|(1\ \text{cm}) = 1.33\ \text{cm}$$

The horizontal dimension w is found by using a geometric proportion evident in Figure 6-6(b):

$$\left(\frac{w}{v-x}\right) = \frac{CL}{v}$$

or

$$w = \left(\frac{v-x}{v}\right)CL = \left(\frac{16.7 - 8.35}{16.7}\right)(1\ \text{cm}) = 0.5\ \text{cm}$$

Thus the light pattern (blur) on the screen is rectangular, with height h of 1.33 cm and width w of 0.5 cm. ∎

6-2 POWER CROSS

Up to this point we have been dealing with cylindrical lenses whose axes are horizontal or vertical. Of course, the cylinder axis could be at any angle. An astigmatic eye, for example, while it possesses predominantly spherical optics, might have a cylindrical lens component whose axis could be horizontal, vertical, or some angle in between. To deal with cylindrical lenses and astigmatism in a general way, then, we must be able to determine the effect of combining cylindrical lenses having arbitrary axis orientations with each other and with spherical lenses. As we shall see, two cylindrical lenses can produce the same effect as a sphero-cylindrical lens. Lens prescriptions for vision correction are, in fact, expressed in terms of combinations of spherical and cylindrical surfaces.

The usual convention for designating cylinder axis orientations is adopted from the standard practice in trigonometry: Angles are measured counterclockwise from the positive x-axis. In applications to the eye we must further specify that the coordinate system is the one seen by the examiner, not by the patient. Looking into the patient's eyes, then, the examiner specifies angular orientation of the cylinder axis according to the scheme shown in Figure 6-7.

Since the power (refracting) axis is perpendicular to the cylinder (non-refracting or zero-power) axis, both power and cylinder-axis orientations can be designated symbolically in a *power cross*. The examples in Figure 6-8, showing both the written and symbolic specifications, as well as the lens orientation, should make this usage clear. For example, the first power cross, indicated as

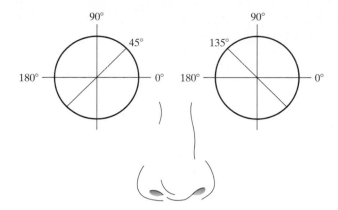

Figure 6-7 In optometric practice, the coordinate system for angular orientation of axes is chosen looking into the patient's eye.

+4.00 × 90 or +4 axis 90, defines a cylinder axis (marked "0" for zero power) at 90°, and perpendicular to it, a horizontal power axis of +4.0 D. This designation is read "plus four axis ninety," where × stands for axis and signifies a *cylinder axis* at 90°. Note in the second example of Figure 6-8 that a horizontal cylinder axis is specified as 180 degrees rather than 0 degrees, to avoid confusion between 0 degrees and 0 power. For other orientations, the smaller of the two possible positive angles is used. An alternative to specifying the orientation of the cylinder axis in this way is to specify the orientation of the power axis instead. When this is done, the "at" symbol, @, is used for the power axis. Such designations for the examples shown in Figure 6-8 would read, in order, +4.00 @ 180, +3.00 @ 90, −2.00 @ 120, +6.50 @ 60.

6-3 COMBINING CYLINDRICAL POWERS

Given two thin cylindrical lenses *in contact*, powers along parallel axes simply add algebraically, and powers along perpendicular axes remain independent of one another. It follows that

1. if two thin cylindrical lenses in contact have their cylinder axes aligned, their cylinder powers add. This is the case in Figure 6-9, where cylinder axes

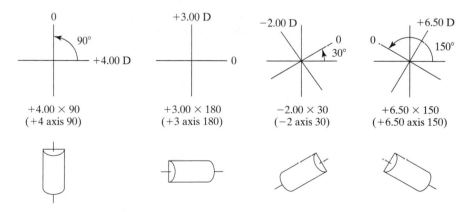

Figure 6-8 Examples of power-cross specification. Corresponding orientation of the cylinder lens is shown in each case.

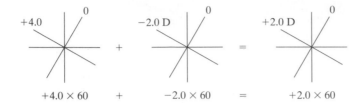

Figure 6-9 Power crosses for a combination of cylindrical lenses in contact, with their cylindrical axes aligned.

of two lenses of different powers are aligned at 60°. The resultant, or equivalent, lens is also aligned at 60°, and its power axis is the sum of the individual powers, or +2.0. The designation for the combination is +2.0 × 60.

2. if two thin cylindrical lenses in contact have their cylinder axes at 90° to one another, as in Figure 6-10, powers are added, as before, along parallel orientations. Thus, along the 30° orientation, we have 0 + (+7.0) = +7.0, and along the 120° orientation, we have +6.0 + 0 = +6.0. The result, therefore, has a power of +7.0 D at 30° and a power of +6.0 D at 120°, as shown. It is not reducible to a simple cylindrical lens, since there is no single direction (cylinder axis) for which the power is zero. Notice how the notation is written for the composite lens in this case.

We can better appreciate the behavior of this lens combination by considering the action of the individual cylindrical lenses in forming images of a point object at some distance in front of the composite lens. Let's consider the following case.

A point object is located 50 cm to the left of the lens combination of Figure 6-10, so that object vergence $U = -2.0$ D. For the first lens with a power of 6.0 D, the image vergence ($V = U + P$) is therefore $V = -2.0 + 6.0 = 4.0$ D, and the image distance v is 25 cm to the right of the lens. Similarly, for the second lens of power 7.0 D, $V = -2.0 + 7.0 = 5.0$ D, and $v = 20$ cm. Since the two power axes are perpendicular, they act independently. The first lens (+6.0 × 30) produces a line image of the point, 25 cm away and parallel to its cylinder axis at 30°, while the second lens (+7.0 × 120) produces a line image of the point, 20 cm away and aligned at 120°. Thus at distances of 20 and 25 cm, the composite lens focuses perpendicularly oriented line images. At other distances, when the limiting aperture of the optical system is circular, the image of the point object as seen on a target screen is elliptical, the result of the converging properties of both lenses acting independently. The change in shape of the spot as a function of distance from the lens is suggested in Figure 6-11, seen from the point of view of an observer looking at the target screen from the lens. At a certain distance between the two line images, as the pattern changes from an ellipse with major axis at 120° to an ellipse with major axis at 30°, the pattern becomes circular. This state of convergence is called the *circle of least confusion*, a best compromise between the two line images in which the spot has its smallest dimensions. The circle of least confusion is the nearest approximation to a focused image conjugate to a point object.

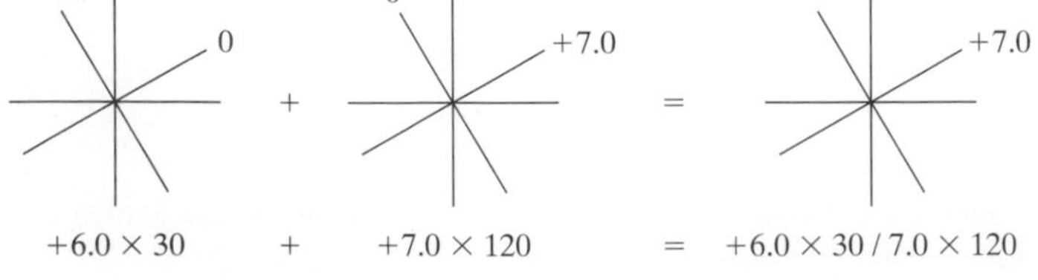

Figure 6-10 Power crosses for a combination of cylindrical lenses in contact, with their cylindrical axes perpendicular.

+6.0 × 30 + +7.0 × 120 = +6.0 × 30 / 7.0 × 120

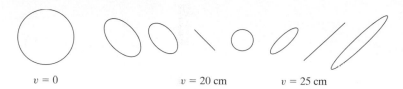

$v = 0$ $v = 20$ cm $v = 25$ cm

Figure 6-11 Variation in spot patterns due to a lens with perpendicular power axes. Line images occur at 20 cm and 25 cm from the lens. Distances in the figure are purposely not shown to scale, in order to better illustrate the evolution of the illumination pattern.

The entire bundle of rays forming the line images is called the *conoid of Sturm* and the optical-axis distance between the two line images is called the *interval of Sturm*. Figures 6-12 and 6-13 may help further to visualize this three dimensional light pattern.[2] Figure 6-12 shows the conoid of Sturm when a square aperture is used at the lens and the two cylinder axes are oriented, respectively, horizontally and vertically. The X forming the top of the pattern represents light rays, while the rectangular planes of cross section indicate the pattern of the illumination intercepted by a screen placed at various positions along the conoid. Figure 6-13 shows two views of Figure 6-12, in part (a), a top view of the outer rays forming the X that includes the horizontal line image H_1H_2, and in part (b), a side view of the rays forming the vertical line image, V_1V_2. Note that, in each case, the line image not shown occurs at the point of intersection of the rays [V in Figure 6-12(a) and H in Figure 6-12(b)] and is oriented perpendicular to the page. The geometry inherent in these figures enables us to calculate the size of both line images and of the light pattern at any distance from the lens. The appropriate geometric proportions are given in the figures. It can be shown (see Problem 6-6) that the distance s from the lens to the circle of least confusion is given by

$$s = \frac{2v_1v_2}{v_1 + v_2} \tag{6-4}$$

and the diameter d of the circle is given by

$$d = \left(\frac{v_2 - v_1}{v_2 + v_1}\right)CL \tag{6-5}$$

where the symbols are those appearing in Figure 6-13.

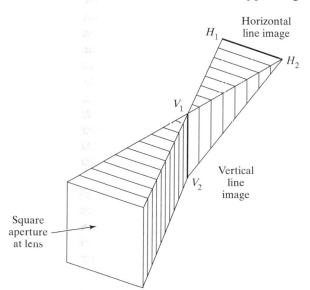

Figure 6-12 Conoid of Sturm for a square aperture. The lens (not shown) has its two power axes oriented in vertical and horizontal directions, each producing its own line image. In this instance, the vertical power axis is the stronger of the two.

[2]See also Figures 7-5(a) and 7-5(b) in Chapter 7.

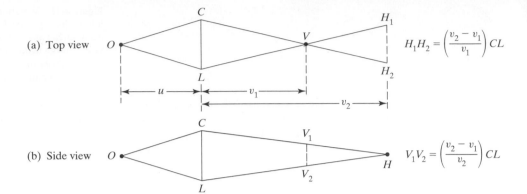

(a) Top view

$$H_1H_2 = \left(\frac{v_2 - v_1}{v_1}\right)CL$$

(b) Side view

$$V_1V_2 = \left(\frac{v_2 - v_1}{v_2}\right)CL$$

Figure 6-13 (a) Top view and (b) side view of the conoid of Sturm shown in Figure 6-12. Each view exhibits one of the perpendicular line images, H_1H_2 and V_1V_2. In each view the other line image, V or H, is perpendicular to the page. The geometry of each figure leads to the expressions shown for the length of the line images.

❑ **Example**

Let a composite cylindrical lens have perpendicular powers of $+6.0 \times 180$ and $+7.0 \times 90$. A point object is situated at 50 cm from the lens. Line images then form as follows (see Figure 6-13): A vertical line image due to the $+7.0$ D power occurs at distance $v_1 = 20$ cm from the lens, and a horizontal line image due to the $+6.0$ D power occurs at distance $v_2 = 25$ cm from the lens. Assume a circular aperture of 4.0 mm diameter.[3] Determine the lengths of the two line images, and describe the circle of least confusion.

Solution The length H_1H_2 of the horizontal line image at 25 cm can be calculated from the proportion given in Figure 6-13(a):

$$H_1H_2 = \left(\frac{v_2 - v_1}{v_1}\right)CL$$

$$H_1H_2 = \left(\frac{25 - 20}{20}\right)4.0 \text{ mm} = 1.0 \text{ mm}$$

The length V_1V_2 of the vertical line image at 20 cm can be calculated from the proportion given in Figure 6-13(b):

$$V_1V_2 = \left(\frac{v_2 - v_1}{v_2}\right)CL$$

$$V_1V_2 = \left(\frac{25 - 20}{25}\right)4.0 \text{ mm} = 0.8 \text{ mm}$$

Thus the conoid of Sturm is bounded in this case by two line images that are 0.8 mm and 1.0 mm in length and separated by a Sturm interval of 5.0 cm. The circle of least confusion is located at the distance

$$s = \frac{2(20)(25)}{(20 + 25)} = 22.2 \text{ cm}$$

[3]Under bright lighting conditions this is roughly the diameter of the pupil of the eye, which serves as the limiting aperture for the eye.

where its diameter is

$$d = \left(\frac{25 - 20}{25 + 20}\right)CL = \frac{1}{9}(4.0 \text{ mm}) = 0.44 \text{ mm}$$

Thus the circle of least confusion has a diameter that is $\frac{1}{9}$ that of the lens aperture and is located 2.2 cm beyond the first of the line images, or 2.2 cm along the interval of Sturm. Notice that it does not fall exactly midway between the two line images. ∎

6-4   TORIC SURFACES

It should be apparent that the combination of cylindrical lenses producing a particular resultant lens is not unique. For example, the lens of the preceding example, with perpendicular axes of 6.0 D and 7.0 D, could result from any two lenses whose powers sum to 6.0 D along one direction and 7.0 D along a perpendicular direction. Figure 6-14(a) illustrates just one example.

As a special case, one of the components might be a spherical lens whose power, necessarily, is the same along any axis. Accordingly, the notation for a spherical lens is simply a number indicating its power. Figure 6-14(b) illustrates a combination involving a spherical lens, but having the same resultant as in (a). Notice also that Figures 6-14(a) and (b), taken together, illustrate what was said earlier: The optical effect of two cylindrical lenses can be accomplished also by a combination of spherical and cylindrical surfaces or lenses. In general, it is possible to find two combinations of a spherical and cylindrical lens—or spherical and cylindrical surfaces of a single lens—that result in a composite lens with specified perpendicular powers. The following example shows how to find the two combinations and demonstrates the characteristics of the two cylindrical choices mentioned above.

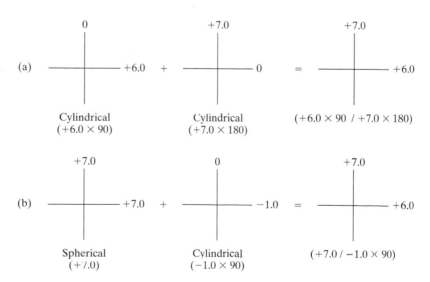

Figure 6-14   Two combinations of lenses that, acting together, produce the same resultant lens. In (a) the component lenses are both cylindrical. In (b) the combination consists of a spherical plus a cylindrical lens. Note the compact notation for the spherical lens.

❏ **Example**

What combination of a spherical and a cylindrical lens produces a lens with vertical power axis of $+10.0$ D and horizontal power axis of $+3.0$ D?

Solution The choice of component lenses is limited by the requirement that the cylinder lens have one axis at zero power, while the spherical lens must have identical powers along both axes. Suppose the zero power axis is chosen as the vertical. Then we have the sum shown in Figure 6-15(a), where we must satisfy the two conditions

$$P_s + 0 = 10.0 \quad \text{or} \quad P_s = 10.0$$

and

$$P_s + P_c = 3.0 \quad \text{or} \quad 10.0 + P_c = 3.0$$

and

$$P_c = -7.0$$

If, on the other hand, we require that the horizontal axis of the cylinder lens have zero power, then Figure 6-15(b) applies and the two conditions are:

$$P_s + P_c = 10.0$$

and

$$P_s + 0 = 3.0$$

which together lead to the solutions $P_s = 3.0$ D and $P_c = 7.0$ D. The two solutions are illustrated in Figure 6-16. In the two combinations, the two choices for the cylindrical surfaces have the same power magnitude but of opposite sign, and the cylinder axes are 90° apart. This is a general result. ∎

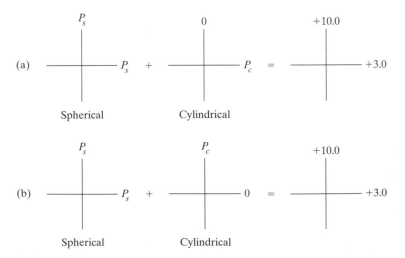

Figure 6-15 An example showing that two combinations of a spherical plus a cylindrical lens can always be found that produce the same (arbitrary) resultant lens. The two solutions are distinguished in that the cylindrical axis is (a) vertical, and (b) horizontal.

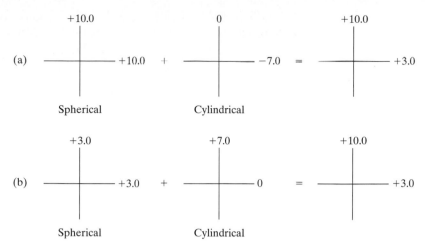

Figure 6-16 Quantitative solutions to the problem illustrated in Figure 6-15.

The resultant power cross of a lens follows, as we have seen, from the sum of the power crosses of its two surfaces. A surface may be spherical (a section of a spherical surface), with a single radius of curvature. It may be cylindrical (a section of a cylindrical surface), having two radii of curvature, of which one is infinite. In the most general case, it may be *toric* (a section of a *toroidal* surface), with two radii of curvature of arbitrary magnitude. In all cases the radius of curvature may be positive or negative, relating to convex or concave shaped surfaces.

A toroidal surface, or *toroid*, has the shape of a doughnut or an inflated inner tube. As shown in Figure 6-17, the surface is characterized by two distinct radii of curvature, r_1 and r_2, exhibited in perpendicular cross sections. If either of

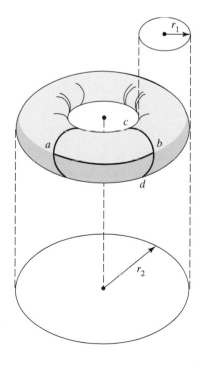

Figure 6-17 A toroid, shaped like a doughnut, showing two circular cross sections of radii r_1 and r_2. On the toroidal surface, the two perpendicular curvatures are shown. Arc *cbd* is of radius r_1 and arc *ab* is of radius r_2.

the radii is allowed to increase at the expense of the other, the shape of the toroidal surface approaches a cylindrical surface. Thus a cylindrical surface is just a special case of a toroidal surface. When two cylindrical surfaces with perpendicular axes are placed in contact, the equivalent surface is a toroidal surface. An example of this was considered in Figure 6-14(a). We have, in fact, been working implicitly with toric lenses. In the most general case, the two surfaces of a lens may be combinations of spherical, cylindrical, and toroidal surfaces.

❑ **Example**

Consider a toric lens whose two surfaces are spherical and toroidal. The spherical side has a power of +6.0 D, while the toroidal side is described by $4.0 \times 45 / -2.0 \times 135$. Determine (a) the power cross of the lens, and (b) the interval of Sturm for a point object at 50 cm from the lens.

Solution Object vergence $U = 1/u = 1/(-0.5) = -2.0$ D.
The power cross is the sum of the two surfaces, spherical and toroidal:

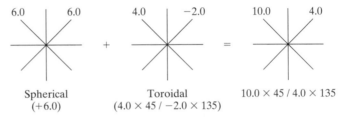

Spherical   Toroidal $10.0 \times 45 / 4.0 \times 135$
(+6.0)   ($4.0 \times 45 / -2.0 \times 135$)

The 4.0 D power axis produces an image vergence of

$$V = P + U = 4.0 + (-2.0) = 2.0 \text{ D}$$

so that the image distance $v = 1/V = 0.5$ m $= 50$ cm. A line image is produced at 50 cm from the lens, tilted at 135° from the horizontal. The other power axis of 10.0 D produces an image vergence $V = 10.0 + (-2.0) = 8.0$ D, so that the image distance is $v = 1/8.0 = 0.125$ m $= 12.5$ cm. Thus the second focused line image occurs at 12.5 cm from the lens, tilted at 45° from the horizontal. The interval of Sturm for this example of astigmatism is then $(50 - 12.5) = 37.5$ cm.

6-5 OCULAR ASTIGMATISM

Astigmatism occurs in the eye because, in addition to the mostly spherical curvature of the cornea, there is a contribution due to an additional cylindrical curvature. In other words, the cornea is not a perfect, spherically shaped "cap." If, for example, astigmatism in the eye is due to a cylindrical component aligned along an axis at 7°, it could be corrected with a cylindrical contact lens of equal but opposite curvature, also aligned at 7°. If correction is accomplished instead by a spectacle lens at some distance from the eye, compensation must be made for the change in curvature of the wave front in passing from the lens to the cornea. This kind of compensation was treated in Section 5-3. In any case, the correction can be expressed in the form of a sphero-cylindrical prescription.

Most of the refraction by the eye occurs at the cornea rather than at the crystalline lens. Furthermore, astigmatism occurs due to refraction at the outer and inner surfaces of the cornea. By far most of this occurs at the outer surface because the refraction at this interface (air to cornea) experiences a greater change in refractive index than at the inner interface (cornea-aqueous). The crystalline lens offers little, if any, contribution to the astigmatism of the eye. For this reason the astigmatic eye can be well-represented by a model consisting of one surface or interface at the cornea, where all the refraction is assumed to take place.

❑ **Example**

Consider a "reduced eye," that is, one whose optical behavior is well-approximated by refraction at a single surface at the cornea, separating air from the vitreous fluid of refractive index 4/3. The axial length of this eye is 24.24 mm and its measured powers along the vertical and horizontal axes are +59 D and +57 D, respectively. What spectacle correction is required for good, distant vision if the correcting lens is to be 12 mm in front of the cornea?

Solution For a distant object, the vergence $U = 0$, so that the power of a perfectly spherical surface is

$$P = V = \frac{n}{v} = \frac{4/3}{24.24 \text{ mm}} = +55.0 \text{ D}$$

The eye in question is myopic since both of its principal powers are greater than 55.0 D, the power required for good focus of a distant object on the retina. Let us adopt a temporary notation to help distinguish carefully between the various powers involved in this problem. The power of 55.0 D is the power required (RP) by a normal eye for a good focus of distant objects. The actual powers (AP) of the astigmatic eye in question are given at 90° and 180°. The power correction (CP) is the difference between the required and actual powers. For this eye, the power corrections at the cornea along the two power axes are calculated as follows:

$$CP_{90} = RP - AP_{90} = 55 - 59 = -4.0 \text{ D}$$

$$CP_{180} = RP - AP_{180} = 55 - 57 = -2.0 \text{ D}$$

Next, the power corrections at the cornea must be adjusted to give the appropriate spectacle power (SP) at the distance x in front of the cornea. Using Eq. (5-3) for effective power, we have

$$SP_{90} = \frac{CP_{90}}{1 - x(CP_{90})} = \frac{-4.0}{1 - (-0.012)(-4.0)} = -4.20 \text{ D}$$

$$SP_{180} = \frac{CP_{180}}{1 - x(CP_{180})} = \frac{-2.0}{1 - (-.012)(-2.0)} = -2.05 \text{ D}$$

Note that the distance x is expressed in meters and is negative because it is measured (upstream) from the cornea backward to the spectacle lens.

Thus, the power cross for the correcting lens for this eye is −4.20 D at 90° (vertical) and −2.05 D at 180° (horizontal). Then, in the spirit of Figure 6-16, we can find two combinations of a spherical and cylindrical surface which result in this correction. The two choices are given, in standard notation, by −2.05/−2.15 × 180 and −4.20/+2.15 × 90. ◼

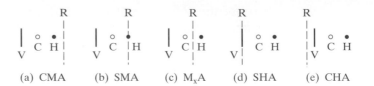

Figure 6-18 Types of astigmatism depend on the position of the elements of the Sturm conoid relative to the retina (R). The cornea is assumed to lie to the left in each case. The vertical line image (V), horizontal line image (H) and circle of least confusion (C) are labeled in (a) and shown in each of the five types. Images are assumed to result from a distant point object in an unaccommodated eye.

In the eye, the location of the Sturm conoid relative to the retina depends on the nature and extent of the astigmatism present. Classification of astigmatism in an unaccommodated eye is illustrated in Figure 6-18. In the figure, we assume for simplicity that the astigmatic line images are along horizontal and vertical axes and that the vertical line image results from the stronger convergence, appearing closer to the cornea. Features of the Sturm conoid are symbolized by a vertical line segment to represent the vertical line image, a solid dot to represent the horizontal line image (perpendicular to the page), and a small, open circle between the line images to represent the circle of least confusion.

The five cases shown in Figure 6-18 result from various positions of the Sturm conoid as it shifts gradually from in front of the retina (a) to behind the retina (e), responding to gradually decreasing converging power of the eye. In (a) *compound myopic astigmatism* (CMA), both line images fall short of the retina. In (b), *simple myopic astigmatism* (SMA), the line image furthest from the cornea falls on the retina. In (c), *mixed astigmatism* (M_xA), the retina falls between the two line images. In (d), *simple hyperopic astigmatism* (SHA), the line image nearest the cornea falls on the retina. In (e), *compound hyperopic astigmatism* (CHA), both line images fall beyond the retina.

What does the eye see in each case? Of course, when one of the line images falls squarely on the retina, the point object is seen as a focused line image, oriented horizontally in case (b) and vertically in case (d). Otherwise, an elliptical spot is seen. In case (a), the blur ellipse is oriented with its long axis along the horizontal, and in case (e), along the vertical. If it happens in case (c) that the circle of least confusion falls on the retina, the blur spot is small and circular (special case of elliptical). Otherwise it has the elliptical orientations suggested in Figure 6-11 for positions along the Conoid interval. These images are summarized in Figure 6-19.

Keep in mind that the images in Figure 6-19 are those of a point object. In practice, of course, we view extended objects. Light is directed to the eye from many points and each produces a set of images like those of Figure 6-19. For example, suppose we are looking at the letter H and we have the type of astigmatism of Figure 6-18(d), for which each point object images as a vertical line on the retina. Consider first the representative object points on the letter H, as shown in Figure 6-20(a). The corresponding images, one vertical line segment for each object point, are shown in Figure 6-20(b). If we then fill in more and more of the object points on H, the large number of line images run together to produce the

Figure 6-19 Blur images on the retina of a point object corresponding to the types of astigmatism shown in Figure 6-18.

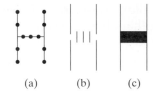

Figure 6-20 Image of the letter H on the retina in case of SHA astigmatism that places a vertical line image on the retina. In (b) are shown the images of the small number of selected object points in (a). In (c) the image seen results from all object points of the letter H.

(a)   (b)   (c)

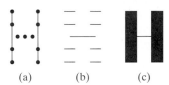

Figure 6-21 Images of the letter H as in Figure 6-20, but with SMA astigmatism that places a horizontal line image on the retina.

(a)   (b)   (c)

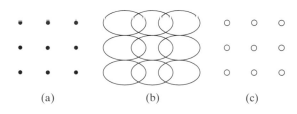

Figure 6-22 Astigmatic image of points of a grid-type object (a). If the elliptical blur images are large enough compared to the size of the grid openings, the images overlap as in (b) and the image loses definition. If the circles of least confusion fall on the retina, the image of the grid as in (c) is recognizable due to the small size of the blur circules on the retina.

(a)   (b)   (c)

image suggested by Figure 6-20(c). This is the astigmatic image seen that needs correction. Notice that in this case the distortion occurs mainly in the horizontal element of H, which appears broader than it should. The vertical limbs of the letter H appear slightly longer than in the object due to imaging of endpoints on the vertical limbs of the object. If, on the other hand, we viewed the same letter with the type of astigmatism of Figure 6-18(b), for which each object point images as a horizontal line, the sketches of Figure 6-20 would appear as in Figure 6-21, where the distortion occurs mainly in the vertical elements of the object.

If the astigmatism instead places an elliptical image on the retina for each object point, both vertical and horizontal elements would appear broadened accordingly. Now imagine an object in which there are no large open spaces, as in the fine mesh of Figure 6-22(a). If the elliptical images of all the point objects at the intersections of the mesh are large enough to overlap, the open spaces in the image are mostly covered, as in Figure 6-22(b). The astigmatic image in this case is unrecognizable and distortion is complete. By comparison, an astigmatism in which the circles of least confusion fall on the retina, the image becomes clearer [Figure 6-22(c)] because of the smaller size of the blur images.

6-6 OBLIQUE ASTIGMATISM

In this chapter, we have been considering astigmatism for axial object points that occurs because the refracting surfaces are not perfectly spherical. However, even when the refracting surfaces are perfectly spherical, *oblique astigmatism* occurs when imaging off-axis object points. This is equivalent to saying that the lens is tilted relative to its optical axis. In such cases, light rays from the object point do not "see" a symmetrical spherical surface. The result is that the usual two line images are formed whose Sturm interval depends on the extent to which the object point is off-axis. The occurrence of oblique astigmatism in the visual image of off-axis objects is especially noticeable in peripheral vision, when the pupil is large enough to allow considerable imaging of off-axis objects. Oblique astigmatism, as well as other aberrations that arise from off-axis object points, are considered in the next chapter.

PROBLEMS

6-1. Light rays emanating in air from a point object on axis strike a plano-cylindrical lens with its convex surface facing the object. Describe the line image by length and location if the lens has a radius of curvature of 5 cm, a refractive index of 1.60, and an axial length of 7 cm. The point object is 15 cm from the lens.

6-2. A plano-cylindrical lens in air has a curvature of 15 cm and an axial length of 2.5 cm. The refractive index of the lens is 1.52. Find the position and length of the line image formed by the lens for a point object 20 cm from the lens. Light from the object is incident on the convex cylindrical surface of the lens.

6-3. A plano-cylindrical lens in air has a radius of curvature of 10 cm, a refractive index of 1.50, and an axial length of 5 cm. Light from a point object is incident on the concave, cylindrical surface from a distance of 25 cm to the left of the lens. Find the position and length of the line image formed by the lens.

6-4. A plano-concave cylindrical lens is used to form an image of a point object 20 cm from the lens. The lens has a refractive index of 1.50, a radius of curvature of 20 cm, and an axial length of 2 cm. Describe as completely as possible the line image of the point.

6-5. Consider the plano-convex cylindrical lens in Problem 6-1. If the point object is only 6.0 cm from the lens, describe the line image.

6-6. Derive Eqs. (6-4) and (6-5), which locate the position and diameter of the circle of least confusion.

6-7. Determine the interval of Sturm for a composite cylindrical lens with perpendicularly oriented powers, given by $+5.00\,D \times 90$ and $+10.00 \times 180$, when the lens is illuminated by rays of light from a point source object at 30.00 cm from the lens. A circular aperture of 1.00 cm diameter is positioned at the lens. Also describe the circle of least confusion.

6-8. A composite cylindrical lens has perpendicular powers of $+4.0 \times 180$ and $+8.0 \times 90$. A point source of light is situated at 40 cm from the lens. Find the two line images and describe the circle of least confusion. Assume a circular aperture at the lens of 8.0 mm diameter.

6-9. What combination of a spherical and a cylindrical lens produces a composite lens with vertical power axis of $+6.00\,D$ and horizontal power axis of $+2.00\,D$?

6-10. What combination of a spherical and a cylindrical lens produces a composite lens with a vertical power axis of $-5.00\,D$ and a horizontal power axis of $+5.00\,D$?

6-11. One side of a lens is spherical, with a power of $+8.00\,D$, and the other side is toric, described by $+3.00 \times 30\,/\,-5.00 \times 120$. Determine the power cross and Sturm interval of the lens for a point object 40 cm from the lens.

6-12. Consider a toric lens whose two surfaces are spherical and toroidal. The spherical side has a power of $+10.0\,D$, while the toroidal side is described by $+8.0 \times 90\,/\,+4.0 \times 180$. Determine the power cross and Sturm interval for a point object at 20 cm from the lens.

6-13. A model of the eye consists of a single refracting surface at the cornea, separating air from the vitreous fluid of refractive index 4/3. The axial length of the eye is 24.24 mm and its measured powers along the vertical and horizontal axes are $+52\,D$ and $+50\,D$, respectively. Determine two combinations of spherical and cylindrical surfaces of a spectacle lens that corrects distant vision. The lenses are worn 15 mm in front of the cornea.

6-14. Solve Problem 6-13 when the measured powers along the vertical and horizontal axes are $+59\,D$ and $+53\,D$, respectively.

6-15. Sketch the image on the retina of the letter L in the case of SMA and SHA astigmatism.

REFERENCES

BENNETT, ARTHUR G. 1984. *Clinical Visual Optics.* Boston: Butterworth-Heinemann.

BYER, ALVIN. 1986. *Theoretical Optics for Clinicians.* Philadelphia: Pennsylvania College of Optometry.

KEATING, MICHAEL P. 1988. *Geometric, Physical and Visual Optics.* Boston: Butterworth-Heinemann.

MICHAELS, D. D. 1980. *Visual Optics and Refraction,* 2d. ed. St. Louis: C. V. Mosby Company.

RUBIN, M. L. 1974. *Optics for Clinicians*, 2d. ed. Gainsville, Fla: Triad Scientific Publishers.

TUNNACLIFFE, ALAN. 1989. *Introduction to Visual Optics.* New York State Mutual Book & Periodical Service, Limited.

WELFORD, W. T. 1986. *Aberrations of Optical Systems.* Boston: Adam Hilger Ltd.

7

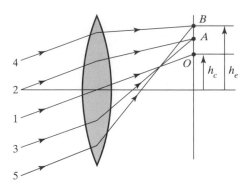

Aberration Theory

INTRODUCTION

The paraxial-ray formulas developed earlier for image formation by spherical reflecting and refracting surfaces are, of course, only approximately correct. In deriving those formulas it was necessary to assume paraxial rays, that is, rays both near to the optical axis and making small angles with it. Mathematically, the power expansions for the sine and cosine functions, given by

$$\sin x = x - \frac{x^3}{3!} + \frac{x^5}{5!} - \cdots$$

$$\cos x = 1 - \frac{x^2}{2!} + \frac{x^4}{4!} - \cdots$$

were accordingly approximated by their first terms. To the extent that these first-order approximations are valid, *Gaussian optics* implies exact imaging. The inclusion of higher-order terms in the derivations, however, predicts increasingly larger departures from "perfect" imaging with increasing angle. These departures are referred to as *aberrations*. When the next term involving x^3 is included in the approximation for $\sin x$, a third-order aberration theory results. The aberrations have been studied and classified by the German mathematician Ludwig von Seidel and are referred to as third-order or *Seidel aberrations*. For monochromatic light there are five Seidel aberrations we need to consider: *spherical aberration, coma, astigmatism, curvature of field*, and *distortion*. An additional aberration, *chromatic aberration*, results from the wavelength dependence of the imaging properties of an optical system. The details of aberration theory are too formidable to treat in this chapter.[1] We include here a qualitative description of each aberration, with typical procedures for its correction.

7-1 RAY AND WAVE ABERRATIONS

The departure from ideal, paraxial imaging may be described quantitatively in several ways. In Figure 7-1 two wave fronts are shown emerging from an optical system. Wavefront W_1 is a spherical wave front representing the Gaussian, or paraxial, approximation that produces a point image at I. Wave front W_2 is an example of an actual wave front, an aspherical envelope whose shape represents a true output of the optical system and which does not come to a focus at any single point in the image plane. This shape could be deduced by precisely tracing a sufficient number of rays through the optical system by the methods described in Section 4-2. Two sample rays from adjacent points A and B, being normal to their respective wave fronts, do not intersect the paraxial image plane at the same point. The "miss" along the optical axis, represented by LI, is called the *longitudinal aberration*, while the miss IS, measured in the image plane, is called the *transverse* or *lateral aberration*. These are *ray aberrations*. Alternatively, the aberration may be described in terms of the deviation of the deformed wave front from the ideal at various distances from the optical axis. At the location of point B, shown in Figure 7-1, the wave aberration is given by the distance AB. Notice that rays from both wave fronts, at their point of tangency O on the optical axis, reach the same image point I. Rays from intermediate points of the actual wave front between O and B intersect the image screen at other points around I producing a blurred image, the result of aberration. The maximum ray aberration thus indicates the size of the blurred image. The ultimate goal of optical design

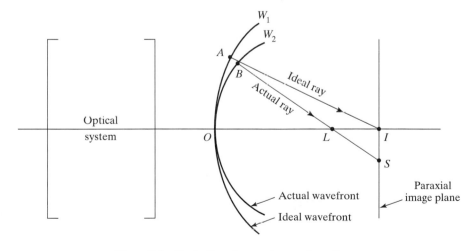

Figure 7-1   Illustration of ray and wave aberrations.

[1]A more quantitative treatment can be found in *Introduction to Optics*, 2nd ed., Chapter 7, Frank L. Pedrotti & Leno S. Pedrotti, Prentice Hall, 1993.

is to reduce the ray aberrations until they are comparable to the unavoidable blurring due to diffraction itself. We discuss the diffraction effect in Chapter 16.

7-2 SPHERICAL ABERRATION

The aberration known as *spherical aberration* is the only third-order aberration that does not depend on the degree of off-axis imaging. Thus spherical aberration exists even for an axial object, as illustrated for a single lens in Figure 7-2(a). The paraxial image point I of object point O is distinct from other axial image points, such as M, for rays from O undergoing refraction by increasingly larger lens apertures. The axial miss distance EI, due to rays from near the extremeties of the lens, provides the usual measure of *longitudinal spherical aberration*, whereas the distance IG in the paraxial image plane measures the corresponding *transverse spherical aberration*. These two aberrations also depend on the object distance for point O. When the image point E is to the left of I, as shown for the case of a positive lens, the spherical aberration is positive. If point O is imaged by a negative lens forming virtual image points, E falls to the right of I, and the spherical aberration is considered negative. At some intermediate point between E and I in Figure 7-2(a), a "best" focus is attained in practice. The broadened image there is called, descriptively, the "circle of least confusion."

Figure 7-2(b) shows spherical aberration for a positive lens when the object is at infinity. Various circular *zones* of the lens about the optical axis produce different focal lengths, so that f is a function of the aperture dimension h. The specified focal length f of the lens is due to the intersection of paraxial rays for which $h \to 0$. This focal length is given by the lensmaker's formula,

$$\frac{1}{f} = (n - 1)\left(\frac{1}{r_1} - \frac{1}{r_2}\right) \tag{7-1}$$

for a thin lens of refractive index n and radii of curvature r_1 and r_2, when used in air. From Eq. (7-1) it appears that a given f may result from different combinations of r_1 and r_2. Various choices of the radii of curvature, while not changing the focal length, may have a large effect on the degree of spherical aberration of the lens. Figure 7-3 illustrates the "bending," or change in shape, of a lens as its radii

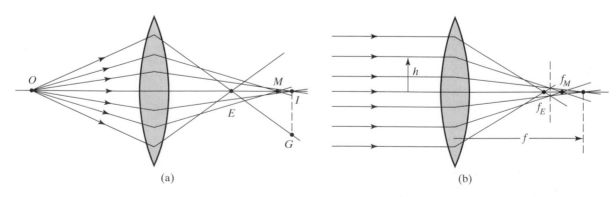

Figure 7-2   Spherical aberration of a lens, producing (a) different image distances, and (b) different focal lengths, depending on the lens aperture.

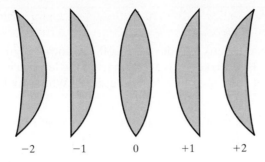

Figure 7-3 "Bending" of a single lens into various surface shapes with focal lengths unchanged. The Coddington shape factor below each version serves to classify them.

−2   −1   0   +1   +2

of curvature vary but its focal length remains fixed. A measure of this bending is the *Coddington shape factor* σ, defined by

$$\sigma = \frac{1 + (r_1/r_2)}{1 - (r_1/r_2)} \tag{7-2}$$

where the usual sign convention for r_1 and r_2 is assumed. For example, a thin lens of $n = 1.50$ and $f = 10$ cm may result from an equiconvex lens of $\sigma = 0$ ($r_1 = 10$, $r_2 = -10$ cm); a plano-convex lens of $\sigma = +1$ ($r_1 = 5$ cm, $r_2 \to \infty$); a meniscus lens of $\sigma = +2$ ($r_1 = 3.33$, $r_2 = 10$ cm). These shapes, as well as their mirror images with negative shape factors, are shown in Figure 7-3.

❏ **Example**

Find the radii of curvature for a thin lens of index $n = 1.50$ and focal length $f = 10.0$ cm, such that it has a Coddington factor $\sigma = 2$.

Solution   The parameters of the lens must satisfy both the lensmaker's equation and the Coddington shape factor formula:

$$\frac{1}{f} = (n - 1)\left(\frac{1}{r_1} - \frac{1}{r_2}\right) \quad \text{and} \quad \sigma = \frac{1 + r_1/r_2}{1 - r_1/r_2}$$

Solving simultaneously (see Problem 7-1) for r_1 and r_2, there results

$$r_1 = \frac{2f(n - 1)}{\sigma + 1} \quad \text{and} \quad r_2 = \frac{2f(n - 1)}{\sigma - 1}$$

In this case,

$$r_1 = \frac{2(10)(0.5)}{2 + 1} = +3.33 \text{ cm} \quad \text{and} \quad r_2 = \frac{2(10)(0.5)}{2 - 1} = +10.0 \text{ cm}$$

The requisite lens is of the form shown for $\sigma = 2$ in Figure 7-3, with radii of curvature as calculated.   ∎

One can show that minimum (but not zero!) spherical aberration results when the bending is such that

$$\sigma = -\frac{2(n^2 - 1)}{n + 2} \frac{(v/u) + 1}{(v/u) - 1}$$

where u and v are object and image distances and n the refractive index of the lens.

❑ **Example**

A positive, thin lens of index $n = 1.50$ and focal length $f = 30$ cm is to image small objects placed on axis at 50 cm from the lens. To minimize spherical aberration, what Coddington factor is required? What should be the radii of curvature of the lens?

Solution The required image distance is found from the thin-lens equation: $1/v - 1/u = 1/f$, or

$$v = \frac{uf}{u + f} = \frac{(-50)(30)}{-50 + 30} = 75 \text{ cm}$$

so that $v/u = 75 / -50 = -1.50$. Then, by Eq. (7-3),

$$\sigma = -\frac{2(n^2 - 1)}{n + 2} \frac{v/u + 1}{v/u - 1} = -\frac{2(1.50^2 - 1)}{1.50 + 2}\left(\frac{-1.50 + 1}{-1.50 - 1}\right) = -0.143$$

The radii of curvature of the lens, as in the previous Example, are then found from

$$r_1 = \frac{2f(n - 1)}{\sigma + 1} = \frac{2(30)(0.5)}{-0.143 + 1} = 35.0 \text{ cm}$$

$$r_2 = \frac{2f(n - 1)}{\sigma - 1} = \frac{2(30)(0.5)}{-0.143 - 1} = -26.2 \text{ cm} \qquad ∎$$

Notice that for an object at infinity ($v/u \rightarrow 0$), $\sigma \cong 0.7$ for a lens of refractive index $n = 1.50$. This shape factor is close to that of the plano-convex lens with $\sigma = +1$. Accordingly, optical systems often make use of plano-convex lenses (with the convex side facing the parallel incident rays) to reduce spherical aberration. In general, a minimum in spherical aberration is associated with the condition of equal refraction by each of the two surfaces, similar to the case of minimum deviation in a prism (discussed in Chapter 9). When lenses are used in combination, the possibility of canceling spherical aberration arises from the fact that positive and negative lenses produce spherical aberration of opposite sign. A common application of this technique is found in the cemented "doublet" lens.

7-3 COMA

Coma is an off-axis aberration that is nonsymmetrical about the optical axis and increases rapidly as the third power of the lens aperture. Because of coma, an off-axis object point images as a blurred shape that resembles a comet with a head and a tail. Hence, the name *coma* is given to this aberration. Figure 7-4 illustrates the way in which the comatic image is formed. Figure 7-4(a) shows four parallel rays, 1, 2, 3, and 4, from a distant, below-axis object point being refracted by a convex lens. These rays all pass through a thin annular region of the lens equidistant from its optical axis. We shall refer to such a region as a *zone*. Thus the lens can be considered a series of concentric zones, extending from its center to its outer edge. One such zone is shown. All rays from the distant object point passing through this zone image as the *comatic circle* shown. Rays 1 and 2 are in a vertical plane and pass through points A and B of the zone, while rays 3 and 4

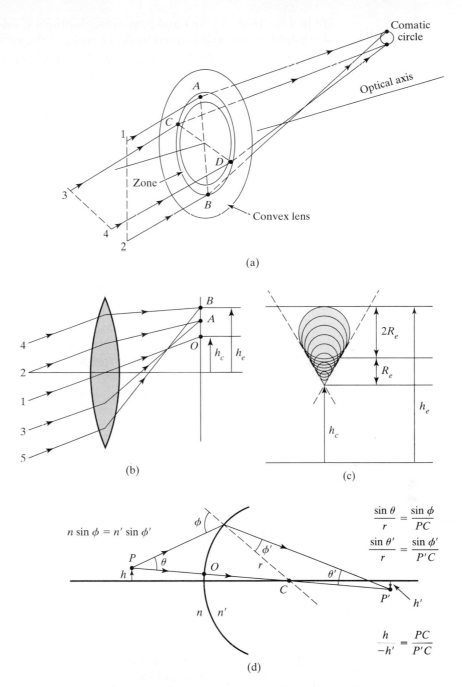

Figure 7-4 (a) and (b) Coma due to a tangential fan of parallel rays. When all such azimuthal fans are considered, each image point in the figure becomes the top of a comatic circle of image points. (c) Formation of a comatic image from a series of comatic circles. The shape of the comatic image is such that its maximum extension is three times the radius of the comatic circle formed by rays from the outer zone of the lens. The angle between the dashed lines is 60°. (d) Nonparaxial rays from object point P near the axis form an image at P', subject to the Abbe sine condition. The condition follows from Snell's law and the geometric relationships given in the figure.

are in a horizontal plane and pass through points C and D of the zone. The top of the comatic circle is formed by rays 1 and 2 of the vertical plane, while the bottom of the comatic circle is formed by rays 3 and 4 of the horizontal plane. The entire comatic circle is formed by all rays passing through the same zone. Now each such zone of the lens produces its own comatic circle, whose diameter increases as the radius of the zone increases. In Figure 7-4(b), a vertical fan of rays is shown passing through the center and two outer zones of the convex lens. The central ray appears at point O in the image plane. Rays through the intermediate zone form the top of the comatic circle for this zone at point A. Rays through the outer zone form the top of the comatic circle associated with this zone at point B. The height above the optical axis of the bottom-most point O is shown as h_c, while the distance to the top of the outermost comatic circle at point B is shown as h_e. A sketch of these and several other of the comatic circles is shown in Figure 7-4(c). Of course, since the lens zones are continuous, so are their associated comatic circles. Thus the inner divisions of these comatic circles are not seen in the image. All that is seen is the outer comatic shape. Figure 7-4(b) shows that each zone produces a different magnification, so that h_c due to central rays is not equal to h_e due to extreme rays. Coma, like spherical aberration, may occur as a positive quantity ($h_e > h_c$) or a negative quantity ($h_e < h_c$). Notice [Figure 7-4(c)] that the maximum extent of the comatic image—$R_e + 2R_e$—is three times the radius R_e of the largest comatic circle.

Without the usual paraxial approximation—restricting rays to those making small angles with the axis—one can show that for a small object near the axis, any ray from an object point that is refracted at a spherical interface must satisfy the *Abbe sine condition*,[2]

$$nh \sin \theta + n' \, h' \sin \theta' = 0 \qquad (7\text{-}3)$$

Here h and h' are object and image size, respectively, and the angles θ and θ' are the slope angles[3] of the rays in optical media n and n', respectively. These quantities are illustrated in Figure 7-4(d). When Eq. (7-3) is rearranged to express the lateral magnification, the Abbe sine condition can be written

$$m = \frac{h'}{h} = -\frac{n \sin \theta}{n' \sin \theta'}$$

To prevent coma, the lateral magnification resulting from refraction by all zones of a lens must be the same. Thus coma is absent when, for all values of θ,

$$\frac{\sin \theta}{\sin \theta'} = \text{constant}$$

The bending of a lens, found useful in reducing spherical aberration, is also useful in reducing coma. The Coddington shape factor, Eq. (7-2), which results in minimum spherical aberration, is close to that producing zero coma, so that both aberrations may be significantly reduced in the same lens by proper bending. One can show that coma is absent in a lens when

$$\sigma = -\left(\frac{2n^2 - n - 1}{n + 1}\right)\left(\frac{v/u + 1}{v/u - 1}\right) \qquad (7\text{-}4)$$

---

[2]This relation can be derived by combining the equations given in Figure 7-4(d), resulting from Snell's law, the sine law for a triangle, and the proportionalities inherent in similar triangles.

[3]For small objects near the axis, the angles θ and θ' shown in Figure 7-4 are close to the slope angles the rays make with the axis.

For the example of the lens considered previously, with $n = 1.50$ and object at infinity, Eq. (7-4) gives a value of $\sigma = 0.8$, quite close to the value of $\sigma = 0.7$, which yielded minimum spherical aberration. A lens or optical system free of both spherical aberration and coma is said to be *aplanatic*.

7-4 ASTIGMATISM AND CURVATURE OF FIELD

Aplanatic optics is still susceptible to two closely related aberrations, *astigmatism* and *curvature of field*. Both aberrations increase similarly with increasing off-axis distance of the object and with increasing aperture of the spherical surface. Unlike astigmatism, however, curvature of field is symmetrical about the optical axis.

In the preceding chapter astigmatism was discussed as an optical aberration of the eye. There we were interested in astigmatism resulting from nonspherical components in the eye, even though objects were on-axis. Thus it is useful to distinguish between *oblique* (or *marginal*) *astigmatism* due to off-axis object points and *refractive astigmatism* due to lack of sphericity in one or more refracting components in the eye or optical system itself. In this chapter, we address primarily oblique astigmatism, though refractive astigmatism also enters into the discussion.

Figures 7-5(a) and (b) illustrate the astigmatic images of an off-axis point P due to a tangential fan of rays through the section tt' and a sagittal fan of rays through the section ss' of a single lens. Since these perpendicular fans of rays focus at different distances from the lens, the two images are line images, shown as T and S for the tangential and sagittal fans, respectively. The focal line T lies in the sagittal plane, and the focal line S falls in the tangential plane. If a screen held perpendicular to the principal ray is moved from S to T, intermediate images will be elliptical in shape. Approximately midway between S and T the focus is circular, the *circle of least confusion*. The locus of the line images S and T for various object points P are paraboloidal surfaces with cross sections symmetric about the optical axis, as illustrated in Figure 7-5(c). The deviation between the two surfaces along any principal ray from a given object point P measures the magnitude of the astigmatism for this object, approximately proportional to the square of the perpendicular distance from the optical axis. When the T surface falls to the left of the S surface, as shown, the astigmatic difference is taken as positive; otherwise it is negative. The T and S surfaces in Figure 7-5(c) are sometimes referred to as the "teacup and saucer" diagram, with the T surface resembling the cup and the S surface its saucer.

If points like P in Figure 7-5(a) fall along a circle in an object plane perpendicular to the optical axis, the corresponding focused line images in the T surface merge along the circumference of the image circle, making the circle sharply focused. In the S surface, however, each image corresponding to an object point P is perpendicular to the circumference of the image circle. The image of the circle due to such object points is therefore not sharp, but has everywhere the width of the focused S line. See Figure 7-6(a). On the other hand, object points along radial lines in the object circle produce sharp radial images only in the S surface, where the elongated radial images merge to produce sharply focused radial lines. See Figure 7-6(b). In the T surface, the line images together produce a line whose width is the length of the individual object point images. Thus if the object plane contains both circular and radial elements, the image distance for a good focus

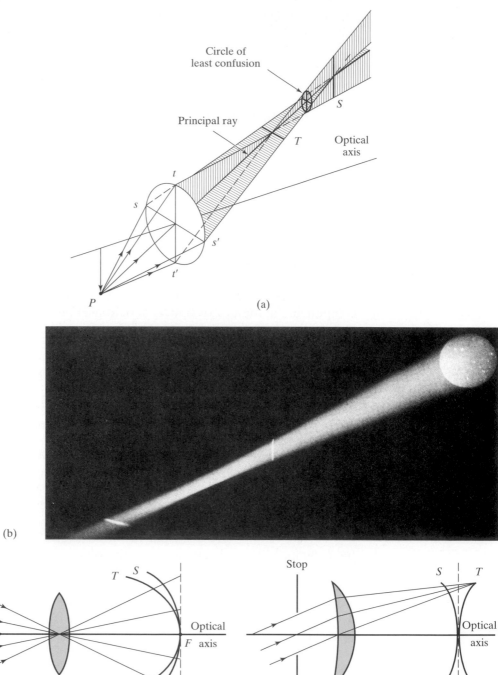

(a)

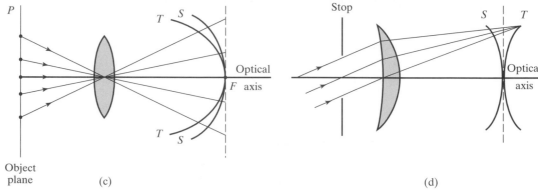

(b)

(c)

(d)

Figure 7-5 (a) Astigmatic line images T and S of an off-axial point P due to tangential (tt') and sagittal (ss') fans of light rays through a lens. (b) Photograph of astigmatic images formed by a lens, as illustrated in Figure 7-5(a). The separated line images T and S are revealed as sections of the beam by fluorescent screens. (From M. Cagnet, M. Francon, and J. C. Thrierr, *Atlas of Optical Phenomenon*, Plate 4, Berlin: Springer-Verlag, © 1962.) (c) Astigmatic surfaces in the field of a lens. (d) Use of a stop to artificially "flatten" the field of a lens. The compromise surface between the S and T surfaces is indicated by the dashed line.

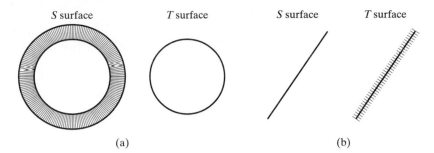

S surface   T surface   S surface   T surface

(a)   (b)

Figure 7-6 Focused astigmatic images of a circle and a radial line in the object plane, as they appear in the T and S surfaces of a lens. (a) A circular object is sharply focused in the T surface but in the S surface, its circumference has the thickness of the S image line of a point object. (b) A radial line object is sharply focused in the S surface, but in the T surface it has the thickness of the T image line of a point object.

will be different for each type of element, with a compromise image somewhere between.

From the point of view of Figure 7-5(c), the elimination of astigmatism requires that the tangential and sagittal surfaces be made to coincide. When the curvatures of these surfaces are changed by altering lens shapes or spacings so that they coincide, astigmatism is eliminated and the resulting image surface is called the *Petzval surface*. In this focal surface, for an aplanatic system, point images are formed. If the surface is curved, then, although astigmatism has been eliminated, the associated aberration called *curvature of field* remains. This aberration causes a plane surface object perpendicular to the optical axis to image as a curved surface. To record sharp images under these conditions, the film must be shaped to fit the Petzval surface. A theoretical Petzval surface exists for any optical system, even when the T and S surfaces do not coincide. Unlike the T and S surfaces, the Petzval surface is unaffected by lens bending or placement and depends only on the refractive indices and focal lengths of the lenses involved. In third-order theory, the Petzval surface is always situated three times farther from the T surface than from the S surface and always lies on the side of the S surface opposite to that of the T surface. Two lenses will have a flat Petzval surface, eliminating curvature of field if

$$n_1 f_1 + n_2 f_2 = 0$$

In general the Petzval surface for a number of thin lenses in air satisfies the condition

$$\sum \frac{1}{n_i f_i} = \frac{1}{R_p} \tag{7-5}$$

where R_p is the radius of curvature of the Petzval surface.

❏ **Example**

A lens doublet is to be made from a cemented positive lens ($n_1 = 1.463$, $f_1 = 5.000$ cm) and a negative lens ($n_2 = 1.632$, $f_2 = 7.000$ cm). Determine (a) the radius of the Petzval surface for this doublet, and (b) the proper focal length of the positive lens that would give a flat Petzval surface.

Solution

(a) The Petzval surface radius for two lenses is found from:

$$\frac{1}{R} = \frac{1}{n_1 f_1} + \frac{1}{n_2 f_2} = \frac{1}{(1.463)5} + \frac{1}{(1.632)7} \quad \text{or} \quad R = 4.46 \text{ cm}$$

(b) When $R = 0$, $n_1 f_1 = -n_2 f_2$ or

$$f_1 = -\frac{n_2 f_2}{n_1} = -\left(\frac{(1.632)(-7)}{1.463}\right) = +7.809 \text{ cm} \qquad \blacksquare$$

Field flattening by the Petzval condition cannot be accomplished for a single lens, but artificial field flattening may be accomplished by the use of an aperture stop positioned as in Figure 7-5(d). In this arrangement oblique chief rays, now determined by the aperture, do not penetrate the lens center. The S and T astigmatic surfaces then appear oppositely curved, and the surface of least confusion, centered around the optical axis, is flat, as shown. This inexpensive method for artificially flattening the field has been used in simple box cameras. In more difficult situations, where the Petzval condition cannot be satisfied without sacrificing other requirements, a low-power lens is sometimes used near the image plane. The lens helps to counteract curvature of field without otherwise seriously compromising image quality. Finally, according to fifth-order aberration theory, the T and S surfaces may actually be made to come together again and intersect at some distance from the optical axis. The result is less average astigmatism over the compromise focal plane. The *anastigmat* camera objective is designed to take advantage of this.

7-5 DISTORTION

The last of the five monochromatic Seidel aberrations, present even if all the others have been eliminated, is *distortion*. Even when object points are imaged as points, distortion shows up as a variation in the lateral magnification for object points at different distances from the optical axis. If the magnification increases with object-point distance from the optical axis, the rectangular grid of Figure 7-7(a), serving as object, will have an image as shown in Figure 7-7(b), descriptively called *pincushion distortion*. On the other hand, if magnification decreases with distance from the axis, the image appears as in Figure 7-7(c), showing so-called *barrel distortion*. The image in either case is sharp but distorted.

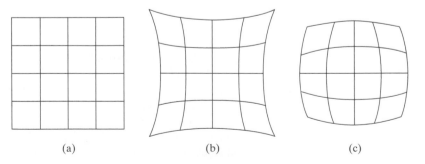

| (a) | (b) | (c) |

Figure 7-7 Images of (a) a square grid, (b) showing pincushion distortion, and (c) barrel distortion, due to nonuniform magnifications.

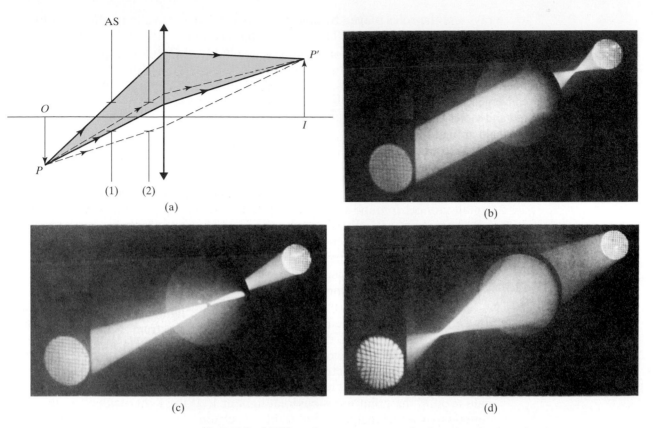

(a)

(b)

(c)

(d)

Figure 7-8 (a) Effect of an aperture stop on the distortion of an image by a lens. The aperture in position (1) produces more barrel distortion than it does in position (2). If object and image are interchanged, the same system produces pincushion distortion. (b) Image of a square grid by a positive lens. With the stop located between object (far right) and lens, barrel distortion occurs in the image. (c) Image of a square grid by a positive lens. With the stop located at the lens, the image is free from distortion. (d) Image of a square grid by a positive lens. With the stop located between lens and image, pincushion distortion occurs in the image. [Figures 7-8(b), (c), and (d) from M. Cagnet, M. Francon, and J. C. Thrierr, *Atlas of Optical Phenomenon*, Plate 5, Berlin: Springer-Verlag, © 1962.]

Such distortion is often augmented due to the limitation of ray bundles by stops or by elements effectively acting as stops. To see this effect, refer to Figure 7-8(a). Shown there is the image P' of an off-axis point P, formed by a single lens. Two pencils of rays are drawn, each limited by the aperture stop (AS) when located (1) at some distance from the lens and (2) near the lens. As the aperture AS approaches the lens, the average distance from point P to the actual lens surface, and from there to P', becomes shorter for the pencil of rays in question. Thus the effective object-to-image distance is greater for position 1 than for position 2, causing the lateral magnification of OP to be smaller for position 1 than for position 2. This decrease in lateral magnification due to aperture position is more noticeable as the object point moves farther off-axis, so that the overall image suffers from barrel distortion.

What if the aperture stop is placed on the image side of the lens? The effect of this arrangement can also be seen from Figure 7-8(a), by simply reversing all rays and the roles played by object and image. Now the ratio of effective object-to-image distance is smaller, lateral magnification of distances such as *OP* increases, and pincushion distortion appears in the overall image. When the aperture stop is placed at the position of the lens, neither distortion occurs. Also, a symmetric doublet with a central stop, combining both effects, is free from distortion for unit magnification. Photographs of the effects of stop location on distortion are reproduced in Figures 7-8(b), (c), and (d).

7-6 CHROMATIC ABERRATION

The final aberration to be discussed is not one of the Seidel aberrations, which are all monochromatic aberrations. Neither our first-order (Gaussian or paraxial) approximations nor third-order theory takes into account an important fact of refraction: the variation of refractive index with wavelength or the phenomenon of *dispersion*. Because of dispersion, an additional *chromatic aberration* (CA) appears, even for paraxial optics, where now images formed by different colors of light are not coincident.

The chromatic aberration of a lens is simply demonstrated by Figure 7-9(a). Since the focal length *f* of a lens depends on the refractive index *n* of the glass, *f* is also a function of wavelength. The figure shows parallel incident light rays passing through the lens and converging to distinct focal points for the red and violet ends of the visible spectrum. Notice that a cone of violet light will form a halo around the point focus of red light at *R*. Similarly, a cone of red light will form a halo around the point focus of violet light at *V*. If the light incident on the lens contains all wavelengths of the visible spectrum, intermediate colors focus between points *R* and *V* on the axis. Just as for a prism, greater refraction of shorter wavelengths brings the violet focus nearer the lens for the positive lens shown. Figure 7-9(b) illustrates chromatic aberration for an off-axial object point *P* and displays both *longitudinal CA* and *lateral CA*. Notice that if longitudinal CA were absent, the residual lateral CA could be interpreted as a difference in magnification for different colors. The longitudinal CA of a convex lens may easily be comparable to its spherical aberration for rays at widest aperture.

Chromatic aberration is eliminated by making use of multiple refracting elements of opposite power. The most common solution is achieved with the *achro-*

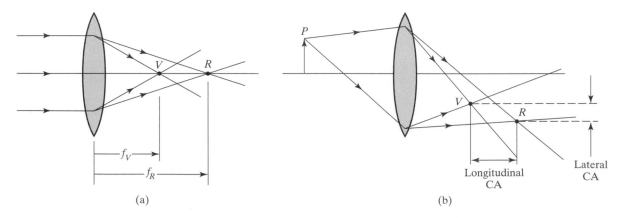

(a)   (b)

Figure 7-9   Chromatic aberration (exaggerated) for a thin lens, illustrating (a) the effect on the focal length and (b) the lateral and longitudinal misses for red (*R*) and violet (*V*) wavelengths.

matic doublet, consisting of a convex and concave lens made of different glasses, such as crown and flint, cemented together. The focal lengths and powers of the lenses differ through shaping of their surfaces to produce a net power of the doublet that may be either positive or negative. The dispersing powers of the components are, through appropriate selection of glasses, in inverse proportion to their powers. The result is a compound lens that has a net focal length but reduced dispersion over a significant portion of the visible spectrum.

Let us consider the quantitative details of this design. The general shape of the achromatic doublet is shown in Figure 7-10. The powers of the two lenses for the yellow center of the visible spectrum, conveniently represented by the *Fraunhofer wavelength*, $\lambda_D = 587.6$ nm, are

$$P_{1D} = \frac{1}{f_{1D}} = (n_{1D} - 1)\left(\frac{1}{r_{11}} - \frac{1}{r_{12}}\right) = (n_{1D} - 1)K_1 \tag{7-6}$$

$$P_{2D} = \frac{1}{f_{2D}} = (n_{2D} - 1)\left(\frac{1}{r_{21}} - \frac{1}{r_{22}}\right) = (n_{2D} - 1)K_2 \tag{7-7}$$

where the radii of curvature are designated in Figure 7-10. Here n_D refers to the refractive index of each glass for the D Fraunhofer line, and we have introduced constants K_1 and K_2 as an abbreviation for the curvatures. It can be shown[4] that the power of a doublet, with lens separation d, is given by

$$\frac{1}{f} = \frac{1}{f_1} + \frac{1}{f_2} - \frac{d}{f_1 f_2} \tag{7-8}$$

or, in terms of power,

$$P = P_1 + P_2 - dP_1P_2 \tag{7-9}$$

For a cemented doublet of thin lenses, $d = 0$, and the powers of the lenses are simply additive:

$$P = P_1 + P_2 \tag{7-10}$$

Incorporating Eqs. (7-6) and (7-7) into Eq. (7-10), we obtain for the D Fraunhofer wavelength

$$P = (n_{1D} - 1)K_1 + (n_{2D} - 1)K_2 \tag{7-11}$$

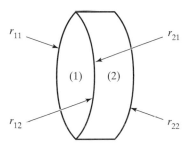

Figure 7-10 Cross section of an achromatic doublet, consisting of (1) crown glass equiconvex lens cemented to (2) a negative flint glass lens. Notation for the four radii of curvature is shown.

---

[4]See, for example, *Introduction to Optics*, 2nd ed., Chapter 4, Frank L. Pedrotti & Leno S. Pedrotti, Prentice Hall, 1993.

Chromatic aberration is absent at the wavelength λ_D if the power is independent of wavelength. This means that if there is a variation in wavelength, there is no corresponding change in power. This requirement, when applied to Eq. (7-11), leads, with the help of some calculus,[5] to the condition

$$V_2 P_{1D} + V_1 P_{2D} = 0 \qquad (7\text{-}12)$$

Here we have economized by introducing a quantity V defined as the *dispersive power*, given by

$$V \equiv \frac{n_D - 1}{n_F - n_C} \qquad (7\text{-}13)$$

where n_F and n_C are the refractive indices at the red and blue Fraunhofer wavelengths of $\lambda_C = 656.3$ nm and $\lambda_F = 486.1$ nm, respectively. Solving simultaneously Eq. (7-12) together with Eq. (7-10) in the form, $P_D = P_{1D} + P_{2D}$, the powers of the individual elements may be expressed in terms of the desired power P_D of the combination:

$$P_{1D} = P_D \frac{-V_1}{V_2 - V_1} \quad \text{and} \quad P_{2D} = P_D \frac{V_2}{V_2 - V_1} \qquad (7\text{-}14)$$

The K curvature factors expressed in Eqs. (7-6) and (7-7) may then be calculated using

$$K_1 = \frac{P_{1D}}{n_{1D} - 1} \quad \text{and} \quad K_2 = \frac{P_{2D}}{n_{2D} - 1} \qquad (7\text{-}15)$$

Finally, from the values of K_1 and K_2, the four radii of curvature of the lens faces may be determined. For simplicity of construction, a crown glass lens (left lens in Figure 7-10) may be chosen to be equiconvex. In addition the curvature of the two lenses must match at their interface. The radii of curvature thus satisfy

$$r_{12} = -r_{11}, \qquad r_{21} = r_{12}, \qquad \text{and} \qquad r_{22} = \frac{r_{12}}{1 - K_2 r_{12}} \qquad (7\text{-}16)$$

where the expression for r_{22} follows from the definition of K_2 in Eq. (7-7).

In the design of an achromatic doublet, the three indices of refraction for each of the glasses to be used are taken from manufacturer's specifications, like those presented in Table 7-1. One also inputs the desired overall focal length of

TABLE 7-1 SAMPLE OF OPTICAL GLASSES

Type	Catalog code $\dfrac{n_D - 1}{10V}$	V $\dfrac{n_D - 1}{n_F - n_C}$	n_C 656.3 nm	n_D 587.6 nm	n_F 486.1 nm
Borosilicate crown	517/645	64.55	1.51461	1.51707	1.52262
Borosilicate crown	520/636	63.59	1.51764	1.52015	1.52582
Light barium crown	573/574	57.43	1.56956	1.57259	1.57953
Dense barium crown	638/555	55.49	1.63461	1.63810	1.64611
Dense flint	617/366	36.60	1.l61218	1.61715	1.62904
Flint	617/380	37.97	1.61564	1.62045	1.63198
Dense flint	689/312	31.15	1.68250	1.68893	1.70461
Dense flint	805/255	25.46	1.79608	1.80518	1.82771
Fused silica	458/678	67.83	1.45637	1.45846	1.46313

[5]See the derivation using calculus in the reference cited in the previous footnote.

the achromat. In the series of calculations leading to the four radii of curvature, a calculation that is easily programmed, Eqs. (7-13), (7-14), (7-15), and (7-16) are employed in sequence. For example, if 520/636 crown glass and 617/366 flint glass are used in designing an achromat of focal length 15 cm, these equations lead to lenses with radii of curvature given by

$$r_{11} = 6.6218 \text{ cm}$$
$$r_{12} = -6.6218 \text{ cm}$$
$$r_{21} = -6.6218 \text{ cm}$$
$$r_{22} = -223.29 \text{ cm}$$

With these values, Eqs. (7-6) and (7-7) permit the calculation of focal lengths for each of the Fraunhofer wavelengths. In this case we find

	f_1 (cm)	f_2 (cm)	f (cm)
λ_D	6.3653	−11.0575	15.0000
λ_C	6.3961	−11.147	15.007
λ_F	6.2966	−10.8485	15.007

For a thin lens, achromatizing renders focal lengths over the visible spectrum (nearly) equal, eliminating longitudinal and lateral CA at the same time. In a thick lens or an optical system of lens combinations, however, the second principal planes for different wavelengths may not coincide as they do in a thin lens. When this is the case, equal focal lengths for two wavelengths, measured as they are from their respective principal planes, do not lead to a single focal point on the axis and some longitudinal CA remains [Figure 7-11(a)]. Lateral CA, however, is eliminated because equal focal lengths for the red and blue wavelengths ensures equal lateral magnifications. If, on the other hand, the focal lengths for red and blue light are made unequal such that they produce a single focus [Figure 7-11(b)], longitudinal CA is absent, but unequal magnifications and lateral CA remain. Thus the condition for removing lateral CA is the equivalence of focal lengths for the two wavelengths, while the condition for removing longitudinal CA is the coincidence of their secondary focal points.

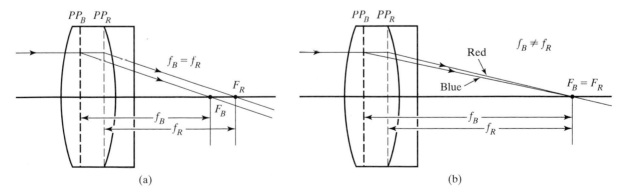

Figure 7-11 Doublet with the second principal planes separated for red and blue light. (a) Equal focal lengths result in residual *longitudinal* chromatic aberration. (b) Equal foci result in residual *lateral* chromatic aberration (⊥ to page).

Another solution for zero lateral CA results if one uses two separated lenses ($d \neq 0$) of the same glass ($n_1 = n_2 = n$). The requirement that the power be independent of the wavelength now leads to the condition

$$d = \frac{f_1 + f_2}{2} \tag{7-17}$$

Thus two lenses of the same material, separated by a distance equal to the average of their focal lengths, exhibit zero lateral CA for the particular wavelengths at which the focal lengths are calculated. This follows from the constancy of the two focal lengths. Significant longitudinal CA, however, remains.

7-7 ABERRATIONS IN VISION

As might be expected, the five monochromatic aberrations and chromatic aberration are not all equally important in the optics of the eye itself or in the spectacle lenses used to correct vision. Among the monochromatic aberrations, image quality is affected by spherical aberration, coma, and astigmatism, while curvature of field and distortion act to deform the image. All of these except spherical aberration are off-axis aberrations. Some steps taken to reduce or eliminate various aberrations in an optical system are not practical when the optical system is the eye. Fortunately, steps taken to reduce or eliminate a particular aberration often help to reduce another aberration as well. In the following paragraphs we describe some of the more obvious conditions leading to impaired vision due to these aberrations and some of the steps taken to correct them.

Spherical Aberration. As we have seen, spherical aberration is a greater problem in large-aperture optical systems. Because of the small pupil diameter, the bundle of light rays entering the eye is quite limited. Thus the eye is a rather small-aperture optical system. Spherical aberration is relatively unimportant in the eye, although it can become more important when the pupil is dilated. Visual acuity or resolving power of the eye is affected when the pupil is larger than about 2.5 mm in diameter. When larger, vision is hampered by spherical aberration; when smaller, vision is hampered by the effects of diffraction.[6]

Other factors enter as well to reduce the effects of spherical aberration in the eye. Outer portions of the cornea depart from a spherical shape to become flatter and refract less than the center portion. The central part of the crystalline lens also refracts more than its outer portions because of a slightly higher refractive index at its center. Also, light entering the peripheral cornea contributes less to the brightness sensed in the retinal image, thus giving greater weight to the action of the central cornea.

Bending can be applied to a spectacle lens to reduce its spherical aberration by equalizing the amount of refraction done by both surfaces of the lens. In general, spherical aberration is not important in ophthalmic lenses unless lenses reach higher powers of $+10.0$ D or more. In such cases aspherical surfaces, most easily made by molding plastic, can be used to reduce the aberration.

Astigmatism. Oblique astigmatism due to off-axis rays entering the eye is relatively unimportant for various reasons. Its effect is reduced by the flattening and decreased resolving power of the peripheral retina. The curvature of

---

[6]The effects of diffraction on resolution of the eye is considered in Chapter 16.

the retina, too, puts it near the circle of least confusion where image quality is optimal. Of course the eye can also suffer from refractive astigmatism due to a lack of sphericity of the cornea and, to a lesser extent, the crystalline lens. When both types of astigmatism are present correction is accomplished using bending of the spectacle lens with a toric front surface.

Peripheral rays through outer portions of a spectacle lens lead to the most important of all aberrations. Oblique astigmatism is usually of the greatest concern in the design of ophthalmic lenses and is usually controlled by modifying the form of the lens through the technique of bending.

Coma. Like spherical aberration, coma is also an aperture-dependent aberration and benefits from the effort to reduce oblique astigmatism. It is not important in the eye for the same reasons listed in our discussion of astigmatism.

Curvature of Field. This aberration is reduced in the effort to correct oblique astigmatism using a meniscus lens. It is not an important aberration of the eye. The curvature in the retinal surface itself helps to correct for curvature in the image. As stated above, the reduced resolution available at the peripheral retina helps to overcome the effect of deterioration in image quality. Some curvature of field can simply be compensated by accommodation in the hyperopic eye.

Curvature of field can be important in peripheral vision through a spectacle lens, owing to rays making large angles with the optical axis. It is responsible for the curvature present in astigmatic images. Next to oblique astigmatism, curvature of field is of greatest importance in the design of ophthalmic lenses and is improved together with the reduction of astigmatism by modifying the form of the lens.

Distortion. There is little distortion in the eye itself. This aberration is not important for the same reasons given above for ocular astigmatism. However distortion may show up when viewing through the periphery of spherical lenses of higher power. Positive spectacle lenses show some pincushion aberration while negative lenses exhibit barrel distortion. When present, it can be contained by rather drastic bending of the lens. Such lenses are both difficult to make and awkward in appearance, so that distortion is usually ignored as an aberration in the design of ophthalmic lenses. Furthermore, distortion is naturally reduced in the process of lens bending to correct oblique astigmatism.

Chromatic Aberration. As discussed above, chromatic aberration results when incident light on the eye is polychromatic, which is almost always the case. The separation of the red and blue ends of the spectrum on the optical axis amount to approximately 1.5 to 2.0 D when the yellow-green or most sensitive part of the spectrum is focused on the retina. Usually considered a disadvantage, chromatic aberration in the eye can also be put to practical use. Because blue light is focused closer to the cornea than red light, its presence can help to get a qualitative assessment of the presence of myopia or hyperopia. If a filter before the eye, such as cobalt glass, admits only blue and red light into the eye, a hyperope sees a blue center in a red field, while a myope sees a red center in a blue field.

Chromatic dispersion is not a major consideration in low power lenses, but in higher power lenses may lead to the observance of lateral colored fringes in the image when vision occurs through outer portions of the lens. This condition exists when high index glasses such as flint are used in certain types of bifocals.

The degree of chromatic dispersion is determined by lens material and so is unaffected by lens bending. It can be checked by the use of a doublet, as described earlier, but the final thickness and weight of the spectacle lens makes this impractical in an ophthalmic lens.

PROBLEMS

7-1. Show that the radii of curvature of a thin lens of refractive index n, focal length f, and Coddington shape factor σ, are given by

$$r_1 = \frac{2f(n-1)}{\sigma+1} \quad \text{and} \quad r_2 = \frac{2f(n-1)}{\sigma-1}$$

which is the result of the simultaneous solution for r_1 and r_2 of the lens maker's equation and the Coddington formula.

7-2. A positive lens of index 1.50 and focal length 30 cm is "bent" to produce Coddington shape factors of 0.70 and 3.00. Determine the corresponding radii of curvature.

7-3. Determine the radii of curvature corresponding to Coddington shape factors of 0, 1, and 2 for a thin lens of refractive index 1.50 and focal length +20 cm.

7-4. Determine the appropriate radii of curvature in bending a thin lens of refractive index 1.52 and focal length −25.0 cm, such that its Coddington shape factors are $\sigma = 2$ and 4.

7-5. Using the results of Problem 7-1, show that **(a)** when the Coddington shape factor is zero, the lens is equiconvex or equiconcave; **(b)** when the Coddington shape factor is plus or minus 1, one surface of the lens is plane.

7-6. A positive thin lens of focal length 20 cm is designed to have minimal spherical aberration in its image plane, for an object 30 cm from the lens. If the lens index is 1.60, determine its radii of curvature.

7-7. Work Problem 7-6 when the object is 40 cm from the lens.

7-8. A thin, plano-convex lens with 1 m focal length and index 1.60 is to be used in an orientation that produces less spherical aberration while focusing a collimated light beam. Prove that the proper orientation is with light incident on the spherical side by comparing the Coddington shape factor for each orientation with the value giving minimum spherical aberration.

7-9. A positive lens is needed to focus a parallel beam of light with minimum spherical aberration. The required focal length is 30 cm. If the glass has a refractive index of 1.50, determine **(a)** the required Coddington factor and **(b)** the radii of curvature of the lens. **(c)** If the lens is to be used instead to produce a collimated beam, how do these answers change?

7-10. Answer Problem 7-9 when the lens is designed to eliminate coma.

7-11. A 20-cm focal length positive lens is to be used as an inverting lens, that is, it simply inverts an image without altering its size. What radii of curvature lead to minimum spherical aberration in this application? The lens refractive index is 1.50.

7-12. Answer Problem 7-11 when the lens is designed to eliminate coma.

7-13. It is desired to reduce the curvature of field of a lens of 20-cm focal length, made of crown glass ($n = 1.5230$). For this purpose a second lens of flint glass ($n = 1.7200$) is added. What should be its focal length? Refractive indices are given for sodium light of 589.3 nm.

7-14. A doublet telescope objective is made of a cemented positive lens ($n_1 = 1.5736$, $f_1 = 3.543$ cm) and negative lens ($n_2 = 1.6039$, $f_2 = 5.391$ cm).

(a) Determine the radius of their Petzval surface.

(b) What focal length for the negative lens gives a flat Petzval surface?

7-15. Design an achromatic doublet of 517/645 crown and 620/380 flint glasses that has an overall focal length of 20 cm. Assume the crown glass lens to be equiconvex. Determine the radii of curvature of the outer surfaces of the lens, as well as its resultant focal length for the *D*, *C*, and *F* Fraunhofer lines.

7-16. Design an achromatic doublet of 5-cm focal length using 638/555 crown and 805/255 flint glass. Determine **(a)** radii of curvature; **(b)** focal lengths for *D, C,* and *F* Fraunhofer lines; **(c)** powers of the individual elements. **(d)** Is Eq. (7-12) satisfied?

7-17. Design an achromatic doublet of −10 cm focal length, using 573/574 and 689/312 glasses. Assume the crown glass lens to be equiconcave. Determine **(a)** radii of curvature of the lens surfaces; **(b)** individual focal lengths for the Fraunhofer *D* line; **(c)** overall focal lengths of the lens for the Fraunhofer *D, C,* and *F* lines.

REFERENCES

Brouwer, William. 1964. *Matrix Methods of Optical Instrument Design*. New York: W. A. Benjamin.

Conrady, A. E. 1957. *Applied Optics and Optical Design*. New York: Dover Publications.

Fincham, W. II. A., and M. H. Freeman. 1980. *Optics*, 9th ed. Boston: Butterworth Publishers. Ch. 18.

Guenther, Robert D. 1990. *Modern Optics*. New York: John Wiley and Sons. Appendix 5-B.

Hopkins, R. E. 1988. "Geometrical Optics." In *Geometrical and Instrumental Optics*, edited by Daniel Malacara. Boston: Academic Press.

Jenkins, Francis A., and Harvey E. White. 1976. *Fundamentals of Optics*, 4th ed. New York: McGraw-Hill Book Company. Ch. 9.

Keating, Michael P. 1988. *Geometric, Physical and Visual Optics*. Boston: Butterworth-Heinemann. Ch. 20.

Kingslake, Rudolf. 1978. *Lens Design Fundamentals*. New York: Academic Press.

Martin, L. C., 1960. *Technical Optics*, Vol. 2, 2d ed. London: Sir Isaac Pitman & Sons. Ch. V, Appendix V.

Nussbaum, Allen. 1968. *Geometric Optics: An Introduction*. Reading, Mass.: Addison-Wesley Publishing Company. Ch. 7, 8.

Smith, F. Dow. 1968. "How Images Are Formed." *Scientific American* (Sept.): 59-70.

Smith, Warren J. 1978. "Image Formation: Geometrical and Physical Optics." In *Handbook of Optics*, edited by Walter G. Driscoll and William Vaughan. New York: McGraw-Hill Book Company.

Welford, W. T. 1962. *Geometrical Optics*. Amsterdam: North-Holland Publishing Company. Ch. 6.

Welford, W. T. 1986. *Aberrations of Optical Systems*. Boston: Adam Hilger Ltd.

8

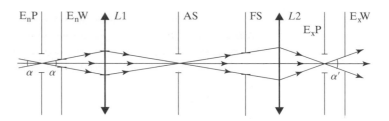

Controlling Light through Optical Systems

INTRODUCTION

Optical instruments are designed principally to deliver clear, sharp images of high quality. We have seen, so far, how mirrors and lenses are shaped and positioned to reduce various aberrations and to produce well-defined images free of distortion. We have learned also how to trace rays through optical systems using the step-by-step application of Gaussian formulas and ray-tracing techniques to locate the position and size of the final image.

Optical instruments are designed also to capture an appropriate field of view that contains the "target" scene of interest to the viewer and, at the same time, to bring as much light as possible from the target scene, point by point, onto the final image scene. It is this *field of view* and *image brightness* that concern us in this chapter. Each can be adjusted by suitably controlling light through optical systems.

The desirable field of view may be large, as in wide-angle cameras, or small, as in telephoto zoom lenses. Regardless of the chosen field of view, the more light that can be made to navigate successfully through an optical system, the brighter will be the final image. Since a final image of excellent definition but inadequate brightness is unacceptable, image brightness is a significant concern in optical design.

In this chapter we examine those elements, which by virtue of their rim sizes, locations, and refractive powers, inevitably control the field of view and brightness delivered by the optical systems of which they are a part. This analysis traditionally involves the study of so-called *stops*, *pupils*, and *windows*. In chapters that follow, where we examine in detail the operation of well-known optical instruments—including the remarkable human eye—we shall see how these elements often affect the overall operation of the system.

8-1 DEFINITION OF TERMS

The analysis of *brightness* and *field of view* in optical systems involves the use of terms that can be confusing to one embarking on a study of the subject. We begin, therefore, with a brief definition of the general terms encountered in this chapter. Their meaning will become clearer as we meet them, one by one, in typical optical systems.

Aperture. This is any optical element, such as a lens, mirror, or prism, which by virtue of its finite rim limits the bundle of light rays passing through it. It also refers to any physical stop, which except for a "porthole" cut into it also effectively blocks the passage of light. The word *aperture*, then, will be used to signify any opening defined by rim size or porthole that limits the passage of light.

Stop. When we use the word *stop* we refer to an opaque plane—a physical stop—with a fixed or variable opening. Such stops are introduced in an optical system, generally to limit the light passing through it at a given location.

Aperture stop. A particular element in an optical system, such as a lens or stop, whose rim effectively limits the maximum cone of rays that move successfully through the system from object point to conjugate image point. It is the key element controlling brightness of image.

Field Stop. A particular element in an optical system, such as a lens or stop, whose rim determines the extent of the object scene we see, and consequently, the extent of the final image. It is the key element controlling field of view.

Pupil. In the context of this chapter, pupils are generally "optical constructs," derived from knowledge of the field stop, which enable us to track the limiting cones of rays from object point to conjugate image point, through the system. They are not dissimilar in function to the role played by the pupil in the eye.

Window. Again, in the context of this chapter, windows are general optical constructs, derived from knowledge of the field stop, which enable us to see how the field of view initially limits the object scene and, consequently, the final image scene.

Limiting apertures, regardless of their origin, play a central role in optical systems. In Chapter 7, we saw how apertures can be used to modify the effects of spherical aberration, astigmatism, curvature of field, and distortion. In other applications, apertures are also used to frame an image, like the sharp outline we see looking into the eyepiece of an optical instrument. Apertures are used also to shield the image from undesirable light scattered from other optical components.

In the section that follows, we concentrate on the effects of various apertures in limiting spatially the cones or bundles of light that successfully navigate through an optical system, from object point to conjugate image point. As such, we will be dealing with aperture stops and pupils. Because real optical systems are varied and complex, we shall confine our analysis to relatively simple systems that, nevertheless, clarify the roles of such stops and pupils.

8-2 CONTROLLING IMAGE BRIGHTNESS: APERTURE STOPS AND PUPILS

As one can see by examining Figure 8-1, the angular extent of the cone of rays passing through the two-lens system, from point O to point I, is determined by the *first* lens in Figure 8-1a, and the *second* lens in Figure 8-1b. Note that the optical systems are identical; only the object position has been changed. Note also that a larger cone of rays (α_2) gets through the system in Figure 8-1b, than the cone of rays (α_1) in Figure 8-1a, thereby increasing brightness of the image in the second system.

Knowledge of image brightness depends on knowledge of angles such as α_1 and α_2. The element in an optical system that determines the maximum cone of light passing through the system is called the *aperture stop*. In Figure 8-1a, it is the first lens; in Figure 8-1b, the second lens. In Figure 8-1a, the limiting cone angle, α_1, is easy to calculate. In Figure 8-1b, the limiting cone angle, α_2, is not so obvious. To obtain α_2, we shall need to introduce optical constructs called *pupils*. Let us now consider several simple optical systems in more detail, focusing on aperture stops, entrance pupils, and exit pupils, and the role each plays.

As we have seen, the aperture stop (AS) is that particular element in an optical system—always a real element, such as a physical stop or a lens rim—that limits the angular size of the largest cone of light rays that move successfully from object point to conjugate image point. Some well-known aperture stops include the variable iris in the human eye and the adjustable diaphragm in a camera. In Figure 8-2, a simple optical system is shown consisting of a front stop and a lens. Either element is a candidate for the aperture stop. But as we can see, with the object OO' located as shown, it is the front stop, not the lens, which limits the

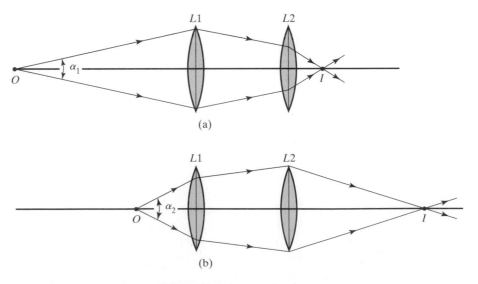

Figure 8-1 The size of the light bundle passing through an optical system depends on the relative location of the object. (a) Lens $L1$ determines bundle angle α_1. (b) Lens $L2$ determines bundle angle α_2.

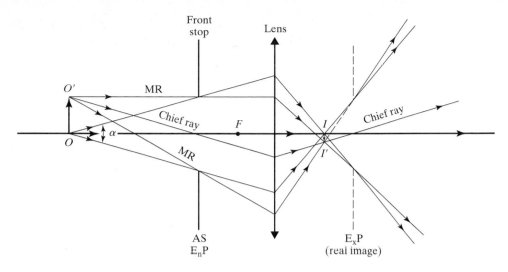

Figure 8-2 An optical system illustrating how the front stop serves as the aperture stop to limit cones of light from object points O and O' to image points I and I'. Note that the aperture stop (AS) is also the entrance pupil (E_nP), while the exit pupil (E_xP) is a real image formed behind the lens.

bundle of rays passing from point O to point I or from O' to I'. Due to the angular limitation imposed by the rim of the front stop, the outer edges of the lens receive no light and therefore do not contribute to the passage of light in this system. A *chief ray* and *marginal rays* (MR) are defined for the cone of rays from O' to I'. We shall say more about these later.

As we implied in Figure 8-1, the front stop may not always serve as the aperture stop if the object OO' is moved. In fact, it is clear from Figure 8-2 that if object OO' is moved closer to the front stop, to a position where the outer rays leaving point O just graze the rim of the front stop, but pass outside the outer rim of the lens, then it is the lens and not the stop that limits the bundle of light from points O and O'. For this new position of the object, and all object points even nearer to the front stop, the lens plays the role of the aperture stop in controlling image brightness. Let us consider next the so-called "pupils," of which there are two, the *entrance pupil* (E_nP) and the *exit pupil* (E_xP).

Entrance Pupil (E_nP). The entrance pupil is the limiting aperture that light rays "see" looking into the optical system from the object. In Figure 8-2, this is simply the front stop (the aperture stop) itself, so in this case, AS and E_nP are identical, as shown. To see that this may not always be the case, look at Figure 8-3. Here, a stop located behind the lens (a *rear stop*)—as in most cameras— serves as the aperture stop. It is this element then, not the lens, which limits rays from O to their smallest angle α relative to the axis. Looking into the optical system *from object space*, one sees the lens directly, but sees the rear stop (AS) only through the lens as a virtual image. In other words, the *effective aperture* due to the aperture stop AS is its image formed by the lens, indicated by the dashed line marked E_nP. Since rays from O, directed toward this *virtual aperture*, make a smaller angle than rays directed toward the lens edge, this virtual aperture E_nP serves as the effective *entrance pupil* of the system. Notice that rays from O, directed toward the edges of E_nP, are in fact refracted by the lens so as to pass through the edges of the real aperture stop—the rear stop. This must be the case, since AS and E_nP are, by definition, conjugate planes: The edges of E_nP are the

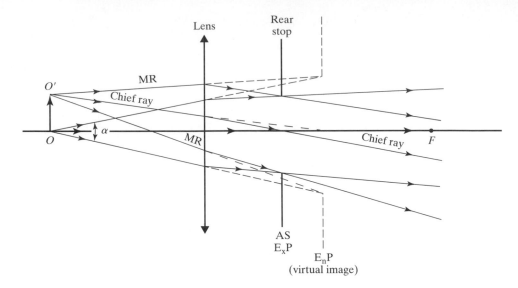

Figure 8-3 An optical system showing how a stop behind a lens serves as the aperture stop (AS) and the exit pupil (E_xP). The entrance pupil (E_nP) is a virtual image formed to the right of the stop.

images of the edges of AS formed by the lens. This example illustrates the general rule: *The entrance pupil is the image of the controlling aperture stop formed by all imaging elements preceding it.*[1] In those instances when the controlling aperture stop is the first such element (a *front stop*), as in Figure 8-2, it serves itself as the entrance pupil.

Another example, in which an aperture placed in front of the lens serves as the AS for the system, is shown in Figure 8-4. It is different from Figure 8-2 in

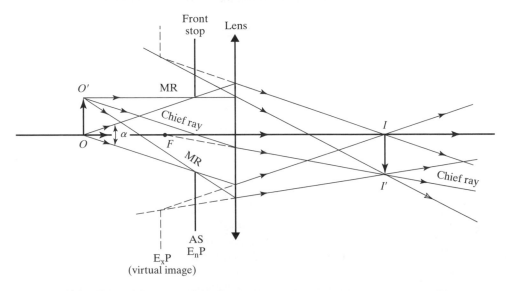

Figure 8-4 An optical system with a front stop placed within the focal length of the lens. The front stop is the aperture stop (AS) and entrance pupil (E_nP), as in Figure 8-2, but the exit pupil (E_xP) is now virtual and located in object space.

[1]"Preceding" is used in the sense that light must pass through those imaging elements first. If we always use light rays directed from left to right, we can simply say, "by all imaging elements to its left."

that the front stop is placed *inside* the focal length of the lens. Nevertheless, the front stop is the AS for the system because it, not the lens, limits the system rays to their smallest angle with the axis. Furthermore, it is the E_nP of the system because it is the first element encountered by light rays from the object.

Exit Pupil (E_xP).   We have described the E_nP of an optical system as the image of the AS one sees when looking into the optical system from object space. If one looks into the optical system from *image space*, beyond the last optical element, another image of the AS can be seen that appears to limit the output beam size. This image is called the *exit pupil* of the optical system. Thus *the exit pupil is the image of the controlling aperture stop formed by all imaging elements following it* (or *to the right* of it). The rear stop in Figure 8-3 is automatically the E_xP for the system because it is the last optical component. In all cases, the exit pupil E_xP is conjugate with E_nP. In Figure 8-2, E_xP shown there is the real image of the entrance pupil E_nP (also the AS). In Figure 8-4, E_xP is the virtual image of E_nP (also the AS), as formed by the lens. Notice that in each case, light rays from axial point O, which just graze the edges of the entrance pupil, also (actually or when extended) graze the edges of the exit pupil.

In a system like that of Figure 8-2, a screen held at the position of the exit pupil displays a sharp, real image of the circular opening of the aperture stop. If the system represents the eyepiece of some optical instrument, the exit pupil of that eyepiece is comfortably matched, in position and diameter, to the pupil of the observer's eye. Notice further that if this same screen is moved closer toward the lens to the image plane, it intercepts a sharp image II' of the object OO'. The exit pupil, whether a real or virtual image, is seen to limit the solid angle of rays affecting the brightness of each point of the image.

Chief Ray.   The *chief* or *principal ray* is a ray from an object point that passes through the *axial point*, in the plane of the entrance pupil. Given the conjugate nature of the entrance pupil with both the aperture stop and the exit pupil, this ray must also pass (*actually or when extended*) through their axial points. This behavior can be verified in each of the three Figures, 8-2, 8-3, and 8-4. Of course, the single lens in those systems could each represent an entire optical system whose ray paths are determined by its cardinal points.

To extend our analysis beyond a single-lens system, look next at a system consisting of two lenses $L1$ and $L2$ with a stop S placed between them, as shown in Figure 8-5. The first question to be answered is: Which element serves as the effective AS for the whole system? The answer to this question is not always obvious, but it can be determined by a straightforward analysis. We need to determine which of the actual elements in the given system—in this case, S, $L1$, or $L2$—defines an entrance pupil that limits oncoming rays to their smallest angle with the axis, as measured from the axial object point O. To decide which candidate presents the limiting aperture, it is necessary then to find the entrance pupil for $L1$, $L2$, and S, by imaging each element through that part of the optical system lying to its left.

$L1$: Since lens $L1$ is the first element, it acts as its own entrance pupil.

S: By ray diagram or by calculation, the image of the stop S, formed by $L1$ (as if light went from right to left), is shown as S_1'.

$L2$: Similarly, by ray diagram or calculation, the image of lens $L2$, formed by $L1$, is $L2'$. Both its location and size (magnification) are shown. It is a real image.

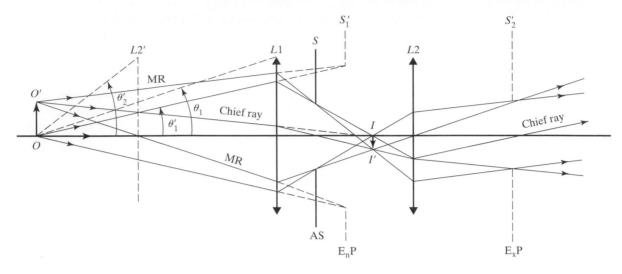

Figure 8-5 Limitation of light rays in an optical system consisting of two positive lenses and a stop. The limiting angles θ_2', θ_1', and θ_1, determined for the three candidates for the aperture stop are shown, with θ_1' as the smallest angle.

The three elements, $L2'$, $L1$ and S_1', are next viewed from the axial point O. Their rims are seen, in Figure 8-5, to subtend angles θ_2', θ_1, and θ_1' (respectively) at object point O. Since S_1' subtends the smallest angle θ_1' at O, it is the entrance pupil E_nP, and, necessarily, its conjugate object, stop S, is the aperture stop of the system. It is so labeled in Figure 8-5.

Once the AS is identified, it is imaged through the optical elements to its right to find the exit pupil. In this case, stop S is imaged through $L2$ to form S_2'. The chief ray and two marginal rays (MR) form a bundle drawn from point O of the object. Notice that the chief ray passes (*actually or when extended*) through the center of AS and its conjugate planes E_nP and E_xP. The chief ray then intersects the optical axis in planes at S, S_1', and S_2'. Note that at S_1' it is the extended ray (virtual) that intersects the optical axis. The role of the marginal rays (MR) is clear, since these rays (*actually or when extended*) just graze the edges of the aperture stop AS and the two pupils, E_nP and E_xP. The image formed by $L1$ is shown as II'; the final image (not shown) for the particular system is virtual, since the rays from either O or O' diverge on leaving $L2$. We can now repeat the discussion above for the bundle of rays shown leaving point O, tracing chief and marginal rays through the system. Note that either bundle of rays, from O or O', limited in size by the edges of the entrance pupil S_1', navigates successfully through the exit pupil S_2'.

8-3 CONTROLLING FIELD OF VIEW: FIELD STOPS AND WINDOWS

In describing the limitations of a cone of rays from an axial object point, we have seen that entrance and exit pupils are related to the aperture stop and so govern the brightness of the image. Particular optical elements also determine the *field of view* handled by the system. The controlling element in this connection is called the *field stop*, and it is related to an *entrance window* and an *exit window* in the same way that the aperture stop is related to entrance and exit pupils.

The simplest experience of a limitation in the field of view is that of looking out through an opening such as a window. The edges of the window determine how much of the outdoors we can see. This field of view can be described in terms of the lateral dimensions of the object viewed, or in terms of the angular extent of the window, relative to the line of sight. Again, one can talk about the field of view in terms of the original object or in terms of the final image (on the retina, in this example).

To see how an aperture restricts the field of view—using diagrams that could well be applied to the case of window and eye lens just discussed—look at Figure 8-6. In part (a), the optical system is a single stop S placed in front of a single lens. Object and image planes are also shown. Rays from an axial point O are limited in angle by stop S and focused by the lens at point O'. Stop S is thus the aperture stop. The same is true for the off-axis point T and its image T'. In both cases, the lens is large enough to intercept the entire cone of rays. If the object plane is uniformly bright and the aperture in stop S is a circular hole, then a circle of radius $O'T'$ is uniformly illuminated in the image plane.

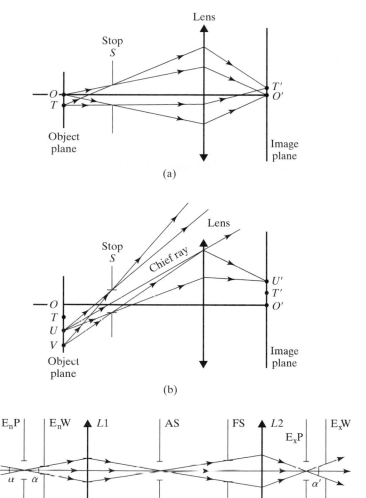

Figure 8-6 Referring to the same optical system, diagrams (a) and (b) illustrate both the way in which an aperture limits the field of view and the process of *vignetting*. Diagram (c) is an example of a more complex optical system, showing the angular field of view in object and image space.

However, if one considers object points below T, the top rays from such points, passing through S, miss the lens. Such a point, U, is shown in part (b) of the figure, the same optical system as (a), but redrawn for clarity. Point U is chosen such that the chief or central ray of its bundle of rays passes just below the top of the lens. About half of the beam is thus lost, so that image point U' receives only about half as much light as points O' and T'. Thus the circular image begins to dim as its radius increases. This partial shielding of the outer portion of the image by stop S for off-axis object points is called *vignetting*. Excessive vignetting may make the image of a point object appear astigmatic. Finally, object point V is chosen such that nearly all of its rays through the stop miss the lens entirely. The field of view (object plane) processed by this optical system is at most a circle of radius OV. It is often defined as the smaller circle of radius OU if one considers the usable field of view as that consisting of all object points that produce image points having at least half the maximum irradiance found at the center of the image.

In this same, simple figure, it turns out that the lens serves conjointly as the field stop, entrance window, and exit window. Let us then examine and describe these in turn, before we consider a more complex system, such as the one in Figure 8-6c.

Field Stop (FS).   The field stop is the optical element whose aperture limits the size or angular width of the object that can be imaged by the system. As seen from the center of the entrance pupil, the field stop (or its image)—by definition—subtends the smallest angle there. When the edge of the field of view is to be sharply delineated, the field stop should be placed in an image plane so that it is sharply focused along with the final image. A simple example of such a field stop is the opening directly in front of the film that outlines the final image in a camera. Intentional limitation of the field of view using an aperture is desirable when either far-off axis imaging is of unacceptable quality due to aberrations or when vignetting severely reduces the illumination in the outer portions of the image.

Entrance Window (E_nW).   The entrance window is the image of the field stop formed by all the optical elements preceding it (in the same way that the entrance pupil is related to the aperture stop). The entrance window delineates the lateral dimensions of the object to be viewed, as in the viewfinder of a camera, and its angular diameter determines the angular field of view. When the field stop is located in an image plane, the entrance window lies in the conjugate object plane, where it defines the lateral dimensions of the object field imaged by the optical system.

Exit Window (E_xW).   The exit window is the image of the field stop formed by all optical elements following it (in the same way that the exit pupil is related to the aperture stop). To an observer in image space, the exit window seems to limit the area of the image just as an outdoor scene appears limited by the window of a room.

In Figure 8-6c, the role of field stop, entrance window and exit window is shown in a more complex optical system consisting of two lenses and two stops. The stop between the lenses is the AS of the system and, as we have seen, is related to the entrance pupil E_nP, its image in $L1$, and an exit pupil E_xP, its image in $L2$. The second stop, to the left of the second lens L_2, is the field stop FS. The entrance window E_nW is its image formed by L_1 to its left. The exit window E_xW is the image formed by L_2 to its right. The angular field of view in object space is then defined by the angle α, the angle subtended by the rim of the entrance window at the center of the entrance pupil. Similarly, the angular field of view in image space can be described by angle α', the angle subtended by the rim of the exit window at the center of the exit pupil. We see that the size of the field imaged by the optical system is effectively determined by the entrance window and, of course, by the size of the field stop. Notice that since E_nW and E_xW are both images of the FS, they are conjugate planes. Thus the beam that fills the entrance window also fills the field stop and the exit window.

The *Summary of Terms* given in Table 8-1 is provided as a convenient reference to a subject that requires patience, as well as experience gained by working through examples.

TABLE 8-1 SUMMARY OF TERMS

Brightness	
Aperture stop:	The *real* aperture in an optical system that limits the size of the cone of rays accepted by the system from an axial object point.
Entrance pupil:	The image of the aperture stop formed by all refracting optical elements (if any) that *precede* it.
Exit pupil:	The image of the aperture stop formed by all refracting optical elements (if any) that *follow* it.

Field of view	
Field stop:	The *real* aperture that limits the angular field of view formed by an optical system.
Entrance window:	The image of the field stop formed by all refracting optical elements (if any) that *precede* it.
Exit window:	The image of the field stop formed by all refracting optical elements (if any) that *follow* it.

8-4 A COMPREHENSIVE ILLUSTRATIVE EXAMPLE

At this point, let us solidify the concepts presented in this chapter with a comprehensive example. We define the parameters for a simple optical system and then perform step-by-step calculations to determine the position and size of the aperture stop, entrance and exit pupils, field stop, and the entrance and exit windows. We also trace the chief ray and marginal rays through the system, from an off-axis object point to its conjugate point in the final image.

❏ **Example**

Consider the optical system made up of two positive thin lenses $L1$ and $L2$ with a stop S located between them, as shown in Figure 8-7a. For this system our task will be to

(a) Locate the position and size of the final image, by a simple ray trace and also by calculation.

(b) Locate the AS, E_nP, and E_xP for the system.

(c) Locate the FS, E_nW, and E_xW for the system.

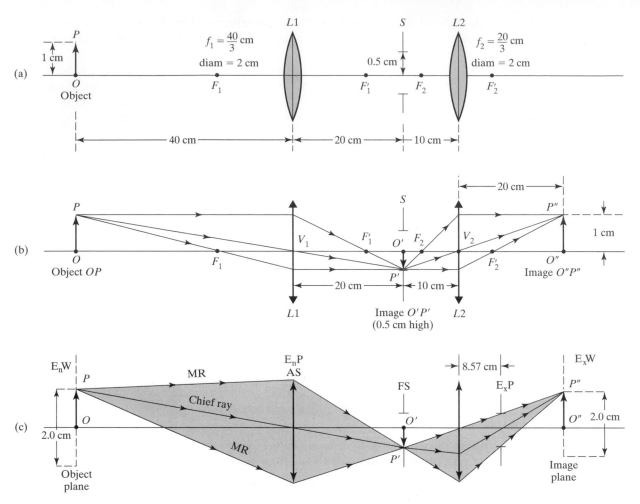

Figure 8-7 Analysis of location and size of stops, pupils, and windows for a given optical system. Diagram (a) defines the optical system. Diagram (b) shows a ray trace that locates the images formed. Diagram (c) shows a superposition of the stops, pupils, and windows on the optical system, and illustrates a cone of light through the system from point P to P' and on to final conjugate image point P''.

(d) Locate the relative positions of the appropriate stops, pupils, and windows on a scale drawing of the optical system and draw the chief ray and marginal rays for the bundle of light from object point P to conjugate image point P''.

Solution

(a) In Figure 8-7b, three rays from object point P are traced through the vertex and focal points of lens $L1$, in accordance with the ray-tracing principles presented in Chapter 3. These rays converge at point P', forming the first image $O'P'$ (inverted) in the plane of the stop S. Three more rays from image point P' are then traced through the vertex and focal points of lens $L2$ and found to converge at P'', there to locate the final (erect) image $O''P''$.

By calculation, we locate $O'P'$ and $O''P''$ as follows. With the thin-lens equation $1/v - 1/u = 1/f$, and proper sign conventions, we find for $O'P'$, using $u_1 = -40$ cm and $f_1 = 40/3$ cm,

$$\frac{1}{v_1} - \frac{1}{-40} = \frac{1}{40/3}$$

giving $v_1 = 20$ cm to the right of L_1. From the magnification equation, $m = v_1/u_1$, we find $m = -1/2$, so that $O'P'$ is inverted and 0.5 cm tall, given that OP is 1.0 cm tall. Then, using $u_2 = -10$ cm and $f_2 = 20/3$ cm in the thin-lens equation, we find that $v_2 = 20$ cm. Thus the final image $O''P''$ is located 20 cm to the right of L2, is erect, and has a size equal to 1.0 cm, since $m = 20/(-10) = -2$, and $O'P' = 0.5$ cm.

These calculations agree exactly with the results indicated by the ray trace. The images of correct size, location, and orientation are shown in Figure 8-7b.

(b) To find the AS we must image the rims of $L1$, S, and $L2$ in all optics to their left, then determine the angles that the rims of these images subtend at axial object point O. The smallest of these angles leads us back to the identity of the aperture stop, and, consequently, to the entrance and exit pupils.

Lens $L1$: $L1$ is its own image because there are no optics to its left. Since it is 40 cm from point O and has a half-diameter of 1.0 cm, it subtends a half-angle at O of $\theta_{L1} = \tan^{-1}(1/40) = 1.43°$.

Stop S: When S is imaged back through lens $L1$, using $1/v - 1/u = 1/f$, we find that $1/v - 1/(-20) = 1/(40/3)$, giving $v = 40$ cm. This tells us that the image of S is located in the object plane, is twice the size of the stop, and subtends, therefore, at point O, a half-angle $\theta_S = \tan^{-1}(1/0) = 90°$.

Lens $L2$: Imaging $L2$ back through $L1$ using $u = -30$ cm and $f = 40/3$ cm, gives $v = 24$ cm. Thus, the image for $L2$ is located 24 cm to the left of $L1$, or 16 cm to the right of object OP. The magnification $m = v/u = 24/(-30) = -0.8$ gives an inverted-image diameter of 2(0.8) or 1.6 cm. Thus the half-angle subtended at point O by the rim is $\theta_{L2} = \tan^{-1}(0.8/16) = 2.86°$.

From these calculations, we see that θ_{L1} is the smallest angle subtended at point O. Thus we conclude that lens $L1$ is the aperture stop (AS) and, being the first element in the system, also the entrance pupil (E_nP). Since the exit pupil E_xP is the image of the aperture stop AS through all optics to its right, we can again use the lens equation to find the location and size of the image of $L1$ through lens $L2$. With $u = -30$ cm and $f = 20/3$ cm, we find that E_xP is located 8.57 cm to the right of lens $L2$, with an image diameter of 4/7 cm. AS, E_nP, and E_xP are each shown in Figure 8-7c, roughly to scale.

(c) To find the field stop (FS), we must determine the half-angles subtended at the center of the entrance pupil (lens $L1$) by the images of each of the elements $L1$, S, $L2$, as formed by all optics to their left. Since these images were found in our work to identify the aperture stop, we need now only calculate the half-angles that the rims of these images subtend at the center of E_nP, that is, the center of lens $L1$.

Lens $L1$: Since lens $L1$ was found to be the entrance pupil, it must subtend an angle of $\theta_{L1} = \tan^{-1}(1/0) = 90°$ at its own center.

Stop S: The image of stop S in object space was found to be a 2-cm rim in the object plane. Thus the half-angle this rim subtends at the center of $L1$ is

$$\theta_S = \tan^{-1}(1/40) = 1.43°.$$

Lens $L2$: The image of $L2$ in object space was found to be a 1.6-cm rim, 24 cm to the left of lens L_1. Thus the half-angle this rim subtends at the center of $L1$ is $\theta_{L2} = \tan^{-1}(0.8/24) = 1.90°$.

Examination of the three angles shows that θ_S is the smallest. Thus, since θ_S is related to the element S, the stop S is the field stop (FS). Then the entrance window, E_nW, the image of FS in all optics to its left, as already determined above, is a window of 2-cm diameter, located in the object plane. Furthermore, the exit window, E_xW, the image of FS in all optics to its right, is a window of 2-cm diameter, located in the image plane. All of this is shown in Figure 8-7c.

(d) The bundle of rays, from object point P to intermediate image point P', and on to final image point P'', is shown in Figure 8-7c. Note that the chief ray passes through the centers of the E_nP, AS, and E_xP, as it must. Notice also that the two marginal rays (MR) pass through the edges of the E_nP, AS, and E_xP, as they must. At the same time this bundle of rays, after leaving object point P and entering the optical system, must converge to single image points at P' and P''. See Figure 8-7c. ■

The eye, perhaps the most interesting of optical instruments, will be treated in Chapter 10. There we shall have occasion to model the eye as an optical system. We will revisit the topic of stops, pupils, and windows, as they pertain to the eye, where they control both the brightness and the field of view of images.

PROBLEMS

8-1. Two positive lenses, each of diameter 5 cm and focal length 15 cm, are separated by a distance of 10 cm.
  (a) Draw a sketch of the optical system, roughly to scale, and locate the focal points for each lens.
  (b) For what range of object positions along the optical axis will the first lens control the amount of light passing through the system?
  (c) For what range of object positions along the optical axis will the second lens control the amount of light passing through the system?

8-2. An aperture of opening 5 cm in diameter is positioned 15 cm to the left of a positive lens of rim diameter 10 cm and focal length 12 cm. Each element is centered on the optical axis.
  (a) For an object located 6 cm to the left of the aperture, which element serves as the aperture stop?
  (b) Where are the entrance and exit pupils located?

8-3. Consider the same arrangement of aperture and lens as in Problem 8-2, but with an object now located 24 cm to the left of the aperture.
  (a) Which element is the aperture stop?
  (b) Where is the entrance pupil? What is its size?
  (c) Where is the exit pupil? What is its size?

8-4. An aperture of 5-cm opening is located 8 cm to the right of a positive lens of diameter 10 cm and focal length 12 cm.
  (a) For an object located 6 cm to the left of the lens, which element is the aperture stop?

(b) Where is the entrance pupil? What is its size?

(c) Where is the exit pupil? What is its size?

8-5. Consider an optical system similar to that in Problem 8-4, with these changes: Object is 12 cm in front of a lens of diameter 5 cm and focal length 12 cm. Aperture of diameter 6 cm remains 8 cm to the right of the lens.

(a) Which element is the aperture stop?

(b) Where is the entrance pupil? What is its size?

(c) Where is the exit pupil? What is its size?

8-6. Draw the optical system given in Problem 8-4 roughly to scale, locating the aperture stop, entrance pupil, and exit pupil, also to scale, on the same drawing.

(a) Determine image location, size and orientation for a 2-cm tall object located 6 cm to the left of the lens.

(b) Verify your solution to (a) by locating the image with the methods of ray tracing.

(c) Determine the solid angle of the cone of light passed successfully from the axial object point through the system.

8-7. An object measures 2 cm high above the axis of an optical system of a 2-cm aperture stop and a thin convex lens of 5-cm focal length and 5-cm aperture. The object is 10 cm in front of the lens, and the aperture stop is 2 cm in front of the lens. Determine the position and size of the entrance and exit pupils, as well as the image. Sketch the chief ray and the two marginal rays through the optical system, from top of object to its conjugate image point.

8-8. Repeat Problem 8-7 for an object 4 cm high, with a 2-cm aperture stop and a thin convex lens of 6-cm focal length and 5-cm aperture. The object is 14 cm in front of the lens and the aperture stop is 2.50 cm behind the lens.

8-9. Repeat Problem 8-7 for an object 2 cm high, with a 2-cm aperture stop and a thin convex lens of 6-cm focal length and 5-cm aperture. The object is 14 cm in front of the lens, and the aperture stop is 4 cm in front of the lens.

8-10. An optical system, centered on an optical axis, consists of (left to right)

- object plane,
- thin lens L_1 30 cm from the object plane,
- aperture A 15 cm farther from L_1,
- thin lens L_2 10 cm farther from A,
- image plane.

Lens L_1 has a focal length of 10 cm and a diameter of 6 cm; lens L_2 has a focal length of 5 cm and a diameter of 6 cm; aperture A has a centered, circular opening of 2.0 cm diameter.

(a) Sketch the system, roughly to scale.

(b) Locate the image plane.

(c) Locate the aperture stop and entrance pupil.

(d) Locate the exit pupil.

(e) Locate the field stop, entrance window, and exit window.

8-11. Consider a typical eye to have the following characteristics:

- radius of single surface cornea $R_C = 7.8$ mm
- radius of front (anterior) surface of eye lens $R_{AL} = 10.0$ mm
- radius of rear (posterior) surface of eye lens $R_{PL} = -6.0$ mm
- refractive index of aqueous medium $n_A = 1.336$
- refractive index (average) of lens $n_L = 1.413$
- refractive index of vitreous medium $n_V = 1.336$
- distance from cornea to front surface of lens = 3.6 mm
- distance from front to rear surface of lens = 3.6 mm
- distance from cornea to retina = 24.17 mm

All radii are measured from centers of curvature along the optical axis, and all distances are measured along the same optical axis. Assume that the cornea can be treated as a single surface and that the real pupil of the eye—the *aperture stop*, quite appropriately—is located in a plane tangent to the front surface of the lens.

(a) Sketch a vertical cross section (side view) of the eye and label all parts and dimensions given above.

(b) Determine the location of the entrance pupil, relative to the aperture stop (real pupil), and state whether it is larger or smaller in size than the real pupil.

(c) Repeat (b) for the exit pupil.

REFERENCES

BYER, ALVIN. 1986. *Theoretical Optics for Clinicians*. Philadelphia: Pennsylvania College of Optometry.

HECHT, E., and A. ZAJAC. 1987. *Optics*, 2d. ed. Reading, Mass.: Addison-Wesley Publishing Company.

KEATING, MICHAEL P. 1988. *Geometric, Physical and Visual Optics*. Boston: Butterworth-Heinemann.

KLEIN, MILES V. 1986. *Optics*, 2d ed. New York: John Wiley and Sons.

MÖLLER, K. D. 1996. *Optics*. Sausalito: University Science Books.

RABBETTS, RONALD B., and ARTHUR G. BENNETT. 1990. *Clinical Visual Optics*, 2d ed. Newton Butterworth-Heinemann.

TUNNACLIFFE, A. H. 1989. *Introduction to Visual Optics*. New York: State Mutual Book & Periodical Service, Limited.

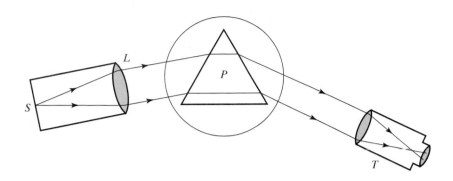

9

Optical Instrumentation

INTRODUCTION

The principles of geometrical optics—enabling us to locate the position and size of images—and the principles of stops, pupils, and windows—enabling us to determine image brightness and field of view—are applied in this chapter to a discussion of several practical optical instruments. We begin with a detailed examination of prisms, since they are present, along with lenses and mirrors, in many different optical instruments. Then, in turn, we look at spectrometers, cameras, magnifiers and eyepieces, microscopes, and telescopes. A discussion of a most remarkable "optical instrument," the human eye, as well as of other ophthalmic instruments, are treated in depth in Chapter 10.

9-1 PRISMS

Prisms are generally not used alone as imaging elements. In combination with lenses and mirrors they find frequent application in redirecting light beams, reordering the orientation of an image, separating out the individual wavelengths (colors) of a polychromatic beam, and enhancing reflection from a "perfect" mirror surface. We begin with an analysis of light through a small, triangular prism.

Angular Deviation of a Prism. The top half of a double-convex, spherical lens can form an image of an axial object point within the paraxial

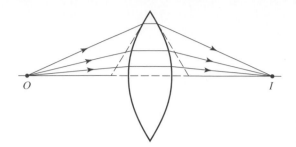

Figure 9-1 Focusing due to half of a convex lens approximates the action of a prism.

approximation, as shown in Figure 9-1. If the lens surfaces are flat, a prism is formed, and paraxial rays can no longer produce a unique image point. It is nevertheless helpful in some cases to think of a prism as functioning approximately like one-half of a convex lens.

In the following we derive the relationships that describe exactly the progress of a single ray of light through a prism. The bending that occurs at each face is determined by Snell's law. The degree of bending is a function of the refractive index of the prism and is, therefore, a function of the wavelength of the incident light. The variation of refractive index and light speed with wavelength is called *dispersion* and is discussed later. For the present we assume the light is monochromatic, with its own characteristic refractive index in the prism medium. The relevant angles describing the progress of the ray through the prism are defined in Figure 9-2. Angles of incidence and refraction at each prism face are shown relative to the normals constructed at the points of intersection of the light ray with the prism faces. The total angular deviation δ of the ray due to the action of the prism as a whole is the sum of the angular deviations δ_1 and δ_2 at the first and second faces, respectively. Snell's law at each prism face requires that

$$\sin \theta_1 = n \sin \theta_1' \tag{9-1}$$

$$n \sin \theta_2' = \sin \theta_2 \tag{9-2}$$

and inspection will show that the following geometrical relations must hold between the angles:

$$\delta_1 = \theta_1 - \theta_1' \tag{9-3}$$

$$\delta_2 = \theta_2 - \theta_2' \tag{9-4}$$

$$B = 180 - \theta_1' - \theta_2' = 180 - A \tag{9-5}$$

$$A = \theta_1' + \theta_2' \tag{9-6}$$

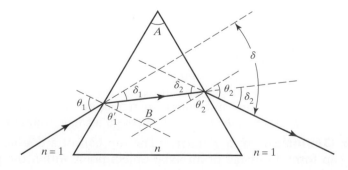

Figure 9-2 Progress of an arbitrary ray through a prism.

The two members of Eq. (9-5) follow because the sum of the angles of a triangle is 180° and because the sum of the angles of a quadrilateral must be 360°. Notice that the angles A and B and the two right angles formed by the normals with the prism sides constitute such a quadrilateral.

Using Eqs. (9-1) through (9-6), a programmable calculator or computer may easily be engaged to perform the sequential operations that finally determine the angle of deviation δ. Assuming that the *prism angle A* and refractive index n are given, then the stepwise calculation for a ray incident at an angle θ_1 is as follows:

$$\theta_1' = \sin^{-1}\left(\frac{\sin \theta_1}{n}\right) \tag{9-7}$$

$$\delta_1 = \theta_1 - \theta_1' \tag{9-8}$$

$$\theta_2' = A - \theta_1' \tag{9-9}$$

$$\theta_2 = \sin^{-1}(n \sin \theta_2') \tag{9-10}$$

$$\delta = \theta_1 + \theta_2 - \theta_1' - \theta_2' \tag{9-11}$$

The variation of deviation with angle of incidence for $A = 30°$ and $n = 1.50$ is shown in Figure 9-3. Notice that a minimum deviation occurs for $\theta_1 = 23°$. Refraction by a prism under the condition of minimum deviation is most often utilized in practice. We may argue rather neatly that when minimum deviation occurs, the ray of light passes symmetrically through the prism, making it unnecessary to subscript angles, as shown in Figure 9-4. Suppose this were not the case, and minimum deviation occurred for a nonsymmetrical case, as in Figure 9-2. Then if the ray were reversed, following the same path backward, it would have the same total deviation as the forward ray, which we have supposed to be a minimum. Hence there would be two angles of incidence, θ_1 and θ_2, producing minimum deviation, contrary to experience. The geometric relations simplify in this case. From Eq. (9-11),

$$\delta = 2\theta - 2\theta' \tag{9-12}$$

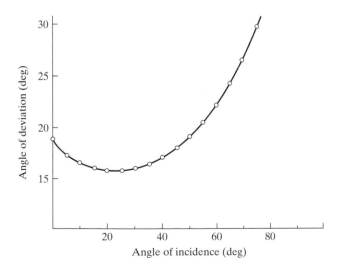

Figure 9-3 Graph of total deviation δ versus angle of incidence θ for a light ray through a prism with $A = 30°$ and $n = 1.50$. Minimum deviation occurs for an angle of 23°.

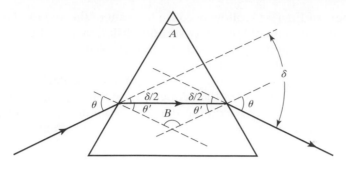

Figure 9-4 Progress of a ray through a prism under the condition of minimum deviation.

and from Eq. (9-6),

$$A = 2\theta' \tag{9-13}$$

Together these allow us to write

$$\theta' = \frac{A}{2} \quad \text{and} \quad \theta = \frac{\delta + A}{2} \tag{9-14}$$

so that Eq. (9-1) becomes

$$\sin\left(\frac{A + \delta}{2}\right) = n \sin\left(\frac{A}{2}\right)$$

or

$$n = \frac{\sin\left(\dfrac{A + \delta}{2}\right)}{\sin\left(\dfrac{A}{2}\right)} \tag{9-15}$$

Eq. (9-15) provides a method of determining the refractive index of a material that can be produced in the form of a prism. Measurement of both prism angle and minimum deviation of the sample determines n. An approximate form of Eq. (9-15) follows for the case of small prism angles and, consequently, small deviations. Approximating the sine of the angles by the angles in radians, we may write

$$n \cong \frac{(A + \delta)/2}{A/2}$$

or

$$\delta \cong A(n - 1), \quad \text{minimum deviation, small } A \tag{9-16}$$

For $A = 15°$, the deviation given by Eq. (9-16) is correct to within about 1%. For $A = 30°$, the error is about 5%.

Ophthalmic Prisms. Small angle prisms are used in ophthalmology to correct those visual anomalies often associated with *double vision*. When the right and left eyes do not aim simultaneously at an object correctly, two nonoverlapping images are perceived, thereby frustrating normal, binocular vision. Prisms with small apex angles can be used to deviate rays so as to bring the two eyes back into normal alignment. The ability of a prism to bend light rays is analyzed in terms of *prism power* measured in *prism diopters*.

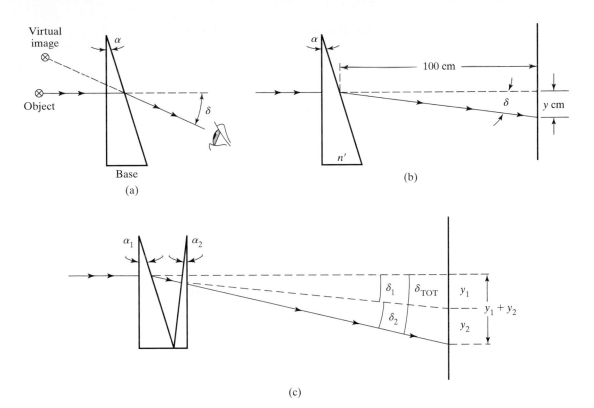

Figure 9-5 Bending power of small-angle prisms. In (a) a prism with base down bends the ray down and virtual image up. In (b) a displacement of y-cm at 100 cm is used to measure prism power. In (c) two small-angle prisms are combined to add their individual bending effects.

Figure 9-5a shows how a small-angle prism of apex angle α, with base down, deviates an object ray down toward the base by an angle δ. The virtual image of the object is seen deviated up toward the apex by an observer looking toward the prism from the right.

Since ophthalmic prisms are thin prisms, we can write, in accordance with Eq. (9-16), the minimum deviation angle δ as

$$\delta \cong (n' - 1)\alpha$$

for a prism of index n' surrounded by air ($n \cong 1$). Since the prism material has an index generally of $n' \cong 1.5$, we see that $\delta \cong 0.5\alpha$. Thus a useful rule of thumb is that the deviation δ is about one-half the prism angle α.

Bending power of the prism (P_{prism}) is given for a small angle prism in prism diopters (pd). It is measured in terms of the displacement y caused by the prism on a screen 1 m from the prism. Figure 9-5b shows a small angle prism of apex angle α, which produces a small deviation angle δ and a displacement of y centimeters at a distance of 100 centimeters. Then we can write tan $\delta = y/100$. If we now define the prism power P_{prism} in terms of the displacement distance y, $P_{prism} \cong y$, we can write

$$P_{prism} = 100 \tan \delta \quad (\delta \text{ in radians}) \tag{9-17}$$

A prism power of one prism diopter (1.0 pd) then causes a displacement of 1 cm at 100 cm, a prism power of 2.0 pd causes a displacement of 2 cm at 100 cm, and so on. From this we see that the deviation angle δ associated with one prism diopter is $\tan \delta = 1/100 = 0.01$ rad, so that $\delta \cong 0.57°$. Combining Eq. (9-17) with Eq. (9-16), and making use of the small angle approximation $\tan \delta \cong \delta$, we can write Eq. (9-17) as

$$P_{prism}(pd) = 100(n' - 1)\alpha \quad (\alpha \text{ in radians}) \tag{9-18}$$

When two thin prisms of small apex angles are used in combination, with both bases down as illustrated in Figure 9-5c, the resultant deviation δ_{tot} is simply the sum of the separate deviations δ_1 and δ_2. The prism power of the combination is then

$$P_{prism}(pd) = 100(\delta_1 + \delta_2) = pd_1 + pd_2 \tag{9-19}$$

If one of the two prisms is base down and the other base up, then the overall deviation δ_{tot} is the algebraic difference of δ_1 and δ_2. For example, if the base-down prism (δ_1) produces a displacement of 3 cm down (pd = −3), and the base-up prism (δ_2) produces a displacement of 4 cm up (pd = +4), the net displacement is 1 cm up. In this instance, then, P_{prism} is equal to 1.0 pd up, obtained from $(pd)_{net} = pd_1 + pd_2 = -3 + 4 = 1$.

❑ **Example**

An ophthalmic prism of prism power pd = 2.5 is desired. If the prism material is of index $n' = 1.56$, what should be its apex angle? How should it be oriented to bend the ray downward from the horizontal?

Solution   From Eq. (9-18), we have

$$\alpha = \frac{P_{prism}}{100(n' - 1)}$$

$$\alpha = \frac{2.5}{100(1.56 - 1)} = 0.0446 \text{ rad} \cong 2.6°$$

So the prism should have an apex angle of about 2.6° and should be oriented base down toward an oncoming ray.   ∎

❑ **Example**

A prism material has an index of 1.50 and an apex angle of 5°.

(a) What is the power of the prism?

(b) By how much does it displace an incident ray at 60 cm?

Solution

(a) $P_{prism} = 100 \tan \alpha \cong 100(n' - 1)\alpha \quad (\alpha \text{ in radians})$
   $P_{prism} = 100 (1.50 - 1) (5/57.3) = 4.36$ pd
   This means that the ray is displaced 4.36 cm at 100 cm.

(b) Setting up a simple ratio,

$$\frac{4.36 \text{ cm}}{100 \text{ cm}} = \frac{x \text{ cm}}{60 \text{ cm}} \quad \text{so that} \quad x = 2.6 \text{ cm}$$

At 60 cm, the ray is displaced 2.6 cm from the incident ray direction. ∎

❏ **Example**

Two small angle prisms, both base down, are used in combination. One has a power of 2.5 pd, the other of 4.4 pd.

(a) What is the deviation angle for the pair?

(b) What is the displacement at 100 cm from the pair?

Solution

(a) $(\text{pd})_{\text{net}} = \text{pd}_1 + \text{pd}_2 = 100\delta_{\text{tot}}$ (δ_{tot} in radians)
$\delta_{\text{tot}} = (4.4 + 2.5)/100 = 0.069 \text{ rad} = 3.95°$

(b) By definition a prism power of 6.9 pd displaces a ray 6.9 cm at 100 cm, when measured from the original direction of the ray. The displacement is downward toward the base of the prisms. ∎

Dispersion. The minimum deviation of a monochromatic beam through a prism is given implicitly by Eq. (9-15) in terms of the refractive index. The refractive index, however, depends on the wavelength, so that it would be better to write n_λ for this quantity. As a result, the total deviation is itself wavelength dependent, which means that various wavelength components of the incident light are separated on refraction from the prism. A typical dispersion curve and the nature of the resulting color separation are shown in Figure 9-6. Notice that shorter wavelengths have larger refractive indices and, therefore, smaller speeds in the prism. Consequently violet light is deviated most in refraction through the prism. The dispersion curve shown is typical, but varies somewhat for different materials. An empirical relation that approximates the curve in Figure 9-6, introduced by Augustin Cauchy, is

$$n_\lambda = A + \frac{B}{\lambda^2} + \frac{C}{\lambda^4} + \cdots \tag{9-20}$$

where $A, B, C, \ldots$ are empirical constants to be fitted to the dispersion data of a particular material. Often the first two terms are sufficient to provide a reasonable fit, in which case experimental knowledge of n at two distinct wavelengths

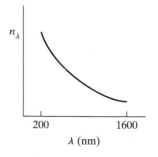

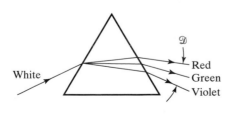

Figure 9-6 Typical dispersion curve and consequent color separation for white light refracted through a prism. The angular spread is a measure of the dispersion 𝒟.

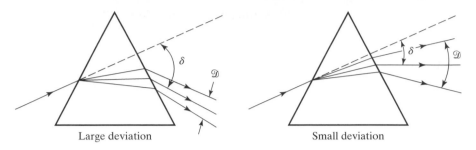

Figure 9-7   Extreme cases showing the effect of dispersion $\mathcal{D}$ for three wavelengths and the deviation δ for the intermediate wavelength.

is sufficient to determine values of A and B that represent the dispersion. The *dispersion* $\mathcal{D}$ is directly related to $\Delta n/\Delta\lambda$, the slope of a dispersion curve of n versus λ, like that of Figure 9-6. From Cauchy's formula [Eq. (9-20)], with the help of some calculus, one can show that, approximately, $\Delta n/\Delta\lambda = -2B/\lambda^3$ for a given λ.

It is important to distinguish dispersion $\mathcal{D}$ from deviation δ. Although prism materials of large n produce a large deviation at a given wavelength, the dispersion or separation of neighboring wavelengths need not be correspondingly large. Figure 9-7 depicts extreme cases illustrating the distinction. Historically, dispersion has been characterized by using three wavelengths of light near the middle and ends of the visible spectrum, called *Fraunhofer lines*. These lines were among those that appeared in the solar spectrum studied by J. von Fraunhofer. Their wavelengths, together with refractive indices, are given in Table 9-1. The F and C dark lines are due to absorption by hydrogen atoms, and the D dark line is due to absorption by the sodium atoms in the sun's outer atmosphere.[1] Using the thin prism at minimum deviation for the D line, for example, the ratio of angular spread of the F and C wavelengths to the deviation of the D wavelength, as suggested in Figure 9-7, is

$$\frac{\mathcal{D}}{\delta} = \frac{\delta_F - \delta_C}{\delta_D} = \frac{(n_F - 1)\alpha - (n_C - 1)\alpha}{(n_D - 1)\alpha} = \frac{n_F - n_C}{n_D - 1}$$

This measure of the ratio of dispersion $\mathcal{D}$ to deviation δ is defined as the *dispersive power* Δ:

$$\Delta = \frac{n_F - n_C}{n_D - 1} \tag{9-21}$$

Using Table 9-1, we may calculate the dispersive power of crown glass to be $\frac{1}{65}$, while that of flint glass is $\frac{1}{29}$, more than twice as great. The reciprocal of the dispersive power is known as the *Abbe number*.

Prism Spectrometers.   An analytical instrument employing a prism as a dispersive element, together with the means of measuring the prism angle and the angles of deviation of various wavelength components in the incident light, is called a *prism spectrometer*. Its essential components are shown in Figure 9-8. Light to be analyzed is focused onto a narrow slit S and then collimated by lens L and refracted by the prism P, which typically rests on a rotatable plat-

---

[1]Because the yellow sodium D line is a doublet (589.0 and 589.6 nm), the more monochromatic d line of helium at 587.56 nm is often preferred to characterize the center of the visible spectrum. The green line of mercury at 546.07, lying nearer to the center of the luminosity curve (Figure 2-5), is also used.

TABLE 9-1 FRAUNHOFER LINES

(nm)	Characterization	n, Crown glass	n, Flint glass
486.1	F, blue	1.5286	1.7328
589.2	D, yellow	1.5230	1.7205
656.3	C, red	1.5205	1.7076

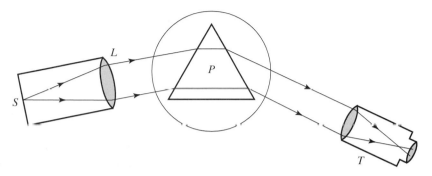

Figure 9-8   Essentials of a spectrometer.

form. Rays of light corresponding to each wavelength component emerge mutually parallel after refraction by the prism and are viewed by a telescope focused for infinity. As the telescope is rotated around the prism table, a focused image of the slit is seen for each wavelength component at its corresponding angular deviation. The deviation δ is measured relative to the telescope position when viewing the slit without the prism in place. When the instrument is used for visual observations without the capability of measuring the angular displacement of the spectral "lines," it is called a *spectroscope*. If means are provided for recording the spectrum, for example, with a photographic film in the focal plane of the telescope objective, the instrument is called a *spectrograph*. When the prism is made of some type of glass, its wavelength range is limited by the ability of the glass to absorb outside the visible region. To extend the usefulness of the spectrograph further into the ultraviolet, for example, prisms made from quartz (SiO_2) and fluorite (CaF_2) have been used. Wavelengths extending further into the infrared can be handled by prisms made of salt (NaCl, KCl) and sapphire (Al_2O_3).

Prisms with Special Applications.   Prisms may be combined to produce achromatic overall behavior, that is, the net dispersion $\Delta\mathscr{D}$ for two given wavelengths λ_1 and λ_2 may be made zero, even though the deviation δ is not zero. On the other hand, the *direct vision prism* accomplishes zero deviation for a particular wavelength while at the same time providing some dispersion $\mathscr{D}$. Schematics showing the behavior of these two prism types are shown in Figure 9-9. The arrangement of prisms in Figure 9-9a, combined so that one prism cancels the dispersion of the other, can also be reversed so that the dispersion is additive, providing double dispersion.

A prism design useful in spectrometers is one that produces a constant deviation for all wavelengths as they are observed or detected. One example is the *Pellin-Broca prism*, illustrated in Figure 9-10. A collimated beam of light enters the prism at face AB and departs at face AD, making an angle of 90° with the incident direction. The dashed lines are merely added to assist in analyzing the operation of the prism, a single structure. Of the incident wavelengths, only one will

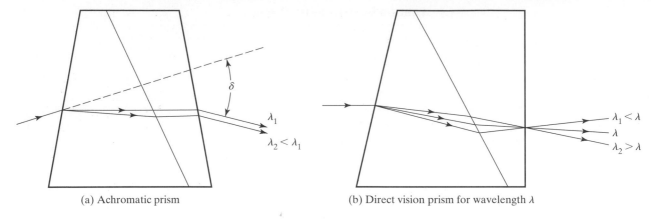

(a) Achromatic prism   (b) Direct vision prism for wavelength λ

Figure 9-9   Nondispersive and nondeviating prisms.

refract at the precise angle that conforms to the case of minimum deviation, as shown, with the light rays parallel to the dashed line AE. At face BC total internal reflection occurs to direct the light beam into the prism section ACD, where it again traverses under the condition of minimum deviation. Since the prism section BEC serves only as a mirror, the beam passes effectively with minimum deviation through sections ABE and ACD, which together constitute a prism of $60°$ apex angle. As shown in Figure 9-10, the spectral line is observed or recorded at F, the focal point of lens L. In use, the observing telescope is rigidly mounted, and, instead, the prism is rotated on its prism table (or about an axis normal to the page). As it rotates, various wavelengths in the incident beam successively meet the condition of incidence angle for minimum deviation, producing a beam along the path indicated, with focus at F. The prism rotation may be calibrated in terms of angle, or better, in terms of wavelength.

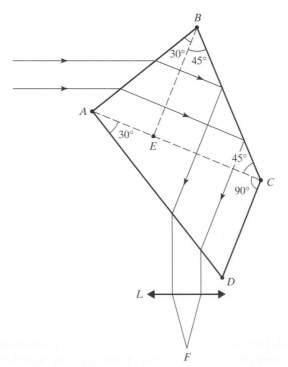

Figure 9-10   Pellin-Broca prism of constant deviation.

Reflecting Prisms. Total internally reflecting prisms are frequently used in optical systems, both to alter the direction of the optical axis and to change the orientation of an image. Of course, prisms alone cannot produce images. When used in conjunction with image-forming elements, the light incident on the prism is first collimated and rendered normal to the prism face to avoid prismatic aberrations in the image. Plane mirrors may substitute for the reflecting prisms, but the prism's reflecting faces are easier to keep free of contamination, and the process of total internal reflection is capable of higher reflectivity. The stability in the angular relationship of prism faces may also be an important advantage in some applications. Some examples of reflecting prisms in use are illustrated in Figure 9-11a–c. The *Porro prism*, Figure 9-11d, consists of two right-angle prisms, oriented in such a way that the faces of each prism are partially

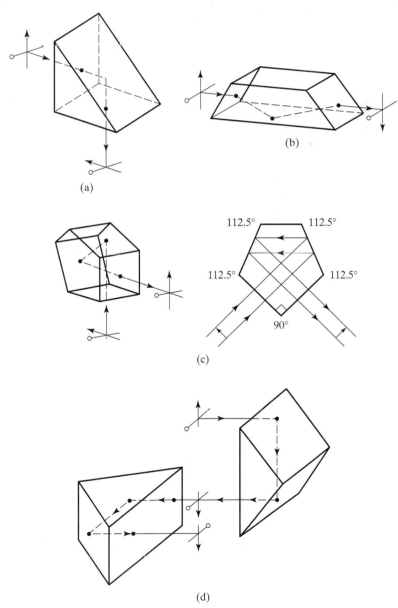

Figure 9-11 Image manipulation by refracting prisms. (a) Right-angle prism. (b) Dove prism. (c) Penta prism, pentagonal cross section. (d) Porro prism.

revealed to input and output the light. In the figure, the prism halves are separated only to clarify the composite action. Notice that images are inverted in both vertical and horizontal directions by the pair; thus, the Porro prism is commonly used in binoculars to produce erect images.

9-2 THE CAMERA

The simplest type of camera is the pinhole camera, illustrated in Figure 9-12a. Light rays from an object are admitted into a light-tight box and onto a photographic film through a tiny pinhole. The pinhole may be provided with any simple means of shuttering, such as a piece of black tape. An image of the object is projected on the back wall of the box, which is lined with a piece of film.

As stated earlier, an image point is determined ideally when every ray from a corresponding object point, each processed by the optical system, intersects at the image point. A pinhole does no focusing and its aperture actually blocks out most of the rays from each object point. Because of the smallness of the pinhole, however, every point in the image is intersected only by rays that originate at *approximately* the same point in the object, as in Figure 9-12b. Alternatively, every object point sends a bundle of rays to the screen, which are limited by the small pinhole and thus form a small circle of light on the screen, as in Figure 9-12a. The overlapping of these circles of light due to every object point maps out an image whose sharpness depends on the diameter of the individual circles. If they are too large, the image is blurred. Thus, as the pinhole is reduced in size, the image improves in clarity, until a certain pinhole size is reached. As the pinhole is reduced further, the images of each object point actually grow larger again due to diffraction, with consequent degradation of the image. Experimentally, one finds that the optimum pinhole size is around 0.5 mm when the pinhole-to-film distance is around 25 cm. The pinhole itself must be accurately formed in as thin an aperture as possible. A pinhole in aluminum foil, supported by a larger aperture, works well. The primary advantage of a pinhole camera (other than its elegant simplicity!) is that, since there is no focusing involved, all objects are in focus on the screen. In other words, the *depth of field* of the camera is unlimited. The primary disadvantage is that, since the pinhole admits so little of the available light, exposure times must be long and stray light minimal. The pinhole camera is not useful in freezing the action of moving objects. The pinhole-to-film distance,

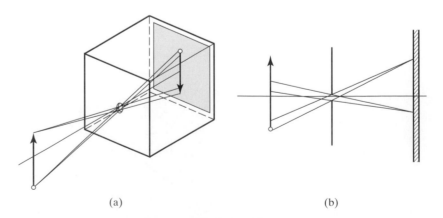

(a)   (b)

Figure 9-12   Imaging by a pinhole camera.

while not critical, does affect the sharpness of the image and the field of view. As this distance is reduced, the angular aperture seen by the film is larger, so that more of the scene is recorded, with corresponding decrease in size of any feature of the scene. Also, the image circles decrease in size, producing a clearer image.

If the pinhole is opened sufficiently to accommodate a converging lens, we have the basic elements of the ordinary camera (Figure 9-13). The most immediate benefits of this modification are (1) an increase in the brightness of the image due to the focusing of *all* the incident rays of light from each object point onto its conjugate image point, and (2) an increase in sharpness of the image, also due to the focusing power of the lens. The lens-to-film distance is now critical and depends on the object distance and lens focal length. For distant objects, the film must be situated in the focal plane of the lens. For closer objects, the focus falls beyond the film. Since the film plane is fixed, a focused image is formed by allowing the lens to be moved farther from the film, that is, by "focusing" the camera. The extreme possible position of the lens determines the nearest distance of objects that can be handled by the camera. "Close-ups" can be managed by changing to a lens with shorter focal length. Thus the focal length of the lens determines the subject area received by the film and the corresponding image size. In general, image size is proportional to focal length. A *wide-angle lens* is a short focal-length lens with a large field of view. A *telephoto lens* is a long focal-length lens, providing magnification at the expense of subject area. The telephoto lens avoids a correspondingly "long" camera by using a positive lens, separated from a second, negative lens of shorter focal length, such that the combination remains positive.

Also important to the operation of the camera is the size of its aperture (the aperture stop, AS), which admits light to the film. In most cameras, the aperture is variable and is coordinated with the exposure time (shutter speed) to determine the total exposure of the film to light from the scene. The light power incident at the image plane (irradiance E_e in watts per square meter) depends directly on (1) the area of the aperture, and inversely on (2) the size of the image. If, as in Figure 9-14, the aperture is circular with diameter D and the energy of the light is assumed to be distributed uniformly over a corresponding image circle of diameter d, then

$$E_e \propto \frac{\text{area of aperture}}{\text{area of image}} = \frac{D^2}{d^2} \tag{9-22}$$

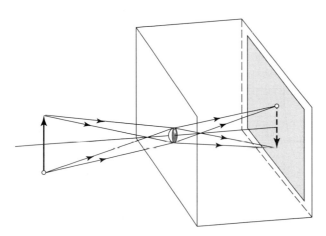

Figure 9-13 Simple camera.

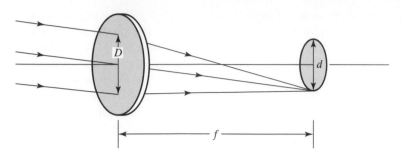

Figure 9-14 Illumination of image. The aperture (not shown) determines the useful diameter D of the lens.

The image size, as in Figure 9-14, is proportional to the focal length of the lens, so we can write

$$E_e \propto \left(\frac{D}{f}\right)^2 \tag{9-23}$$

The quantity f/D is the *relative aperture* of the lens (also called f-number or f/stop), which we symbolize by the letter A,

$$A \equiv \frac{f}{D} \tag{9-24}$$

but is, unfortunately, usually identified by the symbol f/A. For example, a lens of 4-cm focal length that is stopped down to an aperture of 0.5 cm has a relative aperture of $A = 4/0.5 = 8$. This aperture is usually referred to by photographers as $f/8$. The irradiance is now

$$E_e \propto \frac{1}{A^2}$$

Most cameras provide selectable apertures that sequentially change the irradiance at each step by a factor of 2. The corresponding f-numbers then form a geometric series with ratio $\sqrt{2}$, as in Table 9-2.[2] Larger aperture numbers correspond to smaller exposures. Since the total *exposure* (J/m²) of the film depends on the product of irradiance (J/m²-s) and time (s), a desirable total exposure may be met

TABLE 9-2 STANDARD RELATIVE APERTURES AND IRRADIANCE AVAILABLE ON CAMERAS

$A = f$-number	$(A = f\text{-number})^2$	E_e
1	1	E_0
1.4	2	$E_0/2$
2	4	$E_0/4$
2.8	8	$E_0/8$
4	16	$E_0/16$
5.6	32	$E_0/32$
8	64	$E_0/64$
11	128	$E_0/128$
16	256	$E_0/256$
22	512	$E_0/512$

[2]The column of f-numbers shown here are those that appear on cameras and, in some cases, are only close approximations of a geometric series with ratio $\sqrt{2}$. For example, f-11 is really f-11.3 ($8\sqrt{2}$). Although camera apertures are designed correctly, the f-number indications are rounded off.

in a variety of ways. Accordingly, if a particular film (whose *speed* is described by an ASA number) is perfectly exposed by light from a particular scene, with a shutter speed of $\frac{1}{50}$ s and a relative aperture of $f/8$, it will also be perfectly exposed by any other combination that gives the same total exposure, for example, by choosing a shutter speed of $\frac{1}{100}$ s and an aperture of $f/5.6$. The change in shutter speed cuts the total exposure in half, but opening the aperture to the next f/stop doubles the exposure, leaving no change in net exposure.

The particular combination of shutter speed and relative aperture chosen for an optimum total exposure is not always arbitrary. The shutter speed must be fast, of course, to capture an action shot without blurring the image. The choice of relative aperture also affects another property of the image, the *depth of field*. To define this quantity precisely, we use Figure 9-15, which shows an axial object point O at distance u_0 from a lens being imaged at distance v_0 on the other side. All objects in the object plane at O are precisely focused in the image plane, disregarding lens aberrations. Objects both closer to and farther from the lens, however, send bundles of rays that focus farther from and closer to the image plane, respectively. Thus a flat film, situated at distance v_0 from the lens, intercepts *circles of confusion* corresponding to such object points. If the diameters of these circles are small enough, the resultant image is still acceptable. Suppose the largest acceptable diameter is d, as shown, such that all images within a distance x of the precise image are suitably "in focus." The *depth of field* is then said to be the interval in object space conjugate to the interval $v_0 - x$ to $v_0 + x$ in image space, as shown. Notice that although the interval is symmetric about v_0 in image space, the depth of field interval is not symmetric about u_0.

The near-point and far-point distances u_1 and u_2, of the depth of field can be determined once the allowable blurring parameter d is chosen and the lens is specified by focal length and relative aperture. The angle α in Figure 9-15 may be specified in two ways,

$$\tan \alpha \cong \frac{D}{v_0} \quad \text{and} \quad \tan \alpha \cong \frac{d}{x}$$

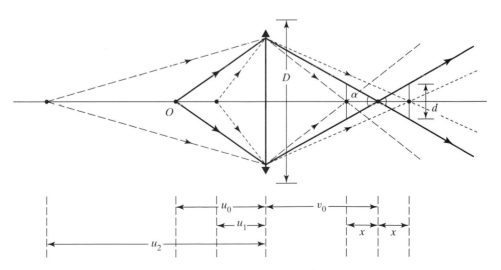

Figure 9-15   Construction illustrating depth of field. Object and image spaces are not shown to the same scale.

so that

$$x \cong \frac{dv_0}{D} \tag{9-25}$$

It is then required to find, from the lens equation, the object distance u_1 corresponding to image distance $v_0 + x$ and the object distance u_2 corresponding to image distance $v_0 - x$. After a moderate amount of algebra, one finds

$$u_1 = \frac{u_0 f(f + Ad)}{f^2 + Adu_0} \tag{9-26}$$

$$u_2 = \frac{u_0 f(f - Ad)}{f^2 - Adu_0} \tag{9-27}$$

where the relative aperture is $A = f/D$. In these equations, the object distances, u_0, u_1, and u_2 are all considered positive. In other words, the sign convention has not been invoked for this limited application. The depth of field, $u_2 - u_1$, can then be expressed as

$$\text{depth of field} = \frac{2Adu_0(u_0 - f)f^2}{f^4 - A^2d^2u_0^2} \tag{9-28}$$

Acceptable values of the circle diameter d depend on the quality of the photograph desired. A slide that will be projected or a negative that will be enlarged requires better original detail and hence a smaller value for d. For most photographic work, d is of the order of *thousandths of an inch*.

❏ **Example**

A camera with a 5-cm focal length lens and $f/16$ aperture is focused on an object 9 ft away. Allowing satisfactory quality of the image when d is 0.04 mm, what are the near and far distances u_1 and u_2, and the depth of field?

Solution   The aperture $A = f/D = 16$, and the product $Ad = 16(.004 \text{ cm}) = 0.064$ cm. The camera is focused on an object 9 ft or ($9 \times 12 \times 2.54 = 274.3$) cm away. The "in focus" objects then extend from a near distance of u_1 to a far distance of u_2. Substituting directly into Eq. (9-26),

$$u_1 = \frac{(274.3)(5)(5 + 0.064)}{5^2 + 0.064(274.3)} = 163 \text{ cm} \cong 5.4 \text{ ft}$$

The calculation for u_2 is the same, except for algebraic signs, and yields the distance $u_2 = 909.3$ cm $\cong 30$ ft. Thus the camera will focus all objects from around 5 ft to 30 ft to an acceptable degree. Eq. (9-28) can also be used to give the depth of field directly, about 25 ft. ∎

Most cameras are equipped with a depth-of-field scale from which values of u_1 and u_2 can be read, once the object distance and aperture are selected. According to Eq. (9-28), depth of field is greater for smaller apertures (larger f-numbers), shorter focal lengths, and longer shooting distances.

The camera lens is called upon to perform a prodigious task. It must provide a large field of view, in the range of 35° to 65° for normal lenses and as large as 120° or more for wide-angle lenses. The image must be in focus and reasonably free from aberrations over the entire area of the film in the focal plane. The aberrations that must be reduced to an acceptable degree are, in addition to chromatic aberration, the five monochromatic aberrations: spherical aberration, coma, astigmatism, curvature of field, and distortion. Since a corrective measure for one type of aberration often causes greater degradation in the image due to another type of aberration, the optical solution represents one of many possible compromise lens designs. The labor involved in the design of a suitable lens that meets particular specifications within acceptable limits has been reduced considerably with the help of computer programming. Human ingenuity is nevertheless an essential component in the design task, since there is more than one optical solution to a given set of specifications. The demands made upon a photographic lens cannot all be met using a single element. Various stages in solving the lens design problem are illustrated in Figure 9-16a, from the single-element meniscus lens, which may still be found in simple cameras, to the four-element Tessar lens. The use of a symmetrical placement of lenses, or groups of lenses, with respect to the aperture is often a distinctive feature of such lens designs. In such placements, one group may reverse the aberrations introduced by the other, reducing overall image degradation due to factors such as coma, distortion, and lateral chromatic aberration. The multiple-element lens in a modern 35-mm camera is shown in the cutaway photo (Figure 9-16b).

9-3 SIMPLE MAGNIFIERS AND EYEPIECES

The simple magnifier is essentially a positive lens used to read small print, (in which case it is often called a *reading glass*), or to assist the eye in examining any small detail in a real object. It is often a simple convex lens but may be a doublet or a triplet, thereby providing for higher-quality images.

Figure 9-17 illustrates the working principle of the *simple magnifier*. A small object of dimension h, when examined by the unaided eye, is assumed to be held at the *near point* of the normal eye—nearest position of distinct vision—position (a), 25 cm from the eye. At this position the object subtends an angle α_0 at the eye. To project a larger image on the retina, the simple magnifier is inserted and the object is moved closer to position (b), where it is at or just inside the focal point of the lens. In this position, the lens forms a virtual image subtending a larger angle α_M at the lens. The *angular magnification*[3] M of the simple magnifier is defined as the ratio α_M/α_0. In the paraxial approximation, the angles may be represented by their tangents, giving

$$M \equiv \frac{\alpha_M}{\alpha_0} = \frac{h/u}{h/-25} = -\frac{25}{u}$$

[3]When viewing virtual images with optical instruments, the images may be at great distances, even "at infinity," when rays entering the eye are parallel. In such cases, lateral magnifications also approach infinity and are not very useful. The more convenient *angular magnification* is clearly a measure of the image size formed on the retina and is used to describe magnification when eyepieces are involved, as in microscopes and telescopes.

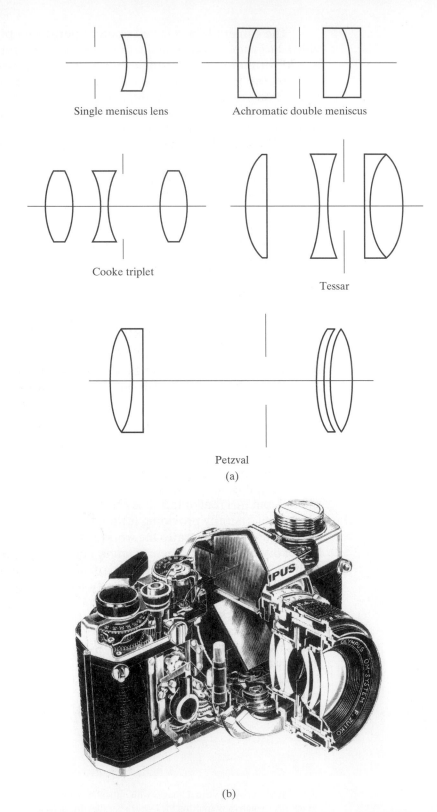

Single meniscus lens

Achromatic double meniscus

Cooke triplet

Tessar

Petzval

(a)

(b)

Figure 9-16 (a) Camera lens design. (b) Cutaway view of a modern 35-mm camera, revealing the multiple element lens. (Courtesy Olympus America, Inc., Melville, N.Y.)

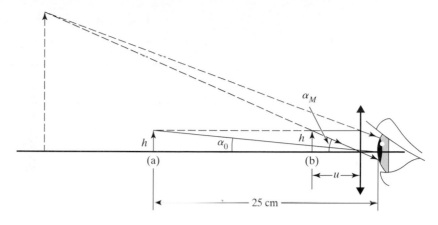

Figure 9-17 Operation of a simple magnifier

If the image is viewed at infinity, $u = -f$ and

$$M = \frac{25}{f} \quad \text{image at infinity} \tag{9-29}$$

At the other extreme, if the virtual image is viewed at the nearpoint of the eye, then $v = -25$ cm, and from the thin-lens equation, we obtain

$$u = \frac{-25f}{25 + f}$$

giving a magnification of

$$M = \frac{25}{f} + 1 \quad \text{image at normal near point} \tag{9-30}$$

The actual angular magnification depends then on the particular viewer who will move the simple magnifier until the virtual image is seen comfortably. For small focal lengths, Eqs. (9-29) and (9-30) do not differ greatly, and in citing magnifications, Eq. (9-29) is most often used. Simple magnifiers may have magnifications in the range of $2\times$ to $10\times$, although the achievement of higher magnifications usually requires a lens corrected for aberrations.

In general, when magnifiers are used to aid the eye in viewing images formed by prior components of an optical system, they are called *oculars* or *eyepieces*. The real image formed by the objective lens of a microscope, for example, serves as the object that is viewed by the eyepiece, whose angular magnification contributes to the overall magnification of the instrument. To provide quality images, the ocular is corrected to some extent for aberrations. In particular, to reduce longitudinal chromatic aberration, two lenses, appropriately separated, are most often used. For example, in an earlier discussion involving Eq. (7-17) in Chapter 7, we stated that if the two lenses are separated by a distance equal to the average of their focal lengths, that is,

$$d = \tfrac{1}{2}(f_1 + f_2) \tag{9-31}$$

then the longitudinal chromatic aberration is reduced to zero for the wavelength at which the focal lengths are determined. This condition is valid, independent of

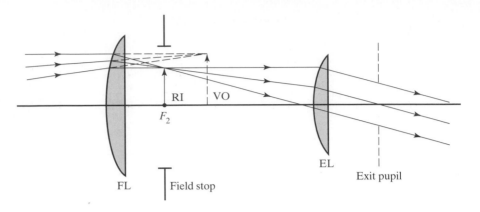

Figure 9-18 Huygens eyepiece.

the lens shapes, leaving the choice of shapes as latitude for compensating other aberrations.

Both the *Huygens* and *Ramsden eyepieces*, Figures 9-18 and 9-19, incorporate Eq. (9-31), requiring that the two plano-convex lenses be separated by half the sum of their focal lengths. In the diagram of Figure 9-18, the focal length of the field lens, FL, is approximately 1.7 times the focal length of the eye lens or ocular, EL. The primary image "observed" by the eyepiece is in this case a virtual object VO for the field lens, since its position falls between the two lenses. The field lens thus forms a real image RI of the virtual object VO, which is then viewed by the eye lens. When the real image falls in the focal plane of the eye lens, the magnified image is viewed at infinity by the eye located at the exit pupil. Note that the Huygens eyepiece cannot be used as an ordinary magnifier, which uses a real object in front of the field lens. When cross hairs or a reticule with a scale is used with the eyepiece to make possible quantitative measurements, to be in focus with the image the cross hairs must be placed in the focal plane of RI, conveniently attached to the field or aperture stop placed there (Figure 9-20). The image of the cross hairs does not share in the image quality provided by the eyepiece as a whole, however, because the eye lens alone is involved in forming the image. This is not a problem in the Ramsden eyepiece, Figure 9-19, in which both the primary and intermediate images are located just in front of the field lens. In this eyepiece, the lenses have the same focal length f and, according to Eq. (9-31), are separated by f. Ideally rays emerge from the eyepiece parallel to

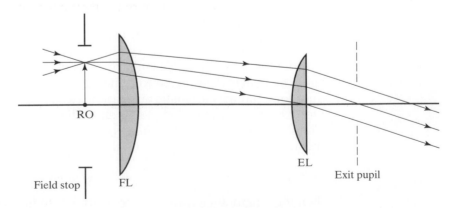

Figure 9-19 Ramsden eyepiece.

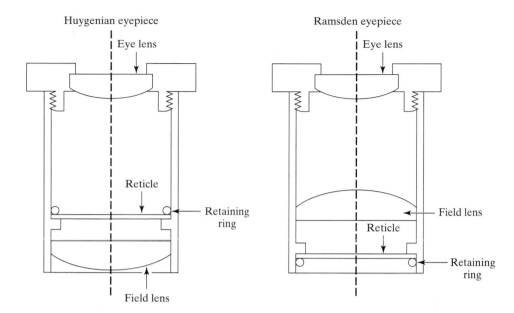

Figure 9-20 Construction of Huygens and Ramsden eyepieces.

one another giving a virtual magnified image at infinity, when the real object RO falls at the position of the first lens. A reticule is placed at this point. A disadvantage of this arrangement is that the surface of the lens is then also in focus, including the dust and smudges it carries. By using a lens separation slightly smaller than *f*, the reticule is in focus at a position slightly in front of the lens, as shown in the ray diagram, Figure 9-19, and in the actual eyepiece, Figure 9-20. With a lens separation somewhat less than *f*, however, the requirement on *L* that corrects for longitudinal chromatic aberration is somewhat violated. A modification of the Ramsden eyepiece that almost eliminates chromatic defects is the *Kellner eyepiece*, which replaces the Ramsden eye lens with an achromatic doublet. Other eyepieces have also been designed to achieve higher magnifications and wider fields.

❏ **Example**

A Huygens eyepiece uses two lenses having focal lengths of 6.25 cm and 2.50 cm, respectively. Determine their optimum separation in reducing longitudinal chromatic aberration, their equivalent focal length, and their angular magnification when viewing an image at infinity.

Solution   The optimum separation is given by Eq. (9-31),

$$d = \tfrac{1}{2}(f_1 + f_2) = \tfrac{1}{2}(6.25 + 2.5) = 4.375 \text{ cm}$$

The equivalent focal length is found from Eq. (7-8),

$$\frac{1}{f} = \frac{1}{f_1} + \frac{1}{f_2} - \frac{d}{f_1 f_2} = \frac{1}{6.25} + \frac{1}{2.50} - \frac{4.375}{(6.25)(2.50)}$$

which gives *f* = 3.57 cm. The angular magnification is then

$$M = \frac{25}{f} = \frac{25}{3.57} = 7$$

■

In designing eyepieces, one usually requires an exit pupil that is not much greater than the size of the pupil of the eye so that image radiance is not lost. Recall that the exit pupil is an image of the aperture stop as formed by all optical elements following it. In instruments such as telescopes, with eyepieces such as we are discussing, the front telescope lens (objective lens) is both the aperture stop and the entrance pupil. The exit pupil is then the image of the entrance pupil formed by the eyepiece, and the magnification is the ratio of entrance to exit pupil diameters. Since the diameter of the objective lens in a simple telescope is that of the entrance pupil, the restriction on the size of the exit pupil places a limit on the magnifying power of the eyepiece and, thus, a lower limit on the focal length. The important specifications of an eyepiece, assuming its aberrations are within acceptable limits for a particular application, include the following:

1. Angular magnification, given by $25/f$, where f is the focal length in centimeters. Available values are $4\times$ to $25\times$, corresponding to focal lengths of 6.25 to 1 cm.
2. Eye relief, that is, the distance from eye lens (rear of eyepiece) to exit pupil. Available eyepieces have eye reliefs in the range of 6 to 26 mm.
3. Field-of-view number, or size of the *primary image* that the eyepiece can cover, is in the range of 6 to 30 mm.

9-4 MICROSCOPES

The magnification of small objects accomplished by the simple magnifier is increased further with the compound microscope. In its simplest form, the instrument consists of two positive lenses, an objective lens of small focal length that faces the object, and a magnifier serving as an eyepiece. The eyepiece "looks" at the real image formed by the objective. Referring to Figure 9-21, where the object lies outside the focal length f_0 of the objective, a real image I is formed within the microscope tube. After coming to a focus, the light rays continue on to the eyepiece, or ocular lens. For visual observations, the intermediate image is made to occur at or just inside the first focal point of the eyepiece. The eye positioned

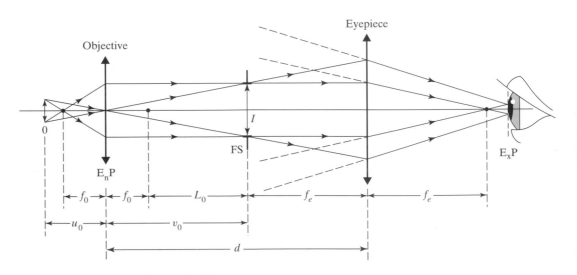

Figure 9-21 Image formation in a compound microscope.

near the eyepiece then sees a virtual image, inverted and magnified, as indicated by the dashed lines in Figure 9-21. The objective lens serves as the aperture stop and entrance pupil of the optical system. The image of the objective formed by the optics to the right—the eyepiece—is then the exit pupil, which locates the position of maximum radiant energy density and thus the optimum position for the entrance pupil of the eye. A special aperture, serving as a field stop, is placed at the position of the intermediate image. The eye then sees both in focus together, giving the field of view a sharply defined boundary. If a camera is attached to the microscope, a real final image is required. In this case the intermediate image must be located outside the ocular focal length, f_e.

Total Magnification. When the final image is viewed by the eye, the magnification of the microscope may be defined as in the case of the simple magnifier. Thus the angular magnification for an image viewed at infinity is

$$M = \frac{25}{f_{\text{eff}}} \tag{9-32}$$

where f_{eff} (in cm) is the effective focal length of the two lenses, separated by a distance d and given by Eq. (7-8),

$$\frac{1}{f_{\text{eff}}} = \frac{1}{f_0} + \frac{1}{f_e} - \frac{d}{f_0 f_e} \tag{9-33}$$

Substituting Eq. (9-33) into Eq. (9-32),

$$M = \frac{25(f_e + f_0 - d)}{f_0 f_e} \tag{9-34}$$

Using the thin-lens equation, however, we may express the ratio of image to object distance for the objective lens by

$$\frac{v_0}{u_0} = \frac{f_0 + f_e - d}{f_0} \tag{9-35}$$

where we have used the fact that $v_0 = d - f_e$, evident in the diagram. Incorporating Eq. (9-35) into Eq. (9-34),

$$M = \frac{25 \, v_0}{f_e \, u_0} \tag{9-36}$$

showing that the total magnification is just the product of the linear magnification of the objective multiplied by the angular magnification of the eyepiece when viewing the final image at infinity. Making use of Newton's formula for the linear magnification of a thin lens, Eq. (3-41),

$$m = -\frac{L_0}{f_0} \tag{9-37}$$

where L_0 represents the distance between the objective image and the second focal point, as shown. The magnification of the microscope may then be expressed, perhaps more conveniently, as

$$M = -\left(\frac{25}{f_e}\right)\left(\frac{L_0}{f_0}\right) \tag{9-38}$$

In many microscopes, the length L_0 is standardized at 16 cm. The focal lengths f_e and f_0 are themselves effective focal lengths of multielement lenses, appropriately corrected for aberrations.

❏ **Example**

In a laboratory experiment, a microscope is made using an objective lens of 3.8-cm focal length and a single eyepiece lens of 5-cm focal length. The lenses are separated by a distance of 16.4 cm. What is the theoretical magnification of the microscope?

Solution

$$L_0 = d - f_0 - f_e = 16.4 - 3.8 - 5 = 7.6 \text{ cm}$$

and

$$M = -\left(\frac{25}{f_e}\right)\left(\frac{L_0}{f_0}\right) = -\left(\frac{25}{5}\right)\left(\frac{7.6}{3.8}\right) = -10 \qquad ■$$

Numerical Aperture. To collect more light and produce brighter images, cones of rays from the object, intercepted by the objective lens, should be as large as possible. As magnifications increase and the focal lengths and diameters of the objective lenses decrease correspondingly, the solid angle of useful rays from the object also decreases. To increase the solid angle, an oil-immersion lens, such as that shown in Figure 9-22, is used for the microscope objective. In this figure, the useful light rays originating at the object point O, passing through a thin cover glass and then air to the first element of the objective lens L, make a half-angle of α_a on the right of the optical axis. Due to critical angle refraction at the glass-air interface, rays making an angle larger than α_a are internally reflected and do not reach the lens. This limitation is somewhat relieved by using a coupling transparent fluid whose index matches as closely as possible that of the glass. On the left of the optical axis in the diagram, a layer of oil is used, and a larger half-angle α_0 is possible. Typically, the cover glass index is 1.522 and the oil index is 1.516, providing an excellent match, so that $\alpha_0 \cong \alpha_0'$. The light-gathering capability of the objective lens is thus increased by increasing the refractive index

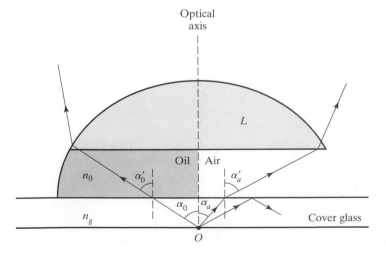

Figure 9-22 Microscope objective, illustrating the increased light-gathering power of an oil-immersion lens.

in object space. A measure of this capability is the product of half-angle and refractive index, called the *numerical aperture*,

$$\text{N.A.} = n \sin \alpha \tag{9-39}$$

The numerical aperture is an invariant in object space, due to Snell's law. That is, in the case of air,

$$\text{N.A.} = n_g \sin \alpha_a = \sin \alpha_a'$$

and when an *oil-immersion objective* is used,

$$\text{N.A.} = n_g \sin \alpha_0 = n_0 \sin \alpha_0'$$

The maximum value of the numerical aperture when air is used is theoretically unity, but when the object space is filled with a fluid of index n, the maximum theoretical numerical aperture may be increased up to the value of n. In practice, the limit is around 1.6. The numerical aperture is an alternative means of defining a relative aperture or of describing how "fast" a lens is. As shown previously, image brightness is *inversely* proportional to the square of the *relative aperture*. In the present context, image brightness is proportional to the square of the *numerical aperture*. The numerical aperture is an important design parameter also because it limits the resolving power and the depth of focus of the lens. The resolving power is proportional to the numerical aperture, whereas the depth of focus is inversely proportional to $(\text{N.A.})^2$. Most microscopes use objectives with numerical apertures in the approximate range of 0.08 to 1.30.

Biological specimens are covered with a cover glass having a thickness of 0.17 or 0.18 mm. For objectives with numerical apertures over 0.30, the cover glass has increasing influence on the image quality, since it introduces a large degree of spherical aberration when oil immersion is not involved. Thus a *biological objective* compensates for the aberration introduced by a cover glass. In contrast, a *metallurgical objective* is designed without such compensation. Objectives may be classified broadly in relation to the corrections introduced into their design. For low magnifications, with focal lengths in the range of 8 to 64 mm, *achromatic objectives* are generally used. Such objectives are chromatically corrected, usually for the Fraunhofer C (red) and F (blue) wavelengths, and spherically corrected, at the Fraunhofer D (sodium yellow) wavelength. For higher magnifications, objective lenses with focal lengths in the range of 4 to 16 mm incorporate some fluorite elements, which, together with the glass elements provide better correction over the visual spectrum. When the correction is nearly perfect throughout the visual spectrum, the objectives are said to be *apochromatic*. Since correction is more crucial at high magnifications, apochromats are usually objectives with focal lengths in the range of 1.5 to 4 mm. For even higher magnifications, the objective is usually designed as an *immersion objective*. Modern techniques and materials have also made possible *flat-field objectives* that essentially eliminate field curvature over the useful portion of the field. With ultraviolet immersion microscopes it is customary to replace the oil with glycerin and the optical glass elements with fluorite and quartz elements because of their higher transmissivity at short wavelengths.

This discussion should make it clear that high-quality microscopes today are designed as a whole and usually for a specific use. The design of an objective or an eyepiece is directly related to the performance of other optical elements in

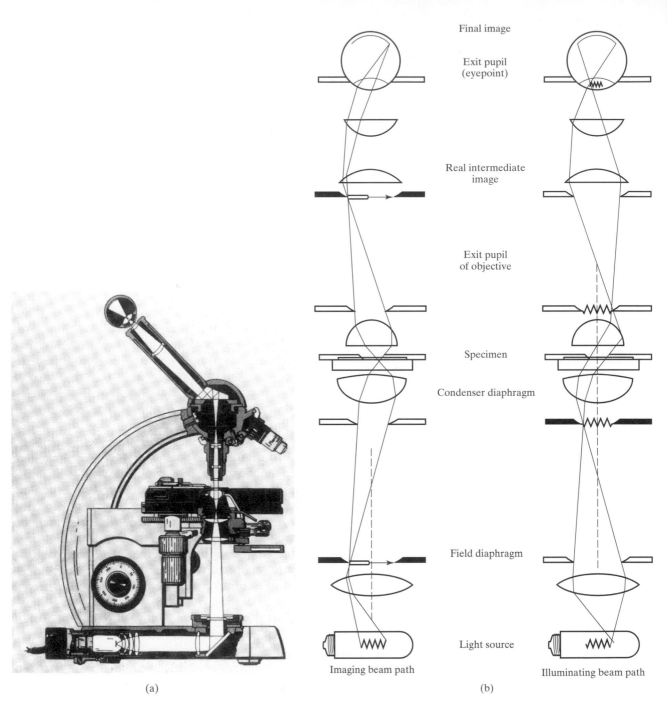

Final image

Exit pupil
(eyepoint)

Real intermediate
image

Exit pupil
of objective

Specimen

Condenser diaphragm

Field diaphragm

Light source

Imaging beam path

Illuminating beam path

(a)

(b)

Figure 9-23 Standard microscope illustrating Koehler illumination. (Courtesy Carl Zeiss, Inc., Thornwood, N.Y.)

the instrument, often including a relay lens within the body tube of the microscope as well. Thus it is generally not possible to interchange objectives and eyepieces between different model microscopes without loss or deterioration of the image.

Figures 9-23a and b illustrate respectively the optical components in a standard microscope and the detailed processing of light rays through the instrument.

Telescopes may be broadly classified as refracting or reflecting, according to whether lenses or mirrors are used to produce the primary image. There are, in addition, *catadioptric* systems that combine refracting and reflecting surfaces. Telescopes may also be distinguished by the erectness or inversion of the final image and by either a visual or photographic means of observation.

Refracting Telescopes. Figures 9-24 and 9-25 show two refracting telescope types, producing, respectively, inverted and erect images. The *Keplerian telescope* in Figure 9-24 is often referred to as an *astronomical telescope* since inversion of astronomical objects in the images produced create no difficulties. The *Galilean telescope*, illustrated in Figure 9-25, produces an erect image by means of an eyepiece of negative focal length. In either case, nearly parallel rays of light from a distant object are collected by a positive objective lens, which forms a real image in its focal plane. The objective lens, being larger in diameter than the pupil of the eye, permits the collection of more light and makes visible point sources such as stars that might otherwise not be detected by the naked eye. The objective lens is usually a doublet, corrected for chromatic aberration. The real image formed by the objective is observed with an eyepiece, represented in the figures as a single lens. This intermediate image, located at or near the focal point of the ocular, serves as a real object RO for the ocular in the astronomical telescope and a virtual object VO in the case of the Galilean telescope. In either case, the light is refracted by the eyepiece in order to produce parallel, or nearly parallel, light rays. An eye placed near the ocular views an image at infinity but with an angular magnification given by the ratio of the angles α'/α, as shown. The object subtends the angle α at the unaided eye and the angle α' at the eyepiece. From the two right triangles formed by the intermediate image and the optical axis, it is evident that the angular magnification is

$$M = \frac{\alpha'}{\alpha} = -\frac{f_0}{f_e} \tag{9-40}$$

The negative sign is introduced, as usual, to indicate that the image is inverted in Figure 9-24, where f_e is positive, and is erect in Figure 9-25, where f_e is negative. In either case, the length L of the telescope is given by

$$L = f_0 + f_e \tag{9-41}$$

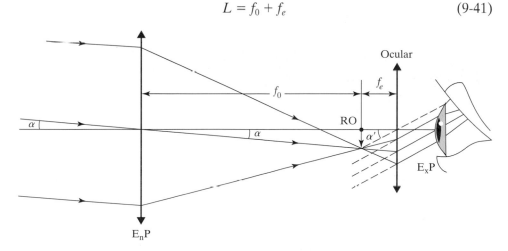

Figure 9-24 Astronomical telescope.

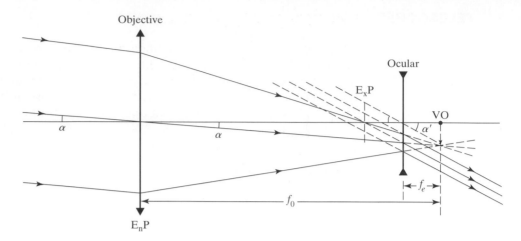

Figure 9-25 Galilean telescope.

permitting a short Galilean telescope, a circumstance that makes this design convenient in the *opera glass*. The astronomical telescope may be modified to produce an erect image by the insertion of a third positive lens whose function is simply to invert the intermediate image, but this lengthens the telescope by at least four times the focal length of the additional lens. Image inversion may also be achieved without additional length, as in binoculars, through the use of inverting prisms, discussed previously.

The objective of either telescope acts as both aperture stop and entrance pupil, whose image in the ocular is then the exit pupil, E_xP, as shown. In the astronomical telescope, the exit pupil is situated just outside the eyepiece and is designed to match the size of the pupil of the eye. A telescope should produce an exit pupil at sufficient distance from the eyepiece to allow a comfortable *eye relief*. Greater ease of observation is also achieved if the exit pupil is a little larger in diameter than the eye pupil, allowing for some relative motion between eye and eyepiece. Notice that in the Galilean telescope the exit pupil falls in front of the eyepiece, where it is inaccessible to the eye. This represents a disadvantage of the Galilean telescope, leading to a restriction in the field of view. Notice also that a field stop with reticle can be employed at the location of the intermediate image in the astronomical telescope, whereas no such arrangement is possible in conjunction with the Galilean telescope. The diameter of the exit pupil D_{ex} is simply related to the diameter of the objective lens D_{obj} through the angular magnification, as follows. Since the exit pupil is the image of the entrance pupil formed by the eyepiece, we may write for the linear, transverse magnification either

$$m_e = \frac{D_{ex}}{D_{obj}} \tag{9-42}$$

or, employing the Newtonian form of the magnification,

$$m_e = -\frac{f}{x} = -\frac{f_e}{f_0} \tag{9-43}$$

where x is the distance of the object (objective lens) from the focal point of the eyepiece, or f_0. Combining Eqs. (9-40), (9-42), and (9-43),

$$m_e = \frac{1}{M} = \frac{D_{ex}}{D_{obj}}$$

Figure 9-26 Cutaway view of binoculars revealing compound objective and ocular lenses and image-inverting prism. (Courtesy Carl Zeiss, Inc., Thornwood, N.Y.)

so that

$$D_{ex} = \frac{D_{obj}}{M} \tag{9-44}$$

Thus the diameter of the bundle of parallel rays filling the objective lens is greater by a factor of M than the diameter of the bundle of rays that pass through the exit pupil. It should be pointed out that the image is not brighter by the same proportion, however, because the apparent size of the image increases by the same factor M. The brightness of the image cannot be greater than the brightness of the object; in fact, it is less bright because of inevitable light losses due to reflections from lens surfaces.

Binoculars (Figure 9-26) afford more comfortable telescopic viewing, allowing both eyes to remain active. In addition, the use of Porro or other prisms to produce erect final images also permits the distance between objective lenses to be greater than the interpupillary distance, enhancing the stereoscopic effect produced by ordinary binocular vision. The designation "6 × 30" for binoculars means that the angular magnification M produced is 6× and the diameter of the objective lens is 30 mm. Using Eq. (9-44), we conclude that the exit pupil for this pair of binoculars is 5 mm, a good match for the normal pupil diameter. For night viewing, when the pupils are somewhat larger, a rating of 7 × 50, producing an exit pupil diameter of 7 mm, would be preferable.

❏ **Example**

Determine the eye relief and field of view for the 6 × 30 binoculars just described. Assume an objective focal length of 15 cm and a field lens (eyepiece) diameter of 1.50 cm.

Solution   The focal length of the ocular is found from

$$f_e = -\frac{f_0}{M} = -\frac{15}{-6} = 2.5 \text{ cm}$$

The eye relief is the distance of the exit pupil from the eyepiece. Since the exit pupil is the image of the objective formed by the eyepiece, the eye relief is the image distance v, given by

$$v = \frac{uf}{u + f} = \frac{-Lf_e}{-L + f_e} = \frac{-(f_0 + f_e)f_e}{-(f_0 + f_e) + f_e} = \frac{-(15 + 2.5)(2.5)}{-(15 + 2.5) + 2.5} = 2.92 \text{ cm}$$

The angular field of view from the objective subtends both the object on one side (height h at distance u) and the field lens of the eyepiece on the other (D_f at distance L). Thus, for objects at a standard distance of $u = 1000$ yd,

$$\theta = \frac{h}{u} = \frac{D_f}{L} = \frac{D_f}{f_0 + f_e}$$

or

$$h = u\theta = \frac{uD_f}{f_0 + f_e} = \frac{(3000 \text{ ft})(1.50)}{15 + 2.5} = 257 \text{ ft at } 1000 \text{ yd} \qquad \blacksquare$$

Reflection Telescopes.   Larger-aperture objective lenses provide greater light-gathering power and resolution. However, large, homogeneous lenses are difficult to produce without optical defects, and their weight is difficult to support. These problems, as well as the elimination of chromatic aberrations, are solved by using curved, reflecting surfaces in place of lenses. The large telescopes, like the Hale 200-inch reflector on Mount Palomar, use such mirrors. Such large reflecting telescopes are usually employed to examine very faint astronomical objects and use the integrating power of photographic plates, exposed over long time intervals, to record an image.

Several basic designs for reflecting telescopes are shown in Figure 9-27. In the *Newtonian* design (a), a parabolic mirror with the help of a flat mirror is used to focus accurately all parallel rays to the same focal point, f_s, near the body of the telescope, where an eyepiece is located to view the image. The use of a parabolic mirror avoids both chromatic and spherical aberration, but coma is present for off-axis points, severely limiting the useful field of view. In the 200-inch Hale telescope, the flat mirror can be dispensed with and the rays allowed to converge at their primary focus f_p. This telescope is large enough so that the observer can sit on a specially built platform situated just behind the primary focus (Figure 9-28). Of course, any obstruction placed inside the telescope reduces the cross section of the incident light waves contributing to the image. In the *Cassegrain* design (Figure 9-27b), the smaller, secondary mirror is hyperboloidal convex in shape, reflecting light from the primary mirror through an aperture in the primary mirror to a secondary focus f_s, where it is conveniently viewed or recorded. The hyperboloidal surface permits perfect imaging between the primary and secondary focal points, which act as the foci of the hyperboloid. Such accurate imaging is also possible when the secondary mirror is concave ellipsoidal, as in the *Gregorian* telescope (Figure 9-27c). The primary and secondary focal points of this telescope are now the foci of the ellipsoid.

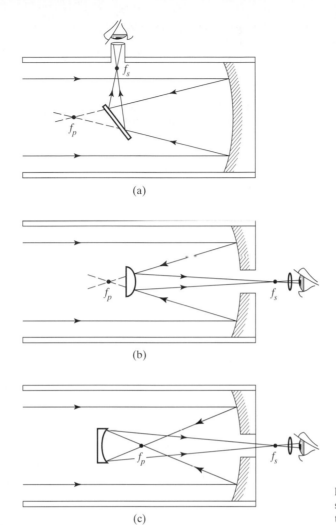

(a)

(b)

(c)

Figure 9-27 Basic designs for reflecting telescopes. (a) Newtonian telescope. (b) Cassegrain telescope. (c) Gregorian telescope.

Figure 9-28 Hale telescope (200-inch) showing observer in prime-focus cage and reflecting surface of 200-inch mirror. (California Institute of Technology.)

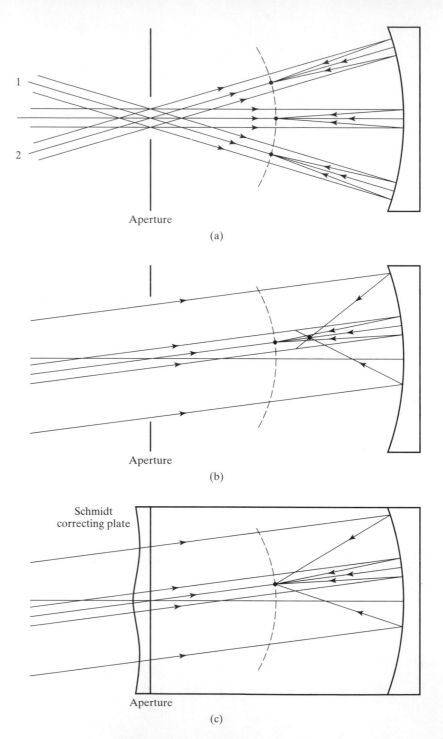

Aperture

(a)

Aperture

(b)

Schmidt
correcting plate

Aperture

(c)

Figure 9-29   The Schmidt optical system.

The Schmidt Telescope.   Perhaps the most celebrated catadioptric telescope is due to a design of Bernhardt Schmidt. He sought to remove the spherical aberration of a primary spherical mirror by using a thin refracting *correcting plate* at the aperture of the telescope. To understand his design, refer to Figure 9-29. A concave primary reflector in (a) receives small bundles of parallel rays from various directions. Each bundle enters at the aperture, which is located at the center of curvature of the primary mirror. Since the axis of any bundle may be considered an optical axis, there are no off-axis points and thus coma and astigmatism do not enter into the aberrations of the system. When the bundles are small, each bundle consists of paraxial rays that focus at the same distance from the mirror, a distance equal to its focal length, or half the radius of curvature of the mirror. The locus of such image points is then the spherical surface indicated by the dashed line. However, when the bundles are large, as shown in (b), spherical aberration occurs, which produces a shorter focus for rays reflecting from the outer zones of the mirror relative to the optical axis of the bundle. Schmidt designed a transparent correcting plate to be placed at the aperture, whose function is to bring the focus of all zones to the same point on the spherical focal surface, as indicated in (c). The shape suggested in the figure is designed to make the focal point of all zones agree with the focal point of a zone whose radius is 0.707 of the aperture radius, the usual choice. The resulting *Schmidt optical system* is, therefore, highly corrected for coma, astigmatism, and spherical aberration. Because the correcting plate is situated at the center of curvature of the mirror, it presents approximately the same optics to parallel beams arriving from different directions and so permits a wide field of view. Residual aberrations are due to errors in the actual fabrication of the correcting plate and also because the plate does not present precisely the same cross section, therefore the same correction, to beams entering from different directions. One disadvantage is that the focal plane is spherical, requiring a careful shaping of photographic films and plates. Notice also that with the correcting plate attached at twice the focal length of the mirror, the telescope is twice as long as the telescopes described previously, in Figure 9-27. Nevertheless, the *Schmidt camera*, as it is often called, has been highly successful and has spawned a large number of variants, including designs to flatten the field near the focal plane.

PROBLEMS

9-1. Plot a curve of total deviation angle versus entrance angle for a prism of apex angle 60° and refractive index 1.52.

9-2. A parallel beam of white light is refracted by a 60° glass prism in a position of minimum deviation. What is the angular separation of emerging red ($n = 1.525$) and blue (1.535) light?

9-3. Ophthalmic prisms of 2.0 and 3.0 pd are to be designed for the left and right lenses, respectively, of a pair of eyeglasses, in order to correct for a 5.0 pd misalignment in the vertical vision of a patient. The prism in the left lens is to produce a downward displacement of the light passing through it, while the prism in the right lens is to produce an upward displacement. The glass used for each lens has a refractive index of 1.523.
(a) What is the prism apex angle for each ophthalmic prism?
(b) What is the orientation of the prism base in each lens?

9-4. While fixating on a real object directly ahead at a distance of 6 meters—one eye uncovered at a time—a certain individual's eyes, each turned slightly outwards, is directed respectively towards positions 30 cm to the left and right of the object.
(a) Based on this geometry of misalignment, how many prism diopters of correction should be prescribed for each eye?
(b) Determine the prism apex angle for each spectacle lens if the refractive index of the glass is 1.562.
(c) The desired effect of the ophthalmic prism in each lens is to bend the light passing through it inward toward the nose, so that the outwardly directed eyes see the virtual image of the object at its displaced positions. What then should be the orientation of the ophthalmic prisms ground into each lens?

9-5. (a) Approximate the Cauchy constants A and B for crown and flint glasses, using the data for the C and F Fraunhofer lines from Table 9-1. Using these constants and the Cauchy relation for n_λ approximated by two terms, calculate the refractive index of the D Fraunhofer line for each case. Compare your answers with the values given in the table.

(b) Calculate the dispersion in the vicinity of the Fraunhofer D line for each glass, using Cauchy's formula for $\Delta n/\Delta\lambda$.

9-6. An equilateral prism of dense barium crown glass is used in a spectroscope. Its refractive index varies with wavelength, as given in the table:

nm	n
656.3	1.63461
587.6	1.63810
486.1	1.64611

(a) Determine the minimum angle of deviation for sodium light of 589.3 nm.

(b) Determine the dispersive power of the prism.

(c) Determine the Cauchy constants A and B in the long wavelength region; from Cauchy's formula for $\Delta n/\Delta\lambda$, find the dispersion of the prism at 656.3 nm.

9-7. A prism of 60° refracting angle gives the following angles of minimum deviation when measured on a spectrometer: C line, 38°20'; D line, 38°33'; F line, 39°12'. Determine the dispersive power of the prism.

9-8. The refractive indices for certain crown and flint glasses are

Crown: $n_C = 1.527$, $n_D = 1.530$, $n_F = 1.536$
Flint: $n_C = 1.630$, $n_D = 1.635$, $n_F = 1.648$

The two glasses are to be combined in a double prism that is a direct vision prism for the D wavelength. The refracting angle of the flint prism is 5°. Determine the required angle of the crown prism and the resulting angle of dispersion between the C and the F rays. Assume that the prisms are thin and the condition of minimum deviation is satisfied.

9-9. An achromatic thin prism for the C and F Fraunhofer lines is to be made using the crown and flint glasses described in Table 9-1. If the crown glass prism has a prism angle of 15°, determine **(a)** the required prism angle for the flint glass, and **(b)** the resulting "mean" deviation for the D line.

9-10. A perfectly diffuse, or *Lambertian*, surface has the form of a square, 5 cm on a side. This object radiates a total power of 25 W into the forward directions that constitute half the total solid angle of 4π. A camera with a 4-cm focal length lens stopped down to $f/8$ is used to photograph the object when it is placed 1 m from the lens.

(a) Determine the radiant exitance, radiant intensity, and radiance of the object. (See Table 2-1.)

(b) Determine the radiant flux delivered to the film.

(c) Determine the irradiance at the film.

9-11. Investigate the behavior of Eq. (9-28), giving the dependence of the depth of field on aperture, focal length, and object distance. With the help of a calculator or computer program, generate curves showing each dependence.

9-12. A camera is used to photograph three rows of students at a distance 6 m away, focusing on the middle row. Suppose that the image defocusing or blur circles due to object points in the first and third rows is to be kept smaller than a typical silver grain of the emulsion, say 1 μm. At what object distance nearer and farther than the middle row does an unacceptable blur occur if the camera has a focal length of 50 mm and is stopped down to $f/4$ setting? (*Hint:* Consider Eqs. (9-26) and (9-27).)

9-13. A telephoto lens consists of a combination of two thin lenses having focal lengths of +20 cm and −8 cm, respectively. The lenses are separated by a distance of 15 cm. Determine the focal length of the combination, distance from negative lens to film plane, and image size of a distant object subtending an angle of 2° at the camera.

9-14. A 5-cm focal length camera lens with $f/4$ aperture is focused on an object 6 ft away. If the maximum diameter of the circle of least confusion is taken to be 0.05 mm, determine the depth of field of the photograph.

9-15. The sun subtends an angle of 0.5° at the earth's surface, where the illuminance is about 10^5 lux at normal incidence. What is the illuminance of an image of the sun formed by a lens with diameter 5 cm and focal length 50 cm?

9-16. (a) A camera uses a convex lens of focal length 15 cm. How large an image is formed on the film of a 6-ft-tall person 100 ft away?

    (b) The convex lens is replaced by a telephoto combination consisting of a 12-cm focal length convex lens and a concave lens. The concave lens is situated in the position of the original lens, and the convex lens is 8 cm in front of it. What is the required focal length of the concave lens such that distant objects form focused images on the same film plane? How much larger is the image of the person using this telephoto lens?

9-17. The lens on a 35-mm camera is marked 50 mm, 1:1.8.

    (a) What is the maximum aperture diameter?

    (b) Starting with the maximum aperture setting, supply the next three f-numbers that would allow the irradiance to be reduced to 1/3 the preceding number at each successive stop.

    (c) What aperture diameters correspond to these f-numbers?

    (d) If a picture is taken at maximum aperture and at 1/100 s, what exposure time at each of the other openings provides equivalent total exposure?

9-18. The magnification given by Eq. (9-29) is also valid for a double-lens eyepiece if the equivalent focal length given by Eq. (7-8) is used. Show that the magnification of a double-lens eyepiece, designed to satisfy the condition for the elimination of longitudinal chromatic aberration, is for an image at infinity

$$M = 12.5\left(\frac{1}{f_1} + \frac{1}{f_2}\right)$$

9-19. A magnifier is made of two thin, plano-convex lenses, each of 3-cm focal length and spaced 2.8 cm apart. Find (a) the equivalent focal length and (b) the magnifying power for an image formed at the near point of the eye.

9-20. The objective of a microscope has a focal length of 0.5 cm and forms the intermediate image 16 cm from its second focal point.

    (a) What is the overall magnification of the microscope when an eyepiece rated at 10× is used?

    (b) At what distance from the objective is a point object viewed by the microscope?

9-21. A homemade compound microscope has, as objective and eyepiece, thin lenses of focal lengths 1 cm and 3 cm, respectively. An object is situated at a distance of 1.20 cm from the objective. If the virtual image produced by the eyepiece is 25 cm from the eye, compute (a) the magnifying power of the microscope and (b) the separation of the lenses.

9-22. Two thin, convex lenses, when placed 25 cm apart, form a compound microscope whose apparent magnification is 20. If the focal length of the lens representing the eyepiece is 4 cm, determine the focal length of the other.

9-23. A level telescope contains a *graticule*—a circular glass on which a scale has been etched—in the common focal plane of objective and eyepiece, so that it is seen in focus with a distant object. If the telescope is focused on a telephone pole 30 m away, how much of the post falls between millimeter marks on the graticule? The focal length of the objective is 20 cm.

9-24. A pair of binoculars is marked 7 × 35. The focal length of the objective is 14 cm, and the diameter of the field lens of the eyepiece is 1.8 cm. Determine (a) the angular magnification of a distant object, (b) the focal length of the ocular, (c) the diameter of the exit pupil, (d) the eye relief, and (e) the field of view in terms of "feet at 1000 yd."

9-25. (a) Show that when the final image is not viewed at infinity, the angular magnification of an astronomical telescope may be expressed by

$$M = \frac{m_{oc}f_{obj}}{v_2}$$

where m_{oc} is the linear magnification of the ocular and v_2 is the distance from the ocular to the final image.

(b) For such a telescope, using two converging lenses with focal lengths of 30 cm and 4 cm, find the angular magnification when the image is viewed at infinity and when the image is viewed at a near point of 25 cm.

9-26. The moon subtends an angle of 0.5° at the objective lens of a terrestrial telescope. The focal lengths of the objective and ocular lenses are 20 cm and 5 cm, respectively. Find the diameter of the image of the moon viewed through the telescope at near point 25 cm.

9-27. An opera glass uses an objective and eyepiece with focal lengths of +12 cm and −4.0 cm, respectively. Determine the length (lens separation) of the instrument and its magnifying power for a viewer whose eyes are focused **(a)** for infinity and **(b)** for a near point of 30 cm.

9-28. An astronomical telescope is used to project a real image of the moon onto a screen 25 cm from an ocular of 5-cm focal length. How far must the ocular be moved from its normal position?

9-29. (a) The Ramsden eyepiece of a telescope is made of two positive lenses of focal length 2 cm each and also separated by 2 cm. Calculate its magnifying power when viewing an image at infinity.

(b) The objective of the telescope is a 30-cm positive lens, with a diameter of 4.50 cm. Calculate the overall magnification of the telescope.

(c) What is the position and diameter of the exit pupil?

(d) The diameter of the eyepiece field lens is 2 cm. Determine the angle defining the field of view of the telescope.

9-30. Show that the angular magnification of a Newtonian reflecting telescope is given by the ratio of objective to ocular focal lengths, as it is for a refracting telescope when the image is formed at infinity.

9-31. The primary mirror of a Cassegrain reflecting telescope has a focal length of 12 ft. The secondary mirror, which is convex, is 10 ft from the primary mirror along the principal axis and forms an image of a distant object at the vertex of the primary mirror. A hole in the mirror there permits viewing the image with an eyepiece of 4-in. focal length, placed just behind the mirror. Calculate the focal length of the convex mirror and the angular magnification of the instrument.

REFERENCES

BAHCALL, J. N., and L. SPITZER, JR. 1982. "The Space Telescope." *Scientific American* (July): 40.

BENFORD, JAMES R., and HAROLD E. ROSENBERGER. 1978. "Microscope Objectives and Eyepieces." In *Handbook of Optics*, edited by Walter G. Driscoll and William Vaughan. New York: McGraw-Hill Book Company.

CREWE, ALBERT. 1971. "A High-Resolution Scanning Electron Microscope." *Scientific American* (Apr.): 26.

GOODMAN, DOUGLAS S. 1988. "Basic Optical Instruments" in *Geometrical and Instrumental Optics*, edited by Daniel Malacara. Boston: Academic Press.

HEAVENS, O. S., and R. W. DITCHBURN. 1991. *Insight into Optics*. New York: John Wiley & Sons.

HORNE, D. F. 1972. *Optical Production Technology*, 2d. ed. New York: Crane, Russak and Company.

HORNE, DOUGLAS F. 1980. *Optical Instruments and Their Applications*. Philadelphia: I. O. P. Publishers.

HOWELLS, MALCOLM R., JANOS KIRZ, and WILLIAM SAYRE. 1991. "X-Ray Microscopes." *Scientific American* (Feb.): 88.

KINGSLAKE. 1978. *Lens Design Fundamentals*. New York: Academic Press, Inc.

KINGSLAKE, R. 1992. *Optics in Photography*. Bellingham: SPIE-International Society for Optical Engineering.

KIRKPATRICK, PAUL. 1949. "X-ray Microscope." *Scientific American* (Mar.): 44.

MALACARA, DANIEL. 1994. *Handbook of Lens Design*. New York: Marcel Dekker, Inc.

McCLAIN, EDWARD JR. 1960. "The 600-Foot Radio Telescope." *Scientific American* (Jan.): 45.

McLAUGHLIN, ROBERT B. 1977. *Special Methods in Light Microscopy*. Chicago: Microscope Publications Ltd.

MULLER, ERWIN W. 1952. "A New Microscope." *Scientific American* (May): 58.

PRICE, WILLIAM H. 1976. "The Photographic Lens." *Scientific American* (Aug.): 72.

QUATE, CALVIN. 1979. "Acoustic Microscope." *Scientific American* (Oct.): 62.

VAN MONCKHOVEN, DESIRE, ROBERT A. SOBIESZEK, and PETER C. BUNNELL. 1979. *Photographic Optics*. North Stratford: Ayer Company Publishers, Incorporated.

WALD, GEORGE. 1950. "Eye and Camera." *Scientific American* (Aug.): 32.

WILSON, ALBERT G. 1950. "The Big Schmidt." *Scientific American* (Dec.): 34.

10

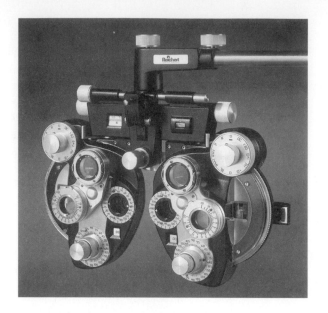

Optics of the Eye

INTRODUCTION

Our previous study of "Geometrical Optics" (Chapter 3), "Vergence and Vision" (Chapter 5), and "Controlling Light Through Optical Systems" (Chapter 8), have provided the necessary foundations for a careful analysis of the performance of the eye as an optical instrument. In this chapter we acquaint ourselves with the optics of the eye. We begin with a brief overview of the biological structure of the eye and its principal functions. Then we examine several schematic models of the eye, developed to represent it and to make calculations somewhat simpler. Following this, we examine common errors of refraction and their correction, building on material presented earlier. We close this chapter with an overview of ophthalmic instruments commonly found in the examination rooms of optometrists and ophthalmologists.

The eyes, in conjunction with the brain, constitute a truly remarkable bio-optical system. Consider briefly the distinctive characteristics of this system. It forms images of a continuum of objects at distances from a foot to infinity. It scans a scene as expansive as the overhead sky or focuses on detail as minute as the head of a pin. It adapts itself to an extraordinary range of intensities, from the barely visible flicker of a candle miles away on a dark night to sunlight so bright that the optical image on the retina causes serious solar burn. It distinguishes between subtle shades of color, from deep blue to deep red. Most importantly for us, its performance as a unique spatial sense organ localizes objects in space, accurately mapping out our three-dimensional world.

10-1 BIOLOGICAL STRUCTURE OF THE EYE

Anatomically, the eyeball is a globe, almost spherically shaped, approximately 23 mm in diameter.[1] Visually, the eyeball can be pictured as a positive optical system that refracts incident light onto its rear surface to form a real image, much as does an ordinary camera.

The basic parts of the eye are shown in Figure 10-1. Let us examine the key biological components of the eye along the optical axis, in the same order they are encountered by light rays in the usual image-forming process. Light first enters the eye through the *cornea*, a transparent tissue devoid of blood vessels but abounding in nerve cells. The corneal cap is roughly 12 mm across and 0.6 mm thick at its center, thickening somewhat further at its edges, with a refractive index of 1.376. Upon entering the eye at the air-cornea interface, where the refractive index changes abruptly from 1.0 to 1.38, light undergoes a significant degree of bending. The corneal surface provides, in fact, about 73% of the total refractive power of the eye. Immediately behind the cornea is the *anterior chamber*, a small space filled with a watery fluid, the *aqueous humor* that provides nutrients for the cornea. This fluid has a refractive index of 1.336, almost equal to that of water (1.333). Because the refractive indices of the cornea and aqueous humor are nearly alike, little additional bending of rays occurs as light moves from the cornea into the anterior chamber. Situated in the aqueous humor is the *iris*, a diaphragm that gives the eye its characteristic color and controls the amount of light that enters. The amount and location of pigment in the iris together determine whether the eye looks blue, green, gray, or brown. The adjustable hole or

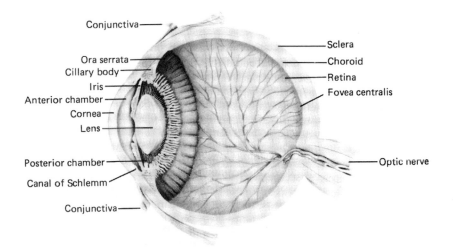

Figure 10-1 Vertical cross section of the eye. (Reproduced with permission of Glaxo Wellcome Inc.)

[1]In fact, since the eye is not a perfect sphere, the vertical dimension is found to be nearer 24 mm, while the transverse dimension, from cornea to retina, is nearer 23 mm.

opening in the center of the iris, through which light passes, is called the *pupil*. The iris contains two sets of delicate muscles that change the pupil size in response to light stimulation, adjusting the diameter, in adults, from a minimum of about 2 mm on a bright day to a maximum of about 8 mm under very dark conditions. While examining the inside of the eye, doctors often use drugs, such as *atropine*, to dilate or enlarge the pupil.

Immediately upon passing through the pupil, light falls on the *crystalline lens*, a transparent structure about the size and shape of a small lima bean. The lens provides the fine tuning required in the image formation process, changing its own shape appropriately to transform an external scene into a sharp image on the retina. The shape of the lens is controlled by the *ciliary muscle*, connected by fibers (*zonules*) to the periphery of the lens. While the muscles are relaxed, the lens assumes its flattest shape, providing the least refraction of incident light rays. In this state, the relaxed eye is focused on distant objects. When the muscles are tensed, the lens bulges, becoming increasingly curved and providing increased refraction of light. In this "strained" state, the eye is focused on nearby objects. The lens is itself a complex, onionlike, layered mass of tissue, held intact by an elastic membrane. Due to the rather intricate laminar structure of fibrous tissue, the refractive index of the lens is not homogeneous. Near the center or core of the lens (on axis), the index is about 1.41; near the periphery it falls to about 1.39.

After its final refraction by the crystalline lens, light enters the *posterior chamber* filled with the *vitreous humor*, a transparent, jelly-like substance whose refractive index (1.336) is again close to that of water. The vitreous humor, essentially structureless, contains small particles of cellular debris that are referred to as *floaters*. They derive their name from the manner in which they are often observed to float across one's field of view as one looks or squints at a white ceiling, for example.

After traversing the vitreous humor, light rays reach their terminus at the inner rear layer of the eye, the *retina*, literally translated as *net*. The retina, or net, is dotted with an overlapping pattern of photoreceptor cells called *rods* and *cones*. The long, thin rods, numbering over 100 million, are located more densely toward the periphery of the retina. They are exceedingly sensitive to dim light, but are unable to distinguish between colors. The wider cones, just under 10 million in number, cluster preferentially near the center of the retina, a 3-mm diameter region called the *macula*. In sharp contrast to the response of the rods, the cones are sensitive to bright light and color but do not function well in dim light. Linked to the photoreceptor cells are three distinct types of nerve cells (*amocrine, bipolar, horizontal*) that transmit the visual impulse to the *optic nerve*. The optic nerve is the main trunkline that carries visual information from the retina to the brain, completing the remarkable process we call *vision*.

In addition to the key optical components encountered by light traveling along the axis of vision, the eye contains other components that should be mentioned. As noted in Figure 10-1, the eye is covered with a tough, white coating called the *sclera* that forms the supporting framework of the eye. Just inside the sclera lies the *choroid*, covering the back four-fifths of the eye and containing most of the blood vessels that nourish the eye. The choroid, in turn, serves as the backing for the retina, the all-important net that houses the rods and cones. At the center of the macula, located somewhere above the optic nerve, is the *fovea centralis*, the region of greatest visual acuity. When it is required that one see sharp and detailed information—while removing a small splinter with a needle, for example—the eyes move continually so that light coming from the area of interest falls precisely on the fovea, a rod-free region about 200 μm in diameter. Quite by contrast, another small region in the retina, located at the point of exit of the optic nerve, is completely insensitive to light. This spot, devoid of receptors, is appropriately called the *blind spot*.

10-2 FUNCTIONS OF THE EYE

To operate as an effective optical system, the eye must form a retinal image of an external object or scene, either distant or nearby, in bright as well as dim light. To achieve efficient operation, the eye takes advantage of special functions. To focus on objects nearby and far away, the eye *accommodates*. To process light signals of varying brightness, the eye *adapts*. To sense the spatial orientation of three-dimensional scenes, the eyes make use of *stereoscopic vision*. To form a faithful, detailed image of the external object, the eye relies on its *visual acuity*. In what follows, we discuss each of these visual functions in somewhat more detail.

Accommodation. Depending on the distance of the object or scene from the eye, the lens accommodates—tenses or relaxes—appropriately to fine-focus the image on the retina. For a distant object, the *ciliary muscle* attached to the lens relaxes and the lens assumes a flatter configuration, increasing its radii of curvature and, consequently, its focal length. As the object moves closer to the eye, the ciliary muscle tenses or contracts, squeezing or bulging the lens and resulting in decreased radii of curvature and a shorter focal length. The smaller the radii of curvature and focal length, the higher the refractive or bending power of the lens, precisely the condition needed to bring closer objects into sharp focus. In a normal eye—and before the normal aging process robs the lens of its elasticity and ability to reshape itself—accommodation permits faithful retinal images of objects anywhere from distant points (at infinity) to nearby points about one foot away. The *near point* (closest point of accommodation) recedes from the eye with advancing age, starting at a position of 7 to 10 cm from the eye for a teenager, increasing to 20 to 40 cm for a middle-aged adult, and extending to as far as 200 cm in later years. For the average far-sighted person, *presbyopia* (loss of accommodation) sets in during the early 40s, signaling the need for reading glasses to restore the near point to a comfortable position near 25 cm or so.

Adaptation. The ability of the eye to respond to light signals that range from very dim to very bright, a range of light intensities that differ by an astonishing factor of about 10^5 (five orders of magnitude), is referred to as *adaptation*. The amount of light (flux or photon number) that enters the eye is regulated first of all by the iris, with its adjustable aperture, the pupil. This adjustment of pupil diameter (from 8 mm down to 2 mm) cannot of itself account for the enormous range of intensities processed by the eye. The remarkable adaptivity of the eye is traced, in fact, to the photoreceptors in the retina, the rods and cones, and to their particular sensitivity to light. The key ingredient seems to be a pigment, called *visual pigment*, contained in both the rods and cones. The rods, stimulated by low-level light signals (*scotopic vision*), contain pigment of only one kind, called *visual purple*. The cones, sensitive to light signals of high intensity and variable color composition (*photopic vision*), each contain one of three different kinds of visual pigment. The numerous, thin rods are multiply connected to nerve fibers, thereby making it possible for any one of 100 or so rods to activate a single nerve fiber. The less numerous, wider cones in the macular region, by contrast, are individually connected to nerve fibers and thus individually activated. The activation of nerve fibers—the very heart of the vision process itself—depends on chemical changes that occur in the visual pigment contained in the rods and cones. When light falls on either type of photoreceptor, the visual pigment changes from a dark state to a clearer state, undergoing a sort of bleaching process. The change in state of the visual pigment in the rods or cones is transformed into an electrical output or nerve fiber impulse. These electrical impulses are sent to the optic nerve and on to the brain, transmitting faithfully the light intensity of the stimulating signal. When the visual purple is fully bleached out in the rods, the photoreceptor cells become insensitive to further light signals. A regeneration of pigment in the rods must occur before they can respond again. Apparently the single type of visual pigment in rods is much more sensitive to light than is any one of the three pigments in cones. Accordingly, rods bleach out completely at much lower light levels than do cones. A change from low-level or scotopic vision to high-level light or photopic vision in the process of adaptation consists of a rapid bleaching out of rod pigment and a resulting insensitivity of the rod receptors. The bright light is then processed efficiently by the less-sensitive cones. Conversely, adaptation from intensely bright light (handled by the cones) to very dim light involves regeneration of pigment in the rods and a restoration of "night vision." In the full process of adaptation, the scotopic response is active over light levels that range from starlight on a clear, moonless night to lunar light from a quarter-moon. The photopic response (rods completely bleached out and inactive) operates between light levels ranging roughly from twilight to bright sunlight. Between light levels of a quarter moon and twilight, rods and cones both receive light and transmit nerve impulses.

Stereoscopic Vision. The ability to judge depth or position of objects accurately in a three-dimensional field is called *stereoscopic vision*. In humans, the optic nerves from the two eyes come together at the *optic chiasma*, near the brain. From the optic chiasma, nerve fibers originating in the right half of *each* retina extend to the right half of the brain. Nerve fibers originating in the left half of each retina terminate in the left half of the brain. Thus, even though each half of the brain receives an image from *both* eyes, the brain forms but a single image. The fusion by the brain of two distinct images into a single image is referred to as *binocular vision*. Nevertheless, the slight differences between the two images from the left and right eyes provide the basis for stereoscopic vision in humans. It should be noted that even monocular vision is not without some depth perception. This is due to visual clues like parallax, shadowing, and the particular perspective of familiar objects.

To have proper binocular vision without *double vision*, the images of an object must fall at corresponding points on each retina. This, of course, is what happens when the eyeballs move appropriately to focus on an object or scene, causing the image to fall on the fovea centralis of each eye. Most individuals are either right-eyed or left-eyed, indicating a dominance of one eye over the other. To determine which is your dominant eye, try the following simple test. Hold a pencil in front of you at eye level. With both eyes open, line the pencil up with the vertical edge of a picture, door or window across the room. Holding the pencil fixed, close one eye at a time. Whichever eye is open when the pencil remains lined up with the reference object is your dominant eye. The brain records the message seen by the dominant eye, while suppressing the other.

Visual Acuity.   Simply stated, visual acuity is the ability to see "clearly." In technical terms, visual acuity depends on the eye's ability to differentiate detail at various distances from the eye. For example, the furthest distance from the eye at which it can still distinguish between adjacent millimeter marks on a meter stick is an indication of its minimum angle of resolution.[2]

Operationally, assessment of the resolving power or visual acuity of the eye is measured in different ways. The ability to discriminate between two closely spaced points is referred to as *minimum separable* resolution; the smallest resolvable angle subtended by a black bar on a white background is called *minimum visible* resolution, and the smallest angle subtended by block letters that can be read (on an eye chart) is called *minimum legible* resolution. Since most of us, at one time or another, are required to read eye charts in a vision test, we limit our discussion of visual acuity to the eye's resolving power associated with minimum legible resolution.

The design of the eye chart owes its existence to a Dutch ophthalmologist, Herman Snellen. According to Snellen, the various letters on the eye chart are constructed so that the overall block size of a specified letter—the line from top to bottom or side to side—subtends an angle of 5′ (five minutes) of arc at some specified test distance, and the width of the detailed lines *within* a letter, such as the vertical bar in the letter T or the horizontal bar in the letter H, subtend an angle of 1′ of arc at the same test distance. See Figure 10-2. The two choices of angle grew out of the best data available to Snellen concerning the minimum legible resolution of normal eyes.

Accordingly, Snellen chose a *standard test distance* of 20 feet and constructed a row of letters labeled 20/20 (See Figure 10-2). At this standard distance

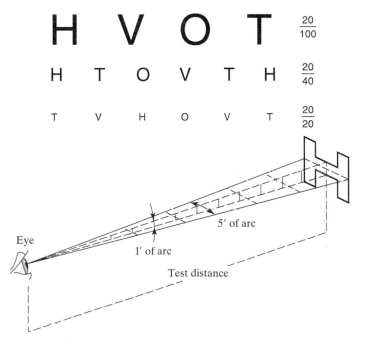

Figure 10-2   Construction of a Snellen eye-chart letter H to measure visual acuity. Notice the measure of the 5′ angle for the entire letter and the measure of the 1′ angle for the width of any member. The top portion of the figure shows a section of an eye chart (reduced) containing the letter H and several other letters, with 20/20, 20/40, and 20/100 rows of letters identified.

---

[2]Ultimately, the limitation in resolution is due to the overlapping of the diffraction patterns in the neighboring images of their conjugate object points. This subject is discussed in Chapter 16, "Fraunhofer Diffraction."

of 20 feet, each of the letters in the 20/20 row subtend the required angles of 5′ and 1′ at the eye of the observer. Snellen then constructed other rows, with successively larger letters in each row, labeling them as 20/40, 20/100, and so on. The larger letters in the 20/40 row were drawn of such a size to subtend the 5′ and 1′ angles when 40 feet distant, the even larger letters in the 20/100 row of such a size to subtend the same angles when 100 feet distant, and so on. Standard charts include rows labeled from 20/10 on up to 20/300 or further.

Thus, a person undergoing an examination for visual acuity views the Snellen chart from a fixed distance of 20 feet—even if in some examination rooms an apparently shorter-than-20-feet distance is lengthened to 20 feet with the help of mirrors. The test subject then attempts to read across the line containing the smallest letters that he or she can identify successfully. An individual with normal vision, who at best correctly identifies each letter in the 20/20 row, is said to have 20/20 vision. An individual in need of spectacles who sees clearly the letters in the 20/100 row, but none below that, is said to have 20/100 vision. This level of acuity, or lack of acuity, indicates that this person can just identify the 20/100 row of letters at a distance of 20 feet, while a person with normal vision (20/20) could back off to a distance of 100 feet and see the letters just as clearly. In general, then, the numerator of the Snellen fraction indicates the fixed testing distance, while the denominator denotes the distance at which each letter in a given Snellen row subtends the standard 5′ and 1′ of arc.

10-3 OPTICAL MODELS OF THE EYE

As we have seen, the normal biological eye is a near spheroid, some 23 mm from cornea to retina. The optical surfaces that provide the bulk of the focusing power are essentially three: the air-cornea interface, the aqueous-lens interface, and the lens-vitreous interface.[3] Overall, the eye can be represented quite simply as a single, thin, positive lens of focal length 17 mm in the relaxed state (emmetropic eye) or 14 mm in the tensed state (near vision). In Chapter 5, "Vergence and Vision," we used this highly simplified model to calculate image positions in emmetropic and ametropic eyes, especially with the addition of corrective spectacles.

While simplified models are convenient for quick results, they are not always adequate for calculations that must approximate more faithfully the true performance of the eye. On the other hand, treating the eye exactly—accounting for all refracting interfaces as well as the varying index of refraction across the lens—introduces many unnecessary complications. From early days, then, attempts have been made to model the eye as a more simplified optical system. Here we shall consider several models, moving consistently from the more exact, so-called *schematic eyes*, to the less exact (but more convenient) *reduced* or *standard* eyes. The models are

1. Helmholtz-Laurance (nearly exact) schematic eye
2. Gullstrand's three-surface, simplified schematic eye
3. Emsley standard reduced 60-diopter eye

---

[3]There are, in fact, four interfaces—the three mentioned here and the cornea-aqueous interface. Since the cornea has an index of refraction of 1.376 and the aqueous humor one of 1.336, minimal refraction occurs here, so the contribution to the focusing power of this interface is ignored.

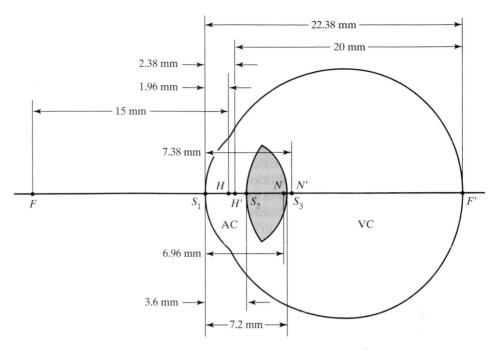

Figure 10-3 Representation of H. V. Helmholtz's schematic eye 1 as modified by L. Laurance. For definition of symbols, refer to Table 10-1. (Adapted with permission from Mathew Alpern, "The Eyes and Vision," Section 12 in *Handbook of Optics*, New York: McGraw Hill, 1978.)

Helmholtz-Laurance Schematic Eye. A schematic eye—after H. V. Helmholtz and L. Laurance—represents a biological eye with fair accuracy. It is shown in Figure 10-3. Relative locations of the refracting surfaces are shown, as are the cardinal points (H, H', N, N', F, F') of interest for the eye as a whole. The schematic eye shown corresponds to its relaxed state. For the fully tensed eye, the front surface of the lens sharpens its curvature from a radius of $R = 10$ mm to $R = 6$ mm. By way of summary and in conjunction with Figure 10-3, Table 10-1 lists the important optical surfaces, their distances from the corneal vertex on the optical axis, several radii of curvature, indices of refraction, and refracting powers of the optical surfaces related to the cornea and lens. Note carefully that the values for the refractive indices of various parts of the eye, as well as radii of curvature of various surfaces, may not agree with values of the biological eye itself. When taken as a whole, however, the optical values that describe the schematic eye do faithfully represent the optical performance of a living, biological eye. Nevertheless, this "nearly exact" model is still far too complicated for calculations carried out by most of the ophthalmic community, so we consider the next simplification.

Gullstrand's Three-Surface Simplified Schematic Eye. Based on an adaptation of A. Gullstrand's model for an exact schematic eye,[4] which can be compared favorably with the Helmholtz-Laurance model described above, a substantial simplification is made in the three-surface schematic eye, shown in Figure 10-4. Here we see that the cornea is treated as a single surface, as it was,

[4]Based on the historically significant work of Allvar Gullstrand (1862–1930), a Swedish ophthalmologist who received the Nobel prize in 1911 for Physiology, specifically for his investigations of the eye as an image-forming optical system.

TABLE 10-1 CONSTANTS OF A SCHEMATIC EYE (HELMHOLTZ-LAURANCE)

Optical surface or element	Defining symbol	Distance from corneal vertex (mm)	Radius of curvature of surface (mm)	Refractive index	Refractive power (diopters)
Cornea	S_1	—	$+8^a$	—	+41.6
Lens (unit)	L	—	—	1.45	+30.5
Front surface	S_2	+3.6	$+10^b$	—	+12.3
Back surface	S_3	+7.2	−6	—	+20.5
Eye (unit)	—	—	—	—	+66.6
Front focal plane	F	−13.04	—	—	—
Back focal plane	F'	+22.38	—	—	—
Front principal plane	H	+1.96	—	—	—
Back principal plane	H'	+2.38	—	—	—
Front nodal plane	N	+6.96	—	—	—
Back nodal plane	N'	+7.38	—	—	—
Anterior chamber	AC	—	—	1.333	—
Vitreous chamber	VC	—	—	1.333	—
Entrance pupil	E_nP	+3.04	—	—	—
Exit pupil	E_xP	+3.72	—	—	—

Source: Adapted with permission from Mathew Alpern, "The Eyes and Vision," Table 1, Section 12 in *Handbook of Optics*, New York: McGraw-Hill Book Company, 1978.

[a]In this model, the cornea is assumed to be infinitely thin. In Gullstrand's exact schematic eye, for example, the cornea is retained as a two-surface element of 0.5 mm thickness.

[b]Value given is for the relaxed eye. For the tensed or fully accommodated eye, the radius of curvature of the front surface is changed to +6 mm.

in fact, in the Helmholtz-Laurance schematic model described above. However, Gullstrand, in his *exact schematic model*[5] (not discussed here) still retained the cornea as two surfaces. The lens in Figure 10-4 is considered to have an average index of refraction of 1.413, and the distance from cornea to macula is fixed in this model at 24.17 mm to guarantee that parallel rays are focused perfectly on the retina for an emmetropic eye. While still not as simple as the standard, reduced model that follows, Gullstrand's three-surface model allows one to change the power of the lens (adjust the spherical curvature of the two surfaces of the lens) to accommodate, as the eye focuses on nearer and nearer objects. Thus, this model is useful for calculating image positions when the natural lens is removed (during cataract surgery), or when an artificial lens of fixed shape (no accommodation) replaces a cataractous lens.

Emsley Standard Reduced 60-Diopter Eye.   The simplest model of the eye, the one used most widely in ophthalmic education, is the one designed by H. Emsley, as a still further simplification to Gullstrand's three-surface eye. It involves a single refracting surface. The essential parameters are shown in Figure 10-5. In this representation, the axial distance of the eye is fixed at 22.22 mm. Further, the two principal points, H and H', and the two nodal points, N and N', shown previously as separate points in Figure 10-3, are now coalesced into single principal points H and N, with H fixed at the point where the corneal surface

[5]More detailed specifications of Gullstrand's eye can be found, for example, in Chapter 2 of Alan Tunnacliffe, *Introduction to Visual Optics*, New York State Mutual Book & Periodical Service, Limited. 1989.

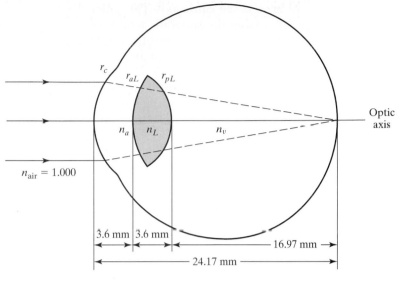

r_c = corneal radius = 7.8 mm
r_{aL} = anterior lens surface = 10 mm
r_{pL} = posterior lens surface = −6 mm
n_a = aqueous humor index = 1.336
n_L = average lens index = 1.413
n_v = vitreous humor index = 1.336

Figure 10-4   Gullstrand's three-surface, simplified schematic eye. The axial length of the eye (24.17 mm) is chosen to ensure the focusing of parallel rays on the retina (emmetropic eye). The radii of curvature r_{aL} and r_{pL} can be adjusted to represent a bulging lens during accommodation.

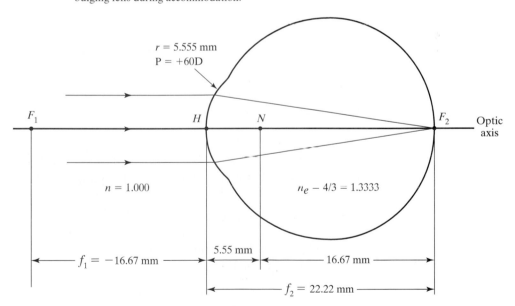

Figure 10-5   Emsley's standard reduced 60-diopter eye. Note that the model of this eye is represented by a single curved surface, where the two principal points H and H' and two nodal points N and N' (shown in Figure 10-3) coalesce into single points H and N, respectively. The focal points F_1 and F_2 are measured relative to the principal point H. Here the axial length of the emmetropic eye, from cornea to retina, is 22.22 mm, thereby ensuring that parallel rays for this model focus on the macula at the retinal surface.

intersects the optic axis. The single, curved corneal surface of 60 D power, encompassing the separate actions of the corneal and lens interfaces found in the biological eye, represents the total focusing power of this model. The index of refraction for the entire medium inside the eye, from cornea to retina, is now simply $n_e = 4/3$.

One can see now that the simplified model of the eye adopted in Chapter 5, with a thin, single, positive lens *in air* of power 60 diopters, and a lens-to-retina distance of 16.67 mm, is just another way to simulate the focusing action of the true biological eye. As long as the complex, biological eye can be treated as a single focusing system, without reference to its internal structure, either the single lens in air ($n = 1.0$), or the single surface eye in a refractive medium of $n = 4/3$ can be used to calculate the effect of spectacle lenses required for the correction of ametropic vision. Before we examine errors of refraction and their correction in the next section, we consider several examples based on Emsley's standard, reduced 60-diopter eye.

❏ **Example**

Given the parameters for Emsley's standard, reduced, 60-diopter eye in Figure 10-5, show by using both $f_1 = -16.67$ mm and $f_2 = 22.22$ mm that the power of the single refracting surface is, indeed, 60 diopters.

Solution   Using the focal point F_1 in air, with $f_1 = -0.01667$ m we see that

$$P = \frac{n_1}{f_1} = \frac{1}{-0.01667 \text{ m}} = 59.988 \text{ D}, \quad \text{or approximately, 60 D}$$

Using the focal point F_2 in the ocular medium, with $n_e = 4/3$ and $f_2 = 0.02222$ m, we see that

$$P = \frac{n_e}{f_2} = \frac{4/3}{0.02222 \text{ m}} = 60.006 \text{ D}, \quad \text{or approximately, 60 D} \quad \blacksquare$$

❏ **Example**

Show that the radius of curvature of the single surface in Emsley's standard, reduced, 60-diopter, emmetropic eye is $r = 5.555$ mm, with correct curvature.

Solution   Using the single-surface imaging formula,

$$\frac{n_2}{v} - \frac{n_1}{u} = \frac{n_2 - n_1}{r}$$

where $u = \infty$ for an emmetropic eye, and $v = 22.22$ mm for perfect focusing of parallel rays on the retina, we can write

$$\frac{4/3}{22.22} - \frac{1}{\infty} = \frac{4/3 - 1}{r}$$

or

$$r = \frac{+(1/3)(22.22)}{4/3} = \frac{+22.22}{4} = +5.555 \text{ mm}$$

In accordance with our adopted sign convention, the positive sign for r indicates that the surface is convex toward the oncoming light. Thus both results are in agreement with Figure 10-5. ∎

❏ **Example**

Suppose an object at the near point, 25 cm from the cornea, is to be viewed by the Emsley standard, reduced eye. Where would the image fall? How should the model eye be modified to represent near-point vision?

Solution With $r = +5.555$ mm, as calculated above, and the imaging formula,

$$\frac{n_2}{v} - \frac{n_1}{u} = \frac{n_2 - n_1}{r}$$

where $u = -250$ mm, we can solve for the image distance v:

$$\frac{4/3}{v} - \frac{1}{-250} = \frac{4/3 - 1}{5.555}$$

$$v = 23.8 \text{ mm}$$

Thus, the Emsley standard reduced 60-D eye, defined for an emmetropic eye, images an object located at the near point at a position 1.58 mm beyond the retina. To focus objects located at the nearpoint correctly on the retina, one would have to increase the power of the model eye. This can be accomplished by decreasing the radius of curvature of the single surface from 5.555 mm to a value that brings the image in by 1.58 mm, thereby focusing perfectly on the retina. If the model is left intact, a lens of appropriate positive power is added in front of the cornea to achieve the same end, as is done in fitting spectacles. ∎

Optical and visual axes. So far we have treated the optical axis as the axis of symmetry for all elements in a general optical system. Accordingly, for the schematic eyes we have modeled to this point, all refracting surfaces are shown coaxially, with the optical axis passing through the center of symmetry of each element. Thus, in Figure 10-6, for a Gullstrand three-surface model, we show the optical axis passing through the principal point H at the front center of the cornea, through the nodal point N at the rear of the lens, and onto the focal point F_2 at the retina.

The fovea centralis, the region of most distinct vision, is not located at F_2, the intersection of the retinal surface and the optical axis. Rather it is found in a region located somewhat temporally (horizontally away from the nose toward the side of the head) and downward (from top of head to bottom). Thus we require a new axis, the *visual axis*, as shown in Figure 10-6. This axis, distinct from the optical axis, is defined as the direction taken by the chief ray in a narrow bundle of rays that enter the pupil and focus on the fovea. As shown in the figure, then, the visual axis coincides with a ray that is headed toward E, the center of the entrance pupil; after refraction by the lens, it leaves the center of the exit pupil E′ as it continues on toward the fovea.

Since the location of the fovea is on the temporal side of the focal point F_2 and somewhat downward, the visual axis *in object space* comes in at an angle $2°$ above the optical axis (thereby heading downward) and $5°$ to the nasal side,

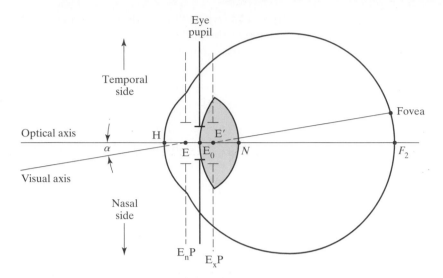

Figure 10-6 Relative locations of the visual and optical axes, the eye pupil, and its entrance and exit pupils. The drawing depicts a horizontal section of the eye with the nasal and temporal sides of the eye located as indicated. The points E, E_0, and E' represent the centers of the entrance pupil, eye pupil, and exit pupil, respectively, and all lie on the optical axis. The visual axis, in object space, makes an angle α with the optical axis and passes through the center of the entrance pupil, terminating at the fovea.

(thereby heading outward). Such a direction essentially connects the center of the entrance pupil and the center of the fovea. While it is important to be aware of the distinction between the visual and optical axes and to take this into account in certain analyses of uncoordinated movements of the left and right eyes, for example, for most applications we shall deal only with the optical axis and consider the fovea and focal point F_2 to be identically located.

Entrance and Exit Pupils. In Figure 10-6, we show the location of the real pupil of the eye, with center at E_0 relative to the single corneal surface and lens of the eye, in accordance with Gullstrand's three-surface schematic eye. For this model of the eye, the real pupil is taken to be a distance of 3.6 mm from the corneal surface (see Figure 10-4), located in a plane tangent to the front (anterior) surface of the lens. Based on the procedure for locating the entrance and exit pupils, relative to the position of the real pupil (aperture stop), as presented in Chapter 8, we can now determine the positions of the points E (center of entrance pupil) and E', (center of exit pupil), as measured from pupil point E_0. The following examples show how this is done.

❏ **Example**

Given the data and drawings in Figures 10-4 and 10-6, determine the location of the entrance pupil E_nP.

Solution To locate E_nP we image the real pupil in all optics to its left—in this case, the corneal surface of radius 7.8 mm—by using the single-spherical surface equation,

$$\frac{n_{air}}{v} - \frac{n_a}{u} = \frac{n_{air} - n_a}{r_c}$$

Paying careful attention to sign conventions, we have $r_c = -0.0078$ m, $u = -0.0036$ m, $n_{\text{air}} = 1.000$, $n_a = 1.336$, so that

$$\frac{1}{v} - \frac{1.336}{-0.0036} = \frac{1.000 - 1.336}{-0.0078}$$

Solving, we obtain $v = -3.048 \times 10^{-3}$ m $= -3.048$ mm. The negative sign tells us that the entrance pupil is a virtual image located on the same side of the corneal surface as the real pupil, a distance 3.048 mm away. Thus the distance HE in Figure 10.6 is 3.048 mm. This places the entrance pupil, slightly magnified, a distance of 0.552 mm in front of (to the left of) the real pupil. ∎

❏ **Example**

Given the data and drawings in Figures 10.4 and 10.6, determine the location of the exit pupil, E_xP.

Solution To locate E_xP, we image the real pupil in all optics to its right. Since the real pupil is tangent to the anterior surface of the lens, it will be imaged only by the posterior surface of the lens, a distance of 3.6 mm away. Thus we use the formula,

$$\frac{n_v}{v} - \frac{n_L}{u} = \frac{n_v - n_L}{r_{pL}}$$

where, according to Figure 10-4, $n_v = 1.336$, $n_L = 1.413$, $r_{pL} = -0.006$ m, and $u = E_0N = -0.0036$ m. Accordingly, we have

$$\frac{1.336}{v} - \frac{1.413}{-0.0036} = \frac{1.336 - 1.413}{-0.006}$$

which, after solving, gives $v = -3.52 \times 10^{-3}$ m $= -3.52$ mm. According to the sign convention, the exit pupil is a virtual image, slightly magnified, located 3.52 mm to the left of the posterior lens surface. Thus the segment $E'N = 3.52$ mm and $E_0E' = 0.08$ mm. ∎

10-4 ERRORS OF REFRACTION AND THEIR CORRECTION

The errors of refraction of the eye lead to three well-known defects in vision: nearsightedness (*myopia*), farsightedness (*hyperopia*), and *astigmatism*. The first two are traceable, for the most part, to an abnormally shaped eyeball, axially too long or too short. As we have already learned, these defects are referred to as ametropic conditions. Either deviation from normal eyeball length impairs the ability of the combined refracting elements, cornea and lens, to form a clear image of objects situated at both remote and nearby positions. The third defect, astigmatism, is due to unequal or asymmetric curvatures in the corneal surface, thereby rendering impossible the simultaneous focusing of all light incident on the eye. Whether the errors of refraction occur singly or in some combination (as they usually do), they are generally correctable with appropriately shaped, external optics (eyeglasses).

As a point of reference for judging the departure of defective vision from the norm, refer to the *normal* eye depicted to the left in Figure 10-7. With accommodation, the normal eye can form a distinct image of objects located anywhere between its far point (FP) at infinity and its near point (NP), nominally at a distance of 25 cm for the young adult. When the normal eye is focused at infinity (distant objects), parallel light enters the relaxed eye and forms a distinct image on the retina (Figure 10-7a). When focused at the near point, diverging light enters the tensed eye (fully accommodated) and is again brought to sharp focus on the retina (Figure 10-7b).

Myopia. When compared with the normal eye, a myopic eye or near-sighted eye is commonly found to be longer in axial distance—from cornea to retina—than the usual, accepted span of about 23 mm. As a consequence, and as illustrated schematically in Figure 10-7c, the myopic eye forms a sharp image of distant objects in *front* of the retina, and, of course, a blurred image at the retina. Distinct retinal images are not formed with the unaccommodated myopic eye until the object moves in from infinity and reaches the myopic far point, the most distant point for clear vision (Figure 10-7d). From the far point inward, with appropriate accommodation, the myopic eye sees quite clearly, even at points *closer* than the normal near point (Figure 10-7e). Since angular magnification of detail increases with proximity to the eye, the myopic eye enjoys "superior" vision of objects held close to the eye. (Therefore it can be an advantage for a watchmaker to be myopic, at least during working hours!) In short then, the near-sighted person has a contracted, drawn-in field of vision, a less remote far point and a closer near point, than a person with normal vision. While the more proximate near point might serve as an advantage, the less remote far point is a distinct disadvantage and calls for correction.

Myopic vision is routinely corrected with spectacles of *negative* dioptic power (diverging lenses), which effectively move the myopic far point and near point outward to normal positions. Figure 10-7f shows the corrected vision for distant objects. Note that as far as the optics of the eye itself is concerned, light from distant objects—after passing through the spectacle lens—appears to originate at its own myopic far point. Similarly, Figure 10-7g illustrates the situation for corrected near vision under full accommodation. Light from an object at the normal near point, after spectacle correction, appears now to originate at the near point of the myopic eye. To gain some insight into the degree of negative lens power required to correct myopic vision, consider the following example.

❏ **Example**

A myopic person (without astigmatism) has a far point of 100 cm and a near point of 15 cm.

(a) What spectacle correction is required to move the myopic far point out to infinity?

(b) With the correction, how near to the eye can the myope place a book and read the print clearly? Solve using both the standard and vergence equations.

Solution

(a) Refer to Figures 10-7d and f. The required spectacle lens must accept light from infinity and bend it so that it appears to the myopic eye to be com-

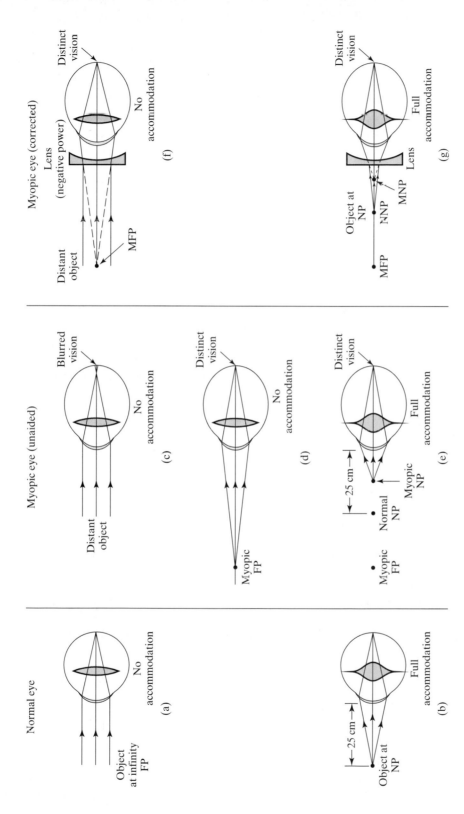

Figure 10-7 A comparison of normal and myopic vision, with optical correction. Note that refraction by the eye lens is not shown. The abbreviations read as follows: MFP = myopic far point; NNP = normal near point; MNP = myopic near point.

ing from its myopic far point, 100 cm away. Thus, with the standard thin-lens equation, we have

$$\frac{1}{v} - \frac{1}{u} = \frac{1}{f}$$

where $v = -100$ cm and $u \to -\infty$. Thus,

$$\frac{1}{-100} - \frac{1}{-\infty} = \frac{1}{f}$$

or $f = -100$ cm. The power $P = 1/f = 1/(-1.0 \text{ m}) = -1.0$ D.

With the vergence equation, we write

$$U + P = V \qquad \text{or} \qquad P = V - U$$

where $V = 1/v = 1/(-1.0 \text{ m}) = -1.0$ D and $U = 1/(-\infty) = 0$. So, $P = -1.0 \text{ D} - 0 = -1.0$ D

Either solution requires a spectacle lens with a correction of -1.0 D.

(b) Refer to Figure 10-7g. The book can be moved into a position near the eye such that the spectacle lens bends the light just enough to make it appear as if it were coming from its myopic near point of 15 cm. Then, with the standard thin-lens equation, $v = -15$ cm, $f = -100$ cm, we have

$$\frac{1}{v} - \frac{1}{u} = \frac{1}{f} \qquad \text{or} \qquad \frac{1}{-15} - \frac{1}{u} = \frac{1}{-100}$$

giving $u = -17.6$ cm. With the vergence equation $U = V - P$, we have $V = 1/-0.15 \text{ m} = -6.667$ D and $P = -1.0$ D. Then

$$U = V - P = -6.667 \text{ D} - (-1.0 \text{ D}) = -5.667 \text{ D}$$

so $u = 1/U = 1/(-5.667) = -0.176 \text{ m} = -17.6$ cm. Either method indicates that with -1.0 D spectacles this particular myope can see objects as near as 17.6 cm from the eye, several inches closer than the 25-cm position for a person with normal vision. ■

Hyperopia.   The far-sighted, or hyperopic, eye is commonly shorter than normal. Whereas the longer-than-normal myopic eye has too much convergence in its optical system and requires a diverging lens to correct its over-refraction, the shorter-than-normal hyperopic eye has too little convergence and requires a converging lens to increase its refraction. The drawings in Figure 10-8, in analogy with those in Figure 10-7, illustrate the defects and correction associated with the far-sighted eye. In Figure 10-8a, light from a distant object enters the relaxed eye and tends toward a focus *behind* the retina, causing blurred vision. The focal point behind the retina is considered the hyperopic far point. Figure 10-8b shows that the hyperopic eye must (and can) accommodate to see distant objects clearly. In Figure 10-8c, it is clear that for distinct vision, the hyperopic near point is farther away than the normal near point. Consequently, objects located closer than the hyperopic near point would be out of focus, even with full accommodation. The far- and near-point corrective measures, with the appropriate positive spectacles lens in place, are indicated in Figure 10-8d and 8e. The corrected eye now sees distant objects clearly without accommodation, and at the normal near point it sees objects clearly with full accommodation. Let's see how an op-

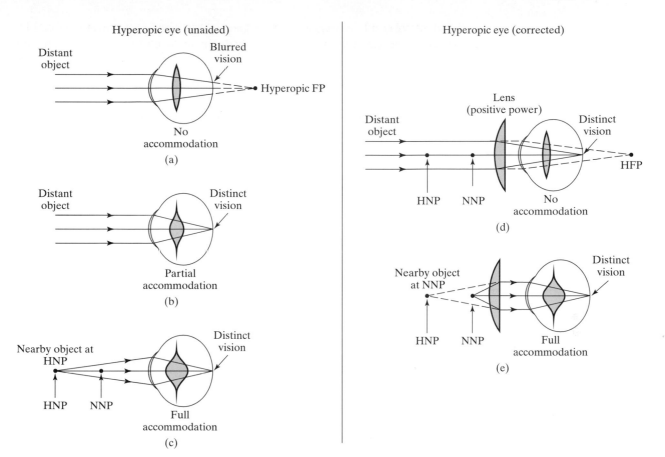

Figure 10-8 Hyperopic vision with correction. The abbreviations read as follows: HNP = hyperopic near point; NNP = normal near point; HFP = hyperopic far point.

tometrist might calculate the spectacle power required to correct hyperopic vision. Consider the following example.

❏ **Example**

A hyperope is diagnosed as having a near point at 85 cm. What corrective power is required for spectacles to enable this far-sighted person to see objects placed at the normal near point, 25 cm from the eye?

Solution Refer to Figure 10-8e. A corrective lens with positive power will bend the light rays coming from the normal near point just enough—under full accommodation—to make it appear as if the light is coming from the hyperope's near point. With the standard thin-lens equation $1/v - 1/u = 1/f$, where $u = -25$ cm and $v = -85$ cm, we obtain

$$\frac{1}{-85} - \frac{1}{-25} = \frac{1}{f}$$

which yields $f = 35.4$ cm, or $P = 1/(0.354 \text{ m}) = +2.82$ D. Using the vergence equation, we have $P = V - U$, where $V = 1/v = 1/(-0.85 \text{ m}) = -1.176$ D, and $U = 1/u = 1/(-0.25 \text{ m}) = -4.0$ D. So, $P = -1.176$ D $- (-4.0$ D$) = +2.82$ D, the identical result, as expected. With spectacles shaped to a power of +2.82 D, the hyperope can see clearly objects brought in as close as 25 cm from the eye. ∎

Astigmatism.[6]   The astigmatic eye suffers from a mixture of spherical and nonspherical curvatures over the surface of the cornea. When astigmatism is present, the radii of curvature of the corneal surface in two meridional planes (those containing the optic axis) are unequal. Such asymmetry leads to different refractive powers and, consequently, to focusing of light at different distances from the cornea, resulting, of course, in blurred vision. If the two planes are orthogonal to one another, say, one horizontal and the other vertical, the defect is referred to as *regular astigmatism*, a condition that is correctable with suitable spectacles. If the two planes are not orthogonal, a rather rare condition called *irregular astigmatism*, the surface anomaly is not so easily corrected. Regular astigmatism can be treated with cylindrical surfaces ground on the back surface of the required spectacle lens.

Assume, for example, that the refractive power in the vertical meridian of the cornea is greater by 1 diopter than the power in the horizontal meridian. This situation means that the corneal surface is more sharply curved in the vertical meridian and that vertically oriented details in an object are brought to a focus nearer the cornea than are horizontally oriented details. Consider a cylindrical surface with negative power of 1 diopter in the vertical meridian. Since a cylinder has no curvature along its axis, the surface has no power in the horizontal meridian. If this surface is included in the spectacle design, it would cancel exactly the distortion introduced by the cornea and equalize powers in both meridians. As a result, vertical and horizontal details in the object scene would be formed at the same distance from the cornea, and astigmatic blurring would not occur.

For many of us, blurred vision is a result of astigmatism mixed with myopia or hyperopia. If myopic astigmatism is present, for example, vision is faulty on two counts. The myopia itself causes an overall blurring of distant objects; the astigmatism compounds the problem by adding considerably more blurring in one meridian than another. Correcting for both defects is accomplished with *sphero-cylindrical* lenses, spherical surfaces to correct for myopia and cylindrical surfaces to correct for astigmatism.

When optometrists prescribe corrective eyeglasses for conditions of myopic or hyperopic astigmatism, they generally identify three numbers. For myopic astigmatism, the three numbers, written in prescription format, might be

$$R_x: \quad -2.00 \quad -1.00 \quad \times 180$$

For hyperopic astigmatism, the prescription might read

$$R_x: \quad +2.00 \quad -1.50 \quad \times 180$$

The first number refers to the *sphere power*, the required power in diopters of the spherical surfaces on the spectacle lens that correct for the overall myopia or hyperopia. The second number refers to the *cylinder power*, the required power of the cylindrical surface superimposed on the back surface of the spectacle lens to correct for astigmatism. The third number refers to the orientation of the *cylinder axis*, specifying whether the axis of the cylinder is to be vertical, horizontal,

---

[6]Astigmatism is also treated at some length in Chapters 6 and 7.

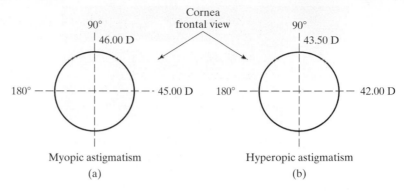

Figure 10-9 Conditions of myopic and hyperopic astigmatism with corrective spectacle prescriptions. (a) Refraction in the 180° meridian yields −2.00 D of myopia. The eyeglass prescription is R_x: −2.00 −1.00 ×180. (b) Refraction in the 180° meridian yields +2.00 D of hyperopia. The eyeglass prescription is R_x: +2.00 −1.50 ×180.

or somewhere in between. In optometric notation, the horizontal axis is referred to as the 180° axis, or simply, ×180, and the vertical axis as ×90.

Figure 10-9 indicates the optical conditions associated with the corrective prescriptions just cited for both myopic and hyperopic astigmatism. For the case of myopic astigmatism, Figure 10-9a, the corneal surface is evidently less sharply curved in the horizontal meridian (power = 45.00 D) than in the vertical (power = 46.00 D). The myopic correction, always measured in the meridian of least refractive power, is found in this instance to be −2.00 D, along the horizontal meridian. The astigmatic correction, with cylinder axis horizontal (×180), is determined to be −1.00 D. With the appropriate cylindrical surface ground on the rear of the spectacle lens, the correction of −1.00 D reduces the power in the vertical meridian from 46.00 D to 45.00 D, thereby equalizing the refracting powers in the two meridians and negating the corneal astigmatism.

Figure 10-9b shows a comparable condition and prescription for hyperopic astigmatism. Note that a sphere power correction of +2.00 D is needed to correct for the hyperopia, and a cylinder power correction of −1.50 D is needed again along the vertical meridian (×180, cylinder axis horizontal) to equalize the refractive power in the two orthogonal meridians.

10-5 COMMON OPHTHALMIC INSTRUMENTS

The anomalies of the eye are many. Whether due to disease, injury, anatomical structure, senescence or simply faulty refraction, the ophthalmological community works to correct the problem and restore optimal vision. A visit to the eye doctor is, more or less, a common occurrence for each of us, whether for a routine checkup or a special problem. Once in the examining room, we come in contact with a variety of optical instruments designed to assess visual acuity and probe the inner parts of the eye, from the cornea to the innermost basal surface, the *fundus.*

In this section, we shall describe the more common of these ophthalmic instruments. We explain briefly their function and, where possible, present simple, schematic diagrams of the optical elements and assemblies that make up the instruments. For our purpose, in this introductory treatment of optics and vision, we shall limit ourselves to an overview of a refractor, direct ophthalmoscope, slit lamp, tonometer, ophthalmometer, and retinal (fundus) camera.

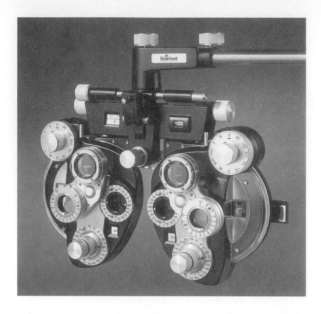

Figure 10-10 A typical refractor used to check for refractive anomalies in vision. (Courtesy Reichert Ophthalmic Instruments, Leica, Inc.: Buffalo, N.Y.)

Refractor. An appointment with an optometrist for a check on visual acuity inevitably brings the patient in contact with a *refractor*. Figure 10-10 provides a close-up of a typical refractor. This instrument contains an assortment of thin, spherical lenses, with powers ranging from −19.00 D to near +17.00 D, variable in steps of 0.25 D, and a set of cylinder powers from 0 to 6.00 D, also in steps of 0.25 D. Combinations of lenses, centered on the patient's visual axes, are presented to each eye by the optometrist, in a sequential fashion, to determine the power and cylinder axis required to correct existing conditions of ammetropia or astigmatism.

An instrument with fully integrated controls, the typical refractor offers choices of prism lenses with powers from 1 to 20 prism diopters, cross cylinder lenses, cylinder axis controls, reading test charts, and green or red lenses for color discrimination. After a routine examination with a refractor, the optometrist prescribes spectacles with corrections such as R_x: +2.00 −1.50 ×180, as described in the previous section.

Direct Ophthalmoscope. Direct ophthalmoscopes are hand-held, self-illuminating optical instruments which enable an examiner to see directly into the patient's eye. This relatively simple instrument, invented by Hermann von Helmholtz in 1851, was first used to view the optic disk, retina, blood vessels, macula, and choroid at the back surface of the eye. Today they are used by observers to see not only the details of the retinal surface but also the ocular regions from the back of the lens to the retina.

Figure 10-11 shows a side view of a direct ophthalmoscope, outlining the bundle of light that moves from the intense tungsten filament source S to illuminate a patch of the retinal surface in the eye. The examiner, through a choice of lenses, looks directly into the eye, at the illuminated fundus.

Light from source S is rendered parallel and directed onto an aperture stop by a condensing lens. The variable aperture stop (AS) presents the examiner with choices of different diameter, circular stops, depending on whether a narrow or wide-angle view of the retina is desired. The focusing lens L_1 and field lens L_2 enable the examiner to vary the size of the spot on the edge of the mirror M and, consequently, the area of the fundus that is to be illuminated. The field of view

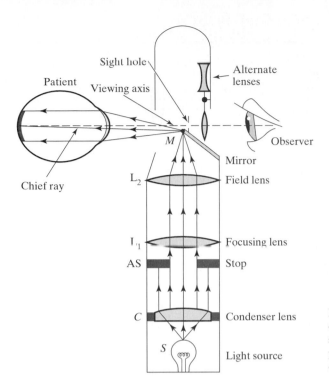

Figure 10-11 Side view of a direct, hand-held ophthalmoscope, showing the light path from source to retina. The examiner looks into the parallel beam illuminating the interior of the eye to see details in the fundus.

seen by the observer is determined by the field stop, which is generally the observer's pupil. The image of the observer's pupil, as formed by the optical system of the patient's eye is, of course, the entrance window, formed usually somewhat behind the patient's retinal surface.

As shown in Figure 10-11, the viewing axis of the examiner along the sight hole and the chief ray of the illuminating light are separated. The separation reduces significantly those reflections from the anterior corneal surface (corneal reflex) that are directed back toward the examiner, thereby potentially obscuring an otherwise clear view of the retina. Interestingly enough, in some ophthalmoscopes, the properties of polarized light are used to reduce this unwanted corneal reflection. In this case, polarized light is used to illuminate the retina. The polarized light reflected from the cornea is removed by an analyzer—located between the sight hole and the examiner—whose transmission axes are adjusted to reject the reflected light. The analyzer, nevertheless, passes about half of the light coming to the observer from the fundus, since internal scattering serves to depolarize the light.

An important outgrowth of the direct ophthalmoscope is the modern *binocular indirect ophthalmoscope*. With the illumination system worn around the forehead of the examiner, it provides stereoscopic observations of large parts of the retina at fairly low magnification. The operation of its illumination and viewing optics is roughly similar to that described for the simpler direct ophthalmoscope.

Slit Lamp. Another workhorse for ophthalmic examination of the eye, from cornea to retina, is the slit lamp. This instrument projects an intensely illuminated, sharply imaged slit over various regions of the eye, at various distances into the eye. In general, slit lamps consist of two main, optical systems, one to develop the bright slit on target, and the other, a microscope to provide a magnified image of the target area. The optics of the slit lamp is shown in Figure 10-12.

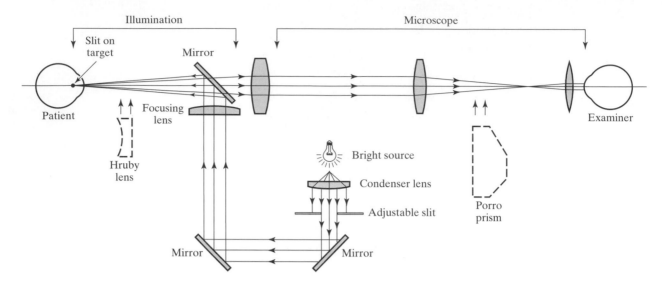

Figure 10-12 Simplified optics of a slit lamp with illumination system, diagnostic microscope and optional Hruby lens.

As can be seen in the figure, a lamp source, condenser lens, adjustable slit, direction-reversing mirror pair, focusing lens, and a semireflecting final eye-alignment mirror all work together to form a sharp, bright image of the slit on the target region in the eye. Light returning from the target area is then processed by lenses in the microscope to provide the observer with a magnified, virtual image (5× to 45×) of the target area. A Hruby lens, shown dotted, is often used in conjunction with the slit lamp, to enable the examiner to view the fundus in detail. In essence, a Hruby lens, often hand-held and movable toward and away from the patient, is used to refocus the distant image of the retina—formed by the optics of the emmetropic eye itself—into a real image on the anterior side of the objective lens of the microscope, near the front focal point of this lens. (Refer back to Figure 9-21, if necessary, to review the optics of a microscope.)

In addition, a Porro prism (Figure 9-11d), also shown dotted, is included in the optical system of the microscope to invert the image, providing the examiner with an upright view, consistent with the view seen when looking into the eye with a direct ophthalmoscope.

A modern version of the slit lamp is shown in Figure 10-13, indicating the binocular microscope head used by the observer and the familiar forehead strap and chin rest for the patient. Slit lamps come replete with sets of objective lenses and eyepieces, filters, and adjustable slits-all included to enhance the versatility of the instrument for diagnostic procedures.

Tonometer. A tonometer is an ophthalmic instrument used to measure the internal pressure within the globe of the eye. Normal pressure is about 15 mm of mercury above atmospheric pressure, which is taken to be 760 mm of mercury at sea level. The rigid shape of the eyeball is maintained by a fluid system that introduces and drains aqueous humor at the base of the iris, at a rate consistent with maintaining correct intraocular pressure. If the pressure rises, due to poor drainage conditions, the consequences can be serious, damaging the retina and leading eventually to varying degrees of blindness. The medical condition describing the blockages that cause abnormal pressure in the eye is called *glaucoma*. Two types of blockage occur, one leading to so-called chronic glaucoma

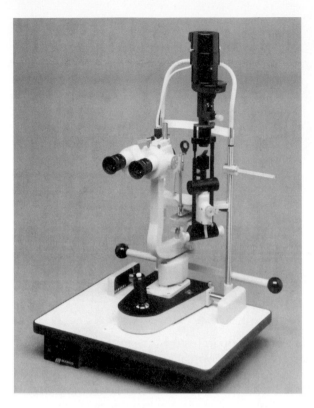

Figure 10-13 A standard slit lamp with a low-power binocular microscope for stereoscopic vision. (Courtesy Marco Ophthalmic, Inc.: Jacksonville, Florida.)

and the other to acute or narrow-angle glaucoma. Both require early detection and treatment, the first with drugs, the second with surgery.

Two types of tonometers are used to measure intraocular pressure. A contact tonometer, used generally in conjunction with a slit lamp, is a mechanical, contact device that produces a flattening of a central area of about 3.00 mm in diameter on the front surface of the cornea. Noncontact tonometers, by contrast, direct puffs of compressed air against the cornea to create the same desired flattening. With either instrument, the degree of mechanical force or strength of air puff required to produce a certain degree of corneal flattening over a given area is correlated with the intraocular pressure. Tonometers, which measure intraocular pressures up to 60 mm of mercury, provide the examiner with images of the flattened cornea. Upon careful interpretation of these images, the examiner determines the existing ocular pressure.

Figure 10-14 shows a compact, modern, noncontact tonometer. This system, through advances in microprocessor control technology, enables the examiner to align the instrument accurately with the corneal surface before "firing" in order to achieve the correct degree of corneal flattening and obtain a direct readout of the intraocular pressure.

Ophthalmometer. The ophthalmic instrument that measures the radius of curvature of the cornea is called an *ophthalmometer*, also referred to as a *keratometer*. Concentrating on a central corneal area of roughly 3.00 mm in diameter, this instrument measures the curvature in any corneal meridian to an accuracy of ±0.01 mm. Consequently, it is useful in ascertaining the degree of astigmatism present in eyes and also in determining the shape of hard contact lenses, which are to be positioned comfortably on the cornea.

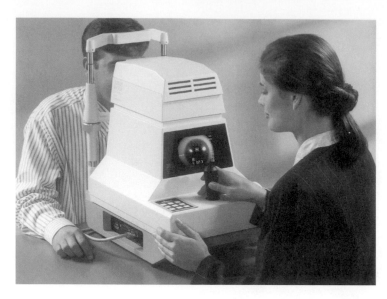

Figure 10-14 A modern, noncontact tonometer used to measure intraocular pressure of a patient's eye. (Courtesy Reichert Ophthalmic Instruments, Leica, Inc.: Buffalo, New York.)

An ophthalmometer operates on the simple principle of projecting a patterned target onto the front, corneal surface and then analyzing the distortions seen in the image reflected from the corneal surface. By comparing the image of the reflected pattern with the image obtained when the pattern is reflected from a normal, perfectly spherical, nonastigmatic cornea, the deviations noted are used to define the degree of astigmatism present in various meridians of the eye. A modern ophthalmometer can measure radii of curvature roughly from 5.5 mm to 12.0 mm in 0.01 mm increments; the corresponding refractive power of the cornea from 28.00 to 60.00 diopters in increments of 0.125 diopters; and the axis of corneal astigmatism from 0° to 180° in 1° increments.

A simplified diagram of the optics involved, without the illumination system, is shown in Figure 10-15a. The patterned objects, called *mires*, are projected onto the cornea of the eye, reflected off the front surface as an image of the pattern, and then processed by a microscope for examination. The two prisms and horizontal apertures, located between the mires and the eyepiece, work together to produce image doubling. Such doubling is required by the examiner to enable measurement of the dimensions of the imaged pattern in perpendicular meridians, and so to determine the details of the astigmatism. Figure 10-15b shows a compact, modern ophthalmometer, containing the elements of the optical system shown in Figure 10-15a, as well as the illumination system.

Retinal camera. A retinal (or fundus) camera is a system that combines the illumination and viewing system typically found in slit lamps with a camera (see Figure 10-12 for the essential optics of a slit lamp). Figure 10-16a is a schematic drawing of the fundus camera system. Imaging of the bright source—a tungsten lamp for viewing and a tungsten arc for photography—is achieved with a double lens (L_1, L_2) focusing/relay system that images the bright source light passing through an aperture onto a mirror M. Light from this mirror is focused on the retinal surface by lens L_3. This same lens captures return light from the illuminated retina and forms a real image of the retina at its focal point F. This image is then relayed by lens L_4 to an observer, looking through eyepiece lens L_5, or onto the film plane of the camera. Suitable mirrors, located in the sys-

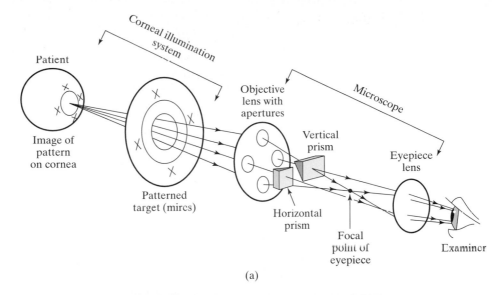

Corneal illumination system

Patient

Image of pattern on cornea

Patterned target (mires)

Objective lens with apertures

Vertical prism

Horizontal prism

Focal point of eyepiece

Microscope

Eyepiece lens

Examiner

(a)

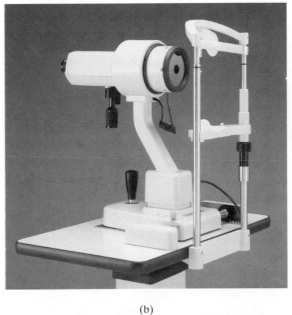

(b)

Figure 10-15 Ophthalmometer. (a) Simplified diagram showing the arrangement of essential optical elements for measuring corneal curvature. (b) Photograph of a mod ern ophthalmometer. (Courtesy Topcon America Corporation: Paramus, New Jersey.)

tem near both the source and observer, facilitate the switching between lamp and arc and between eyepiece and camera.

Figure 10-16b shows a modern retinal camera and Figure 10-16c reproduces a picture of the retinal surface as seen by the observer looking through the eye-piece or as captured on the film plane of the camera. Details of the retinal blood vessels, macular region, and the optic disk are clearly visible in the retinal photo-graph.

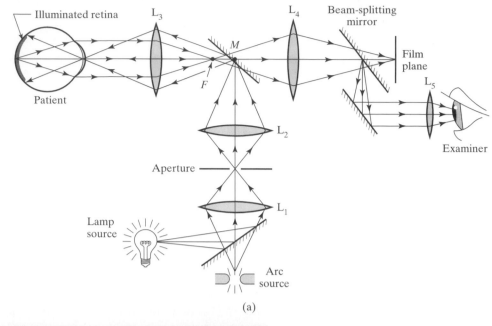

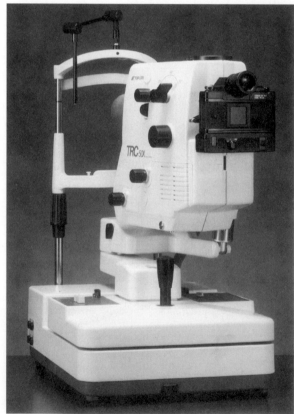

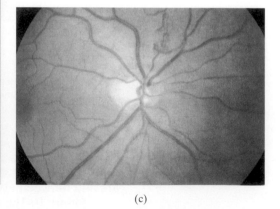

Figure 10-16 Retinal camera. (a) Optical diagram showing the illumination and viewing systems for camera and examiner. (b) Modern retinal camera. (c) View of fundus seen by examiner or camera. (Courtesy Topcon American Coirporation: Paramus, New Jersey.)

PROBLEMS

10-1. Reference to Table 10-1 indicates that the corneal radius of curvature for the unaccommodated schematic eye is 8 mm. Treating the cornea as a thin surface (whose own thickness can be neglected), bounded by air on one side and aqueous humor on the other, determine the refractive power of the corneal surface.

10-2. You have been asked to design a Snellen eye chart for a test distance of 5 ft. The chart is to include rows of letters to test for visual acuities of 20/300 (same as 5/75), 20/100, 20/60, 20/20, and 20/15. Determine the size of the block letter and letter detail (in inches) for each row of letters.

10-3. Consider the unaccommodated crystalline lens of the eye as an isolated unit, having radii of curvature and effective refractive index as given for the schematic eye in Table 10-1.
- **(a)** Calculate its focal length and refracting power as a thin lens in air.
- **(b)** Calculate its focal length and refracting power in its actual environment, surrounded on both sides with fluid of effective index 1.33. Assume a thin lens.
- **(c)** By treating it as a thick lens ($t = 3.6$ mm), calculate the focal lengths f_1 and f_2 relative to the principal planes and the corresponding refracting power of this eye. Refer to Figure 4-3 and Eqs. (4-1) and (4-2).

10-4. Taking values for refractive indices and separation of elements from the Helmholtz-Laurance schematic unaccommodated eye given in Table 10-1 and Figure 10-3, determine the distance behind the cornea where an image is focused for **(a)** an object at infinity, and **(b)** an object at 25 cm from the eye. Use the Gaussian formula for image formation by a spherical surface in a three-step chain of calculations. In (b) assume that the fully accommodated eye differs in the following ways: The front surface of the lens is more sharply curved, having a radius of $+6$ mm, but the back surface remains at -6 mm. As a result, the thickness of the lens along the axis increases to 4.0 mm, and the distance from cornea to the front surface of lens is shortened to 3.2 mm.

10-5. Repeat Problem 10-4 for the Gullstrand three-surface simplified schematic unaccommodated eye, shown in Figure 10-4. Use dimensions given with Figure 10-4.

10-6. Consider the Gullstrand three-surface simplified schematic eye, with geometry and parameters as shown in Figure 10-4, as the model eye for this problem.
- **(a)** Determine the locations of the first and second focal points F_1 and F_2. (*Hint:* Image incident parallel rays forward through the eye in a three-step calculation to determine F_2; image parallel rays similarly backward through the eye to determine F_1.)
- **(b)** Draw a sketch of the Gullstrand simplified eye, roughly to scale, with vertex V_1 at the corneal surface and vertex V_2 at the rear surface of the lens. Locate the focal points F_1 and F_2 along the optical axis.

10-7. Suppose we simplify the Gullstrand eye (Figure 10-4) further by assigning an average refractive index of 1.374 to both the aqueous region and the lens that follows. Then the optical system can be regarded as a thick lens with input plane at the corneal surface ($r = 7.8$ mm), output plane at the rear surface of the lens ($r = -6$ mm), overall thickness of 7.2 mm, and bounded by air ($n = 1.000$) at the front and vitreous ($n = 1.336$) at the rear.
- **(a)** For this thick lens, use Equations (4-1), (4-2) and (4-3), along with Figure 4-3 in Chapter 4, "The Thick Lens," to determine the positions of the four cardinal points $F_1, F_2, H_1,$ and H_2.
- **(b)** Sketch the optical system, roughly to scale, and indicate the approximate locations of $F_1, F_2, H_1,$ and H_2 relative to the input and output planes.
- **(c)** How does the location of F_2 agree with that of the more correct model of the eye in Problem 10-6?

10-8. Given the Emsley standard reduced 60-diopter unaccommodated eye (Figure 10-5), with corneal radius of curvature equal to 5.555 mm, axial thickness of eye from cornea to retina as 22.22 mm, and refractive index of the internal eye medium as 4/3, determine the location of the focal points for this model and compare your results with those shown in the figure.

10-9. How much accommodation, in diopters, must be added to the Emsley standard reduced 60-D unaccommodated eye (Figure 10-5) to focus light from an object at the near point (25 cm from the cornea) clearly on the retina of this eye?

10-10. For the Gullstrand three-surface schematic eye model (Figure 10-4), determine the diameters of the entrance pupil and exit pupil for a dark-adapted eye whose real pupil diameter is 8 mm (See also Figure 10-6).

10-11. A hyperopic person has no astigmatism but has a near point of 125 cm. Correction with glasses requires that this person see objects at the normal near point (25 cm) clearly.
(a) What is the power of the corrective lens?
(b) Will the corrective glasses enable focusing of a distant object on the retina?

10-12. A person has a far point of 50 cm and a near point of 15 cm. What power eyeglasses is needed to correct the far point? Using the eyeglasses, what is the person's new near point?

10-13. Consider each of the following spectacle prescriptions and describe the refractive errors that are involved:
(a) $-1.50, -1.50$, axis 180
(b) -2.00
(c) $+2.00$
(d) $+2.00, -1.50$, axis 180

10-14. Refractors (Figure 10-10) used by optometrists typically contain an assortment of lenses with powers ranging from -19.00 D to $+17.00$ D, in steps of 0.25 D.
(a) Convert the two end point powers for the range in diopters to focal lengths and describe the nature of the lens at each end point.
(b) As the powers of the individual lenses change in steps of 0.25 D, from -19.00 D to $+17.00$ D, the focal lengths for the corresponding lenses do not change by a constant amount. The relationship between a change in dioptric power ΔP and corresponding change in focal length Δf is given by $\Delta f = (-\Delta P/P^2)$. Determine the change in focal length Δf for a change of $\Delta P = 0.25$ D at each of the powers, $P = -19.00$ D, -5.00 D, $+8.00$ D, and $+17.00$ D, to demonstrate this point.
(c) What is the nature of the lens whose power $P = 0$?

10-15. A direct ophthalmoscope (Figure 10-11) is used by an emmetropic observer to examine details of the emmetropic subject's retina. When used without benefit of additional refracting optics, and set for viewing at infinity, the angular magnification provided by the direct ophthalmoscope is given by the usual formula $M = 25/f$, where f is the effective focal length of the optical system of the eye in centimeters, and the 25 refers to the normal near point distance of 25 cm.
(a) For the Emsley standard 60-D eye, what value of M can be expected with the direct ophthalmoscope?
(b) What does this magnification mean so far as the fundus area seen by the observer is concerned?

REFERENCES

ALPERN, MATHEW. 1978. "The Eyes and Vision." In *Handbook of Optics*, edited by Walter G. Driscoll and William Vaughan. New York: McGraw-Hill Book Company.

BENNETT, ARTHUR G., and RONALD B. RABBETTS. 1989. *Clinical Visual Optics*, 2d ed. Boston: Butterworths.

DUKE-ELDER, S., and D. ABRAMS. 1980. *Ophthalmic Optics and Refraction*. Vol. 5 of *Systems of Ophthalmology*, edited by S. Duke-Elder. St. Louis: C. V. Mosby Company.

FEYNMAN, RICHARD P., ROBERT B. LEIGHTON, and MATTHEW SANDS. 1975. *The Feynman Lectures in Physics*, vol. 1. Reading, Mass.: Addison-Wesley Publishing Company. Ch. 35, 36.

Fincham, W. H. A., and M. H. Freeman. 1980. *Optics*, 9th ed. Boston: Butterworth Publishers. Ch. 20.

Michael, Charles R. 1969. "Retinal Processing of Visual Images." *Scientific American* (May): 104.

Michaels, D. D. *Visual Optics and Refraction*, 2d ed. St. Louis: C. V. Mosby Company, 1980.

Milne, L. J., and M. J. Milne. 1956. "Electrical Events in Vision." *Scientific American* (Dec.): 113.

Rubin, M. L. 1974. *Optics for Clinicians*, 2d ed. Gainesville, Fla.: Triad Scientific Publishers.

Sperry, R. W. 1956. "The Eyes and the Brain." *Scientific American* (May): 48.

Tunnacliffe, Alan. 1989. *Introduction to Visual Optics*. New York: State Mutual Book & Periodical Service, Limited.

11

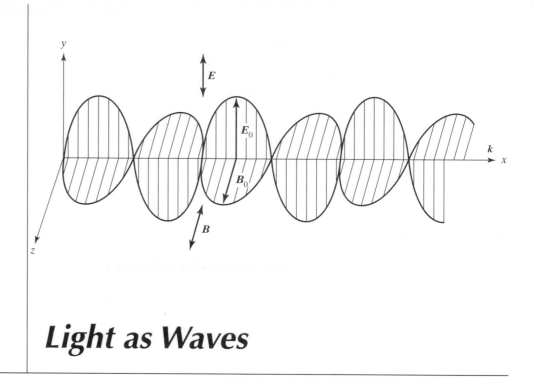

Light as Waves

INTRODUCTION

The fundamentals of wave motion are presented in this chapter as preparation for the study of laser optics, interference, polarization, and diffraction, all of which are treated in the remainder of this book. The wave motion of light is essential to an explanation of much of the subject matter taken up there.

The propagation of light can be described as a wave motion so it is necessary to examine how waves can be represented mathematically. Our most direct notion of wave motion is usually based on experience with clearly visible examples like the waves in a rope or surface waves on a body of water. Other wave motions, like those that transport sound or light energy, are not so clearly visible. Wave motion may be *transverse* or *longitudinal*. The simplest model for a transverse wave is the up and down motion created along a rope when one end is shaken and the other (distant) end is fixed. Each particle of the rope moves up and down in a direction that is perpendicular, or *transverse*, to the direction of the wave moving out along the rope. The apparent direction of the moving wave is the direction in which energy—in this case, mechanical energy—is transported. Light waves, which propagate electromagnetic energy, are of this transverse type. Perhaps the simplest model for a longitudinal wave is the motion observed in a vibrating spring or "slinky." Each particle of the spring moves back and forth in the *same* direction as the wave propagating along the length of the spring. Sound waves are an example of longitudinal wave motion.

Review of units. It may benefit the reader at this point to review the physical units used in this chapter, all in the mks (meter-kilogram-second) system. The coulomb C of electrical charge is the magnitude of the charge carried by 6.242×10^{18} electrons. A current of 1 coulomb of charge per second is one ampere A. The unit of potential difference is the volt V, and the unit of energy is the joule J. Power is the rate of energy flow in joules per second (J/s), which is equivalent to the unit of power, the watt W. The electric field E, described as the force on a stationary unit charge in the presence of other charges, is measured in units of force per unit of charge, or newtons per coulomb (N/C), which is equivalent to volts per meter (V/m). The magnetic field B, described in terms of the force on a moving charge, is measured in teslas (T). For more details relating to the definition and equivalences of these units, an introductory physics textbook should be consulted.

11-1 HARMONIC WAVES

Although a wave motion can be transitory, lasting for a limited time as in the case of a single pulse in a rope, most wave motions of interest to us here are continuous in nature. A continuous wave motion is repetitive, both in space and in time. Of the many shapes a repetitive waveform can take, by far the most useful is the smooth shape of a sine wave. When the waveform has this shape, it is called a *harmonic wave*. Waves produced in nature often are of this harmonic type, such as the sound waves emitted by a vibrating tuning fork or the variations in the electric field of a light wave. More importantly, harmonic waves can be combined to represent more complex waveforms, even a series of rectangular pulses, or "square waves." Such a combination of harmonic wave terms is called a *Fourier series*. An actual waveform can always be represented exactly by an infinite series of harmonic waves. Often, a finite number of such terms approximates the waveform with sufficient accuracy.

The most general mathematical expression of a harmonic wave is given by

$$y = A \sin \left[k(x \pm vt) \right] \tag{11-1}$$

where y is the *displacement*, or whatever physical property is changing harmonically. The speed of the wave is v. The symbols A and k represent constants that can be varied without changing the harmonic character of the wave. The choice of the $(+)$ algebraic sign is appropriate when the wave is moving from right to left (or in the direction of increasing, negative x), while the $(-)$ algebraic sign is correct when the wave is moving from left to right. Since $\sin x \equiv \cos (x - \pi/2)$, the only difference between the sine and cosine functions is a relative translation of $\pi/2$ radians. Thus the cosine function can replace the sine function in the general form of Eq. (11-1). In Figure 11-1, a portion of a sine (or cosine) wave is pictured. In Figure 11-1a, a section of the wave with *amplitude A* is shown at a fixed time, as in a snapshot; in Figure 11-1b, the time variations of the wave are pictured at a fixed point x along the wave. In Figure 11-1a, the repetitive spatial unit of the wave is shown as the *wavelength* λ. Because of this periodicity, increasing all x by λ should reproduce the same wave. This is satisfied by Eq. (11-1) if

$$k = \frac{2\pi}{\lambda}$$

Thus the constant k, called the *propagation constant*, contains information regarding the wavelength. Alternatively, if the wave is viewed from a fixed position,

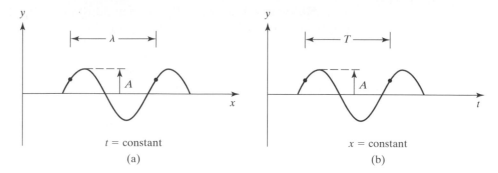

Figure 11-1 A portion of a sine wave, showing its extension in space and time. (a) Sine wave at a fixed time. (b) Sine wave at a fixed point.

as in Figure 11-1b, its motion is periodic in time with a repetitive unit called the *period T*. If all *t* are increased by *T*, the waveform is exactly reproduced. This requirement leads to the relation v = *f*λ where v is the wave speed and *f* is the *frequency* of the wave motion. Now the period represents the time for one complete oscillation, while the frequency is the number of oscillations per second, so that period and frequency are reciprocally related:

$$f = \frac{1}{T}$$

In certain applications, other wave parameters are introduced for convenience. Thus $\omega = 2\pi f$ is called the *angular frequency*, and the reciprocal of the wavelength $\kappa = 1/\lambda$ is called the *wave number*. With these additional relationships, it is easy to show the equivalence of the following common forms for harmonic waves:

$$y = A \sin k(x \pm \mathrm{v}t) \tag{11-2}$$

$$y = A \sin\left[2\pi\left(\frac{x}{\lambda} \pm \frac{t}{T}\right)\right] \tag{11-3}$$

$$y = A \sin (kx \pm \omega t) \tag{11-4}$$

These equations may also be expressed with the cosine function in place of the sine function. The argument of the sine or cosine, an angle that depends on *x* (space) and *t* (time), is called the *phase*, φ. For example, in Eq. (11-2), the phase φ is given by

$$\varphi = k(x \pm \mathrm{v}t)$$

When *x* and *t* change together in such a way that φ remains constant, the displacement $y = A \sin \varphi$ is also constant. The condition of constant phase evidently describes the motion of a fixed point on the waveform, which moves with the speed of the wave.

In any of the wave equations, Eqs. (11-2) to (11-4), notice that under initial conditions *x* = 0 and *t* = 0, the displacement *y* = 0 if the sine function is used and *y* = *A* if the cosine function is used. As pointed out previously, both situations could be handled by either the sine or cosine function, whose only difference is a phase angle of 90° or $\pi/2$ radians. In general, to accommodate any arbitrary initial displacement, some angle φ_0 must be added to the phase. For example, Eq. (11-2) with the sine function becomes

$$y = A \sin [k(x \pm vt) + \varphi_0] \qquad (11\text{-}5)$$

Now suppose our initial boundary conditions are such that $y = y_0$ when $x = 0$ and $t = 0$. Then, substituting these values into Eq. (11-5), we obtain

$$y_0 = A \sin \varphi_0$$

from which the required *initial phase angle* φ_0 can be calculated as

$$\varphi_0 = \sin^{-1}\left(\frac{y_0}{A}\right) \qquad (11\text{-}6)$$

The wave Eqs. (11-2) to (11-4) can be generalized further to yield any initial displacement, therefore, by the addition of an initial phase angle φ_0 to the phase. In many cases, the precise phase of the wave is not of interest. When this is true, φ_0 is set equal to zero for simplicity.

❏ **Example**

A traveling wave propagates along the x-axis according to the expression

$$y(x,t) = 0.35 \sin\left(3\pi x - 10\pi t + \frac{\pi}{4}\right)$$

where x is in meters and t is in seconds. Determine the wavelength, frequency, velocity, and initial phase angle. Also find the displacement at $x = 10$ cm and $t = 0$.

Solution   By direct comparison with Eq. (11-4), $k = 3\pi$, and $\omega = 10\pi$. Thus

$$\lambda = \frac{2\pi}{k} = \frac{2\pi}{3\pi} = \frac{2}{3}\,\text{m} \qquad \text{and} \qquad f = \frac{\omega}{2\pi} = \frac{10\pi}{2\pi} = 5\,\text{Hz}$$

The initial phase ($x = 0$, $t = 0$) is $\pi/4$. The velocity of the wave may be found from $v = \lambda f = (2/3)5 = 3.33$ m/s in the positive *x*-direction (due to the negative sign in the phase). Furthermore, the displacement

$$y(x,t) = y(0.1,0) = 0.35 \sin\left(0.3\pi + \frac{\pi}{4}\right) = 0.35 \sin (0.55\pi) = +0.346\,\text{m} \quad \blacksquare$$

11-2 ELECTROMAGNETIC WAVES

The harmonic wave equations discussed so far can represent any type of wave disturbance that varies in a sinusoidal manner. This includes, for example, waves on a string, water waves, and sound waves. The equations apply to a specific situation as soon as the physical significance of the displacement we have labeled y is identified. The displacement may refer to vertical displacements of a vibrating string or pressure variations due to a sound wave propagating in a gas. For electromagnetic waves that can represent the propagation of light, y stands for

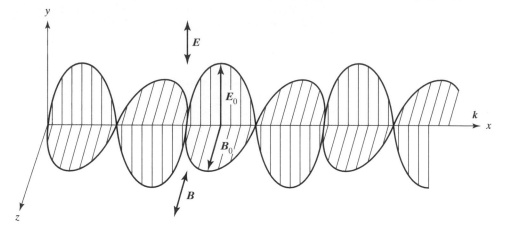

Figure 11-2 Plane eletromagnetic wave. The electric field **E**, magnetic field **B**, and propagation vector **k** are everywhere mutually perpendicular.

either of the varying electric or magnetic fields that together constitute the wave. Figure 11-2 depicts a plane electromagnetic wave of a single wavelength, traveling along the x-axis. From *Maxwell's equations*, which describe such waves, we know that the electric and magnetic fields undergoing harmonic variations are always perpendicular to one another and to the direction of propagation, as suggested by the orthogonal set of axes in Figure 11-2. These variations may be described by the harmonic wave equation in the form

$$\boldsymbol{E} = \boldsymbol{E}_0 \sin{(kx - \omega t)} \tag{11-7}$$

$$\boldsymbol{B} = \boldsymbol{B}_0 \sin{(kx - \omega t)} \tag{11-8}$$

where the electric field vector **E** varies along the y-direction and the magnetic field vector **B** varies along the z-direction. The corresponding amplitudes of the fields are represented in the equations by $\boldsymbol{E}_0$ and $\boldsymbol{B}_0$. Both field components of the wave travel with the same propagation vector **k** and angular frequency ω and thus with the same wavelength λ and the same speed v. Furthermore, electromagnetic theory tells us that the field amplitudes are related by $E_0 = cB_0$, where c is the speed of the wave. At any specified time and place

$$E = cB \tag{11-9}$$

In free space, the velocity c is given by

$$c = \frac{1}{\sqrt{\varepsilon_0 \mu_0}} \tag{11-10}$$

where the constants ε_0 and μ_0 are, respectively, the *permittivity* and *permeability* of vacuum. Measured values of these constants, $\varepsilon_0 = 8.8542 \times 10^{-12}$ (C-s)^2/kg-m^3 and $\mu_0 = 4\pi \times 10^{-7}$ kg-m/(A-s)^2, provide an indirect method of determining the speed of light in free space and yield a value of $c = 2.998 \times 10^8$ m/s.

Such a wave—as shown in Figure 11-2 and described by Eqs. (11-7) and (11-8)—represents the transmission of energy. The energy density u_E (in units of J/m^3) associated with the electric field in free space is

$$u_E = \frac{\varepsilon_0}{2} E^2 \tag{11-11}$$

and the energy density u_B associated with the magnetic field in free space is

$$u_B = \frac{1}{2\mu_0}B^2 \qquad (11\text{-}12)$$

These expressions, easily derived for the static electric field of an ideal capacitor and the static magnetic field of an ideal solenoid, are generally valid. By incorporating Eqs. (11-9) and (11-10) into either Eq. (11-11) or (11-12), one can show that u_E and u_B are equal. For example, starting with Eq. (11-12),

$$u_B = \frac{1}{2\mu_0}\left(\frac{E}{c}\right)^2 = \left(\frac{\varepsilon_0\mu_0}{2\mu_0}\right)E^2 = u_E$$

The energy of an electromagnetic wave is therefore divided equally between its constituent electric and magnetic fields. The total energy density is the sum

$$u = u_E + u_B = 2u_E = 2u_B$$

or

$$u = \varepsilon_0 E^2 = \left(\frac{1}{\mu_0}\right)B^2 \qquad (11\text{-}13)$$

Consider next the power of the electromagnetic wave—that is, the rate at which it transports energy. In a time τ, the energy transported through a cross section of area A (Figure 11-3) is the energy associated with the volume V of a rectangular volume of length $c\tau$. Thus

$$\text{power} = \frac{\text{energy}}{\tau} = \frac{uV}{\tau} = \frac{u(Ac\tau)}{\tau} = ucA \qquad (11\text{-}14)$$

or the power transferred per unit area, S, is

$$S = uc \qquad (11\text{-}15)$$

We now express the energy density u in terms of E and B, as follows, making use of Eqs. (11-10) and (11-13):

$$u = \sqrt{u}\sqrt{u} = \left(\sqrt{\varepsilon_0}E\right)\left(\frac{B}{\sqrt{\mu_0}}\right) = \frac{\varepsilon_0}{\sqrt{\varepsilon_0\mu_0}}EB = \varepsilon_0 cEB$$

Inserting this result into Eq. (11-15),

$$S = \varepsilon_0 c^2 EB \qquad (11\text{-}16)$$

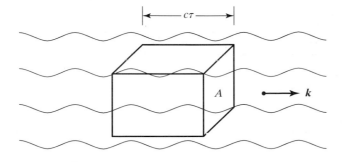

Figure 11-3 Energy flow of an electromagnetic wave. In time τ, the energy enclosed in the rectangular volume flows across the surface A.

The power per unit area, S, when assigned the direction of propagation, is called the *Poynting vector*. Because of the rapid variation of the electric and magnetic fields, whose frequencies are 10^{14} to 10^{15} Hz in the visible spectrum, the magnitude of the Poynting vector in Eq. (11-16) is also a rapidly varying function of time. In most cases a time average of the power delivered per unit area is all that is required. This quantity is called the *irradiance, E_e*. An additional factor of $\frac{1}{2}$ occurs in the expression for irradiance, traceable simply to the fact that the average of the functions $\sin^2 \theta$ or $\cos^2 \theta$ over a period is exactly $\frac{1}{2}$. Thus, the irradiance in free space can be calculated from any of the following alternative forms:

$$E_e = \frac{1}{2} \varepsilon_0 c^2 E_0 B_0$$

$$E_e = \frac{1}{2} \varepsilon_0 c E_0^2 \qquad (11\text{-}17)$$

$$E_e = \frac{1}{2}\left(\frac{c}{\mu_0}\right) B_0^2$$

These equations apply also to a medium of refractive index n if permittivity ε_0 is replaced by $n^2\varepsilon_0$ and speed in vacuum c is replaced by the speed in the medium, c/n.

❏ **Example**

A laser beam of radius 1 mm carries a power of 6 kW. Determine its average irradiance and the amplitude of its E and B fields.[1]

Solution   The average irradiance

$$E_e = \frac{\text{power}}{\text{area}} = \frac{6000}{\pi(10^{-3})^2} = 1.91 \times 10^9 \text{ W/m}^2$$

From Eq. (11-17), using numerical values for ε_0 and c given earlier,

$$E_0 = \left(\frac{2E_e}{\varepsilon_0 c}\right)^{1/2} = \left[\frac{2(1.91 \times 10^9)}{\varepsilon_0 c}\right]^{1/2} = 1.20 \times 10^6 \text{ V/m}$$

and, from Eq. (11-9),

$$B_0 = \frac{E_0}{c} = \frac{1.20 \times 10^6}{c} = 4.00 \times 10^{-3} \text{ T} \qquad \blacksquare$$

11-3 DOPPLER EFFECT

The familiar *Doppler effect* for sound waves has its counterpart in light waves, but with an important difference. Recall that when dealing with sound waves, the apparent frequency of a source increases or decreases depending on the motion

---

[1]Knowledge of the magnitude of the average irradiance in a laser beam, or the strength of its electrical field, are both important when assessing invasive outcomes or cellular damage resulting from laser surgery or intervention in ocular or other human tissue.

of both source and observer along the line joining them. This phenomenon is typified by the familiar example of the change in sound of an automobile horn or passenger train whistle as it passes the listener. The frequency shift due to a moving *source*, as experienced by a fixed observer, is based physically on a change in the transmitted wavelength received by the listener. The frequency shift detected by an *observer*, moving toward or away from a fixed sound source, is based physically on the change in speed of the sound waves relative to the observer. The two effects are physically distinct and described by different equations. They are also essentially different from the case of light waves. The difference between the Doppler effect in sound and light waves is more than the difference in wave speeds. Whereas sound waves propagate through a material medium, light waves propagate in vacuum. As soon as the medium of propagation is removed, there is no longer a physical basis for the distinction between moving observer and moving source. There is only *one* relative motion between them that determines the frequency shift in the Doppler effect for light. The derivation of the Doppler effect for light requires the theory of special relativity, so is not carried out here. The result[2] is expressed by

$$\frac{\lambda'}{\lambda} = \sqrt{\frac{1 - v/c}{1 + v/c}}$$

where λ is the wavelength of the light source at rest relative to the observer, λ' is the Doppler-shifted wavelength, and v is the relative velocity between source and observer. The sign of v is positive when they are approaching one another. When $v \ll c$, this equation is approximated by

$$\frac{\lambda'}{\lambda} = 1 - \frac{v}{c} \tag{11-18}$$

The Doppler effect is especially important when used to determine the speed of astronomical sources emitting electromagnetic radiation. The red shift is the shift in wavelength of such radiation toward longer wavelengths, due to a relative speed of the source away from us. Conversely, if the source were to move toward us, the shift would be a blue shift, a shift toward shorter wavelengths. Doppler broadening of spectral lines represents another important application in which fast-moving atoms of a gas radiate light with both increases and decreases in frequency due to their random motion toward and away from the observer making spectroscopic measurements. The center frequency of the spectral line originates from atoms moving in a direction perpendicular to the line of sight of the observer, neither toward or away.

❏ **Example**

Light from a distant galaxy shows the characteristic lines of the oxygen spectrum, except that the wavelengths are shifted somewhat from their values as measured from radiating oxygen sources in the laboratory. In particular, the line expected at 513 nm shows up at 525 nm, shifted toward the red. What is the speed of the galaxy relative to Earth?

---

[2]See for example Resnick (1972).

Solution   Here, $\lambda = 513$ nm and $\lambda' = 525$ nm. Thus, using Eq. (11-18),

$$\frac{525}{513} = 1 - \frac{v}{c}$$

$$v = -0.0234c = -7013 \text{ km/s}$$

Since the apparent λ is larger (the frequency less), the galaxy is moving away from Earth with a speed of approximately 7013 km/s.   ■

11-4 SUPERPOSITION PRINCIPLE

Quite commonly it is necessary to deal with situations in which two or more waves arrive at the same point in space or exist together along the same direction. Important cases of the combined effects of two or more harmonic waves are treated in the following discussion. The first case deals with the superposition of harmonic waves of differing amplitudes and phases, but with the same frequency. The analysis shows that the resultant is just another harmonic wave having the same frequency. This leads to an important difference between the irradiance attainable from randomly phased and coherent harmonic waves. Another important case of superposition treated here is the standing wave, which results from the addition of a harmonic wave with its reflected counterpart.

To explain the combined effects of waves successfully one must ask specifically: What is the net displacement y at a point in space where two waves with independent displacements y_1 and y_2 exist together? In most cases of interest, the correct answer is given by the *superposition principle*: The resultant displacement is the sum of the separate displacements of the constituent waves:

$$y = y_1 + y_2$$

Using this principle, the resultant wave amplitude and irradiance (W/m²) can be calculated and verified by measurement. In this way, the superposition principle has been determined to be valid for all kinds of waves.

The superposition of two electromagnetic (em) waves may be expressed in terms of their electric or magnetic fields by the equations

$$E = E_1 + E_2 \quad \text{and} \quad B = B_1 + B_2 \tag{11-19}$$

In general, the orientation of the electric or magnetic fields must be taken into account. The superposition of waves at a point where their electric fields are orthogonal, for example, does not yield the same result as the case in which they are parallel. For the present, we treat electric fields as scalar quantities. This treatment is strictly valid for cases in which the individual E vectors are parallel; it is often applied in cases where they are nearly parallel.

Nonlinear effects for which the superposition principle does not predict all the observed results can occur when light of very large amplitude interacts with matter. The possibility of producing high-energy densities, using laser light, has facilitated the study and use of such effects, making *nonlinear optics* an important branch of modern optics.

11-5 SUPERPOSITION OF WAVES OF THE SAME FREQUENCY

The first case of superposition to be considered is the situation in which two harmonic waves of the same frequency combine to form a resultant wave disturbance. We permit the two waves to differ in amplitude and phase. Beginning with a wave in the form

$$E = E_0 \sin (kx + \omega t + \varphi_0)$$

where an initial phase angle φ_0 is added for generality, we set $x = 0$ because we wish to examine waves at a fixed point in space. At any fixed point, kx is constant, so choosing the point at the origin of coordinates is merely the most convenient since the term kx then disappears from the equation. Two such waves, intersecting at the fixed point and differing in amplitude and phase, can then be expressed by

$$E_1 = E_{01} \sin (\omega t + \varphi_{01}) \tag{11-20}$$

$$E_2 = E_{02} \sin (\omega t + \varphi_{02}) \tag{11-21}$$

By the superposition principle, the resultant electric field E_0 at the point is

$$E_0 = E_1 + E_2 = E_{01} \sin (\omega t + \varphi_{01}) + E_{02} \sin (\omega t + \varphi_{02}) \tag{11-22}$$

The addition can be accomplished by considering the two waves as *phasors*, which obey the same addition rules as vectors. The magnitudes of the two phasors are E_{01} and E_{02}, respectively, and their "directions" are determined by the corresponding phase angles, φ_{01} and φ_{02}. The waves are represented graphically in Figure 11-4a. Their sum, the result of superposition, is the phasor of magnitude E_0 in the direction determined by the resultant phase angle φ. The resultant magnitude and phase angle of the sum can be found from Figure 11-4b, which shows the orthogonal components of the individual phasors. Using the Pythagorean theorem,

$$E_0^2 = (E_{01} \cos \varphi_{01} + E_{02} \cos \varphi_{02})^2 + (E_{01} \sin \varphi_{01} + E_{02} \sin \varphi_{02})^2 \tag{11-23}$$

or, equivalently, using the cosine law for a triangle,

$$E_0^2 = E_{01}^2 + E_{02}^2 + 2E_{01}E_{02} \cos (\varphi_{02} - \varphi_{01}) \tag{11-24}$$

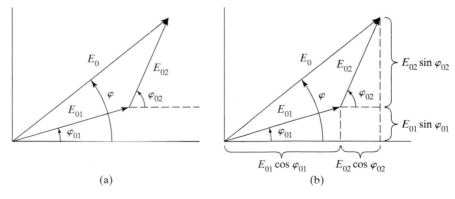

(a) (b)

Figure 11-4 Phasor diagrams for the superposition of two harmonic waves. (a) Adding two harmonic waves. (b) Phasor components.

Either Eq. (11-23) or (11-24) allows the calculation of the resulting magnitude E_0. If more than two phasors are being added, Eq. (11-23) is more useful, since it can be generalized for N phasors by inserting terms for additional phasors within the parentheses, up to $E_{0N} \cos \varphi_{0N}$ in the first parentheses and $E_{0N} \sin \varphi_{0N}$ in the second. The resultant phase angle φ is read from Figure 11-4b:

$$\tan \varphi = \frac{E_{01} \sin \varphi_{01} + E_{02} \sin \varphi_{02}}{E_{01} \cos \varphi_{01} + E_{02} \cos_{02}} \qquad (11\text{-}25)$$

This result can also be generalized to N phasors by adding like terms to numerator and denominator for the additional phasors.

Summarizing, the sum of N harmonic waves of identical frequency is again a harmonic wave of the same frequency, with amplitude given by Eq. (11-23) or Eq. (11-24), and phase given by Eq. (11-25), suitably generalized for N waves.

❏ **Example**

Determine the result of the superposition of the following harmonic waves:

$$E_1 = 7 \sin\left(\omega t + \frac{\pi}{3}\right)$$

$$E_2 = 12 \cos\left(\omega t + \frac{\pi}{4}\right)$$

$$E_3 = 20 \sin\left(\omega t + \frac{\pi}{5}\right)$$

Solution To put all three phasors in the same trigonometric form, first change the cosine wave for E_2 to a sine wave:

$$E_2 = 12 \cos\left(\omega t + \frac{\pi}{4}\right) = 12 \sin\left(\omega t + \frac{\pi}{4} + \frac{\pi}{2}\right) = 12 \sin\left(\omega t + \frac{3\pi}{4}\right)$$

Then, extending Eq. (11-23) to three phasors,

$$E_0^2 = \left[7 \cos\left(\frac{\pi}{3}\right) + 12 \cos\left(\frac{3\pi}{4}\right) + 20 \cos\left(\frac{\pi}{5}\right)\right]^2$$

$$+ \left[7 \sin\left(\frac{\pi}{3}\right) + 12 \sin\left(\frac{3\pi}{4}\right) + 20 \sin\left(\frac{\pi}{5}\right)\right]^2$$

$$E_0^2 = (11.195)^2 + (26.303)^2 \quad \text{and} \quad E_0 = 28.6$$

The units of E_0 are of course the same as those of E_{01} and E_{02}. The phase angle of the resulting harmonic wave is found using an extension of Eq. (11-25). Since the sums forming the numerator and denominator have already been calculated in the first part, we have

$$\tan \varphi = \frac{26.303}{11.195} \quad \text{and} \quad \varphi = 66.9° \text{ or } 1.17 \text{ radians}$$

Thus the resulting harmonic wave can be expressed as

$$E_0 = 28.6 \sin (\omega t + 1.17) \quad \text{or} \quad E_0 = 28.6 \sin (\omega t + 0.372\pi)$$

Notice that the phase, a term in the argument of the sine function, must be dimensionless or in radians. ∎

11-6 STANDING WAVES

Another important case of superposition arises when a given wave exists in both forward and reverse directions along the same medium. This condition occurs most frequently when the forward wave experiences a reflection at some point along its path, as in Figure 11-5a. Let us assume for the moment an ideal situation in which none of the energy is lost on reflection nor absorbed by the transmitting medium. This permits us to write both waves with the same amplitude. Forward and reverse waves are, respectively,

$$E_1 = E_0 \sin (kx - \omega t) \tag{11-26}$$

$$E_2 = E_0 \sin (kx + \omega t) \tag{11-27}$$

The resultant wave in the medium, by the principle of superposition, is

$$E_R(x,t) = E_1 + E_2 = E_0 [\sin (kx - \omega t) + \sin (kx + \omega t)] \tag{11-28}$$

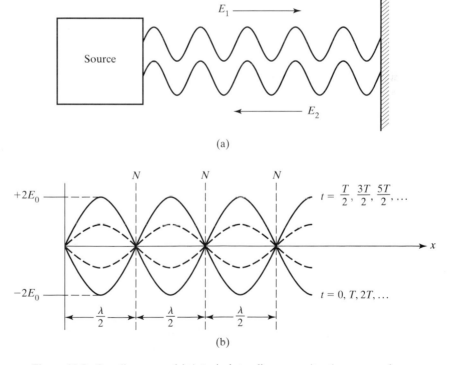

(a)

(b)

Figure 11-5 Standing waves. (a) A typical standing wave situation occurs when a wave and its reflection exist along the same medium. A phase shift (not shown) generally occurs on reflection. (b) Resultant displacement of a standing wave, shown at various instants. The solid lines represent the maximum displacement of the wave. The displacement at the nodes (N) is always zero.

It is expedient in this case to define

$$\alpha \equiv kx + \omega t \quad \text{and} \quad \beta \equiv kx - \omega t$$

and employ the trigonometric identity

$$\sin \alpha + \sin \beta \equiv 2 \sin \tfrac{1}{2}(\alpha + \beta) \cos \tfrac{1}{2}(\alpha - \beta)$$

Applied to Eq. (11-28) this leads immediately to the result

$$E_R(x,t) = (2E_0 \sin kx) \cos \omega t \tag{11-29}$$

which represents a standing wave graphed in Figure 11-5b. Interpretation is facilitated by regarding the quantity in parentheses as a space-dependent amplitude. At any point x along the medium, the oscillations are given by

$$E_R(x,t) = A(x) \cos \omega t$$

where $A(x) = 2E_0 \sin kx$. There exist values of x for which $A(x) = 0$, and thus $E_R(x,t) = 0$ for all t. These values occur whenever

$$\sin kx = 0, \quad \text{or} \quad kx = \frac{2\pi x}{\lambda} = m\pi, \quad m = 0, \pm 1, \pm 2, \ldots$$

or, considering only the positive values for m,

$$x = m\left(\frac{\lambda}{2}\right) = 0, \frac{\lambda}{2}, \lambda, \frac{3\lambda}{2}, \ldots \tag{11-30}$$

Such points are called the *nodes* of the standing wave and are separated by half a wavelength. At various times, the standing wave will appear as sine waves, like those shown in Figure 11-5b. Although their amplitudes vary with time, all pass through zero at the fixed nodal points. The overall displacement $E_R(x,t)$ has its maximum value at all points when $\cos \omega t = \pm 1$, or when

$$\omega t = 2\pi f t = \left(\frac{2\pi}{T}\right) t = m\pi, \quad m = 0, \pm 1, \pm 2, \ldots$$

Thus the outer envelope of the standing wave occurs at times, for positive m values, of

$$t = m\left(\frac{T}{2}\right) = 0, \frac{T}{2}, T, \frac{3T}{2}, \ldots$$

where T is the period. There are also periodic times when the standing wave is everywhere zero, since $\cos \omega t = 0$ for $t = T/4, 3T/4, \ldots$.

Unlike traveling waves, standing waves transmit no energy. All the energy in the wave goes into sustaining the oscillations between nodes, at which points forward and reverse waves cancel. However, since mirrors are not perfect reflectors and the transmitting medium generally absorbs some of the wave energy, wave amplitude decreases with x. Unless the source continues to replace lost energy, the amplitude also decreases with time. In this case, the two waves do not cancel exactly at the nodes nor do they add to the maximum of $2E_0$ at the *antinodes*, points halfway between the nodes. The resultant wave will then be found to include a traveling wave component that carries energy to the mirror and back.

Introduction of a relative phase between the waves of Eqs. (11-26) and (11-27), which would be expected on reflection, leads to a phase angle component in the sine and cosine factors of Eq. (11-29). Nodes will then be displaced from the positions shown in Figure 11-5b, but their separation remains $\lambda/2$. Times at which the form is everywhere zero or everywhere at its maximum displacement also change. The principal features of the standing wave, however, remain unaffected.

11-7 WAVE COHERENCE

The subject of *coherence* in waves can be treated quite rigorously and mathematically, but here we bypass mathematical analysis and describe coherence in a general sense. We strive only for a qualitatively useful understanding of coherence that will come to our aid, especially in studying the laser (Chapter 12). Coherence is the property that most distinguishes the laser from other light sources, such as the sun or a gas discharge lamp.

Coherence, simply stated, is a measure of the degree of *phase correlation* that exists in the radiation field of a light source at different locations and different times. Basically, two simple waves are coherent when their maxima and minima (or peaks and valleys when represented schematically) line up and remain aligned in the region of interest. Coherence is often described in terms of a *temporal coherence* and a *spatial coherence*. Temporal coherence is a measure of the time interval during which the propagating light remains in phase. This turns out to be a measure of how monochromatic the light is. Spatial coherence is a measure of the uniformity of phase across the propagating wave front.

To obtain a qualitative understanding of temporal and spatial coherence, consider the simple analogy of water waves created at the center of a quiet pond by a regular, periodic disturbance. The source of disturbance might be a cork bobbing up and down in regular fashion, creating a regular progression of outwardly moving crests and troughs, as in Figure 11-6. Such a water wavefield can be said to have perfect temporal and spatial coherence. The temporal coherence is perfect because there is but a single wavelength; the crest-to-crest distance remains constant. As long as the cork keeps bobbing regularly, the wavelength remains fixed, and one can predict with great accuracy the location of all crests and troughs on the pond's surface. The spatial coherence of the wavefield is also perfect because the cork is a small source, generating ideal waves, circular crests, and

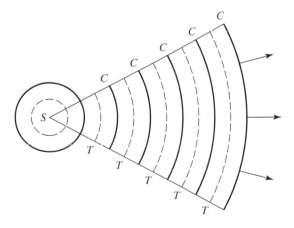

Figure 11-6 Portion of a perfectly coherent water wavefield created by a regularly bobbing cork at S. The wavefield contains perfectly ordered wave fronts, C (crests) and T (troughs), representing water waves of a single wavelength.

troughs of ideal regularity. Along each wavefront then, the spatial variation of the relative phase of the water motion is zero, that is, the surface of the water all along a crest or trough is in step or in phase. Again, one can predict with great accuracy, anywhere on the pond, the vertical displacement of the water surface.

The water wave field described above can be rendered both temporally and spatially *incoherent* by the simple process of replacing the single cork with many corks and causing each cork to bob up and down with a different and randomly varying periodic motion. There would then be little correlation between the behavior of the water surface at one position and another. The wavefronts would be highly irregular geometrical curves, changing shape haphazardly as the collection of corks continued their jumbled, disconnected motions. It does not require much imagination to move conceptually from a collection of corks that give rise to water waves to a collection of excited (radiating) atoms that give rise to light. Disconnected, uncorrelated creation of water waves results in an incoherent water wave field. Disconnected, uncorrelated creation of light waves results, similarly, in an incoherent radiation field.

To emit light of high coherence then, the radiating region of a source must be small in extent (in the limit, of course, of a single atom) and emit light of a narrow spread of wavelengths or *bandwidth* (in the ideal limit, with the spread $\Delta\lambda = 0$). For real light sources, neither of these conditions is attainable. Real light sources, with the exception of the laser, emit light via the uncorrelated action of many atoms, involving many different wavelengths. The result is the generation of incoherent light. To achieve some measure of coherence with a nonlaser source, two modifications to the emitted light are possible. First, a pinhole can be used with the light source to limit the spatial extent of the source. Second, a narrow-band filter can be used to decrease significantly the bandwidth $\Delta\lambda$ (also referred to as *linewidth* in a spectrum) of the light. Each modification improves the coherence of the light given off by the source, but only at the expense of a drastic loss of light energy.

The *coherence time* t_c of a light source is a measure of the average time interval over which one can continue to predict the correct phase of the light beam at a given point in space. The *coherence length* L_c is related to the coherence time by the equation $L_c = ct$, where c is the speed of light. Thus the coherence length is the average length of light beam along which the phase of the wave remains unchanged. The helium-neon laser, for example, has a coherence time of the order of milliseconds, compared with about 10^{-11} second for light from a sodium discharge lamp. The coherence length of the same laser is thousands of kilometers, compared with fractions of a centimeter for the sodium lamp.

PROBLEMS

11-1. A harmonic wave is moving in the negative z-direction with an amplitude (arbitrary units) of 2, a wavelength of 5 m, and a period of 3 s. Its displacement at the origin is zero at time zero. Write a wave equation for this wave (a) that exhibits directly both wavelength and period; (b) that exhibits directly both propagation constant and velocity.

11-2. **(a)** Write the equation of a harmonic wave traveling along the x-direction at $t = 0$, if it is known to have an amplitude of 5 m and a wavelength of 50 m.
   (b) Write an expression for the disturbance at $t = 4$ s if it is moving in the negative x-direction at 2 m/s.

11-3. For a harmonic wave given by $y = 10 \sin (628.3x - 6283t)$, with x and y in centimeters and t in seconds, determine **(a)** wavelength; **(b)** frequency; **(c)** propagation constant; **(d)** angular frequency; **(e)** period; **(f)** velocity; **(g)** amplitude.

11-4. A harmonic wave traveling in the $+x$-direction has, at $t = 0$, a displacement of 13 units at $x = 0$ and a displacement of -7.5 units at $x = 3\lambda/4$. Write the equation for the wave at $t = 0$.

11-5. **(a)** Show that if the maximum positive displacement of a sinusoidal wave occurs at distance x_0 centimeters from the origin when $t = 0$, its initial phase angle φ_0 is given by

$$\varphi_0 = \frac{\pi}{2} - \left(\frac{2\pi}{\lambda}\right)x_0$$

where the wavelength λ is in centimeters.
(b) Determine the initial phase and sketch the wave when $\lambda = 10$ cm and $x_0 = 0, 5/6, 5/2, 5,$ and $-1/2$ cm.
(c) What are the appropriate initial phase angles for (b) when a cosine function is used instead?

11-6. The energy flow to Earth associated with sunlight is about 1.4 kW/m². Find the maximum values of E and B for a wave of this power density.

11-7. A light wave is traveling in glass of index 1.50. If the electric field amplitude of the wave is known to be 100 V/m, find the amplitude of the magnetic field.

11-8. The solar constant is the radiant flux (irradiance) from the sun at Earth's surface and is about 0.135 W/cm². Assume an average wavelength of 700 nm for the Sun's radiation which reaches Earth. Find
(a) the amplitude of the E- and B-fields;
(b) the number of photons that arrive each second on each square meter of a solar panel;
(c) a harmonic wave equation for the E-field of the solar radiation, inserting all constants numerically.

11-9. **(a)** The light from a 220-W lamp spreads uniformly in all directions. Find the irradiance of these optical electromagnetic waves and the amplitude of their E-field at a distance of 10 m from the lamp. Assume that 5% of the lamp energy is converted to light.
(b) Suppose a 2000-W laser beam is concentrated by a lens into a cross-sectional area of about 1×10^{-6} cm². Find the corresponding irradiance and amplitudes of the E- and B-fields there.

11-10. How fast does one have to approach a red traffic light to see a green signal? So that we all get the same answer, let's say that the wavelength of a good red is 640 nm and the wavelength of a good green is 540 nm.

11-11. A quasar near the limits of the observed universe to date shows a wavelength that is 4.80 times the wavelength emitted by the same molecules on Earth. If the Doppler effect is responsible for this shift, what velocity does it determine for the quasar?

11-12. **(a)** Show in a phasor diagram the following two harmonic waves:

$$E_1 = 2 \sin \omega t \quad \text{and} \quad E_2 = 7 \sin\left(\omega t + \frac{\pi}{4}\right)$$

(b) Determine the mathematical expression for the resultant wave.

11-13. Find the resultant of the superposition of two harmonic waves in the form

$$E = E_0 \sin (\omega t + \alpha)$$

with amplitudes of 3 and 4 and phases of 30° and 90°, respectively. Both waves have a period of 1 s.

11-14. Two waves traveling together along the same line are given by

$$y_1 = 5 \sin\left[\omega t + \frac{\pi}{2}\right] \quad \text{and} \quad y_2 = 7 \sin\left[\omega t + \frac{\pi}{3}\right]$$

Write the resultant wave equation.

11-15. Write the equation of the superposition of the following harmonic waves:

$$E_1 = \sin(\omega t - 10°) \quad E_2 = 3\cos(\omega t + 100°) \quad E_3 = 2\sin(\omega t - 30°)$$

where it is understood the angles in degrees must be expressed in radians to make dimensional sense. The period of each wave is 2 s.

11-16. Two plane waves of the same frequency and with vibrations in the z-direction are given by

$$f(x,t) = 4\sin\left[20t + \left(\frac{\pi}{3}\right)x + \pi\right] \quad \text{and} \quad f(y,t) = 2\sin\left[20t + \left(\frac{\pi}{4}\right)y + \pi\right]$$

Write the resultant wave equation expressing their superposition at the point $x = 5$ and $y = 2$.

11-17. Standing waves are produced by the superposition of the wave

$$y = 7\sin 2\pi\left[\frac{t}{T} - \frac{2x}{\pi}\right] \quad (x, y \text{ in cm}; t \text{ in s})$$

and its reflection in a medium whose absorption is negligible. For the resultant wave, find the amplitude, wavelength, length of one loop, velocity, and period.

11-18. A medium is disturbed by an oscillation described by

$$y = 3\sin\left(\frac{\pi x}{10}\right)\cos(50\pi t) \quad (x,y \text{ in cm}; t \text{ in s})$$

(a) Determine the amplitude, frequency, wavelength, speed, and direction of the component waves whose superposition produces this result.
(b) What is the internodal distance?
(c) What is the displacement of a particle in the medium at $x = 5$ cm and $t = 0.22$ s?

11-19. The coherence times for several light sources are given as:
(a) student laboratory He-Ne laser: $\tau = 0.03$ ns
(b) cadmium red light from a discharge lamp: $\tau = 1.0$ ns
(c) research-grade CO_2 laser: $\tau = 36$ μs
Find the corresponding coherence lengths for these sources.

REFERENCES

GARBUNY, MAX. 1965. *Optical Physics*. New York: Academic Press.

GHATAK, AJOY K. 1972. *An Introduction to Modern Optics*. New York: McGraw-Hill Book Company. Ch 1.

HEAVENS, O. S., and R. W. DITCHBURN. 1991. *Insight into Optics*. New York: John Wiley & Sons.

HECHT, EUGENE, and ALFRED ZAJAC. 1974. *Optics*. Reading, Mass: Addison-Wesley Publishing Company. Ch. 2.

PEŘINA, JAN. 1985. *Coherence of Light*. New York: Kluwer Academic Publisher.

RESNICK, ROBERT. 1972. *Basic Concepts in Relativity and Early Quantum Mechanics*. New York: John Wiley and Sons. Ch. 2.

WALDMAN, GARY. 1983. *Introduction to Light*. Englewood Cliffs, N.J.: Prentice Hall.

12

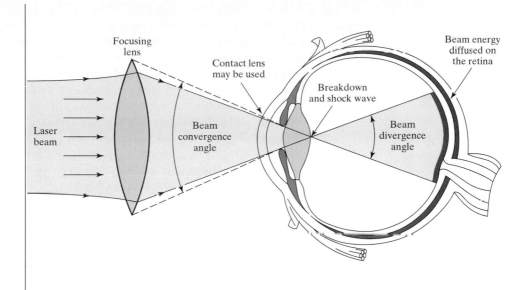

Focusing lens

Contact lens may be used

Beam convergence angle

Laser beam

Breakdown and shock wave

Beam divergence angle

Beam energy diffused on the retina

Lasers and the Eye

INTRODUCTION

The laser is perhaps the most important optical device to be developed in the past 50 years. Since its arrival in the 1960s, rather quiet and unheralded outside the scientific community, it has provided the stimulus to make optics one of the most rapidly growing fields in science and technology today. In fact, by the mid-1990s, lasers, coupling the fields of optics and electronics, had given rise to a dramatic new field in technology called *photonics*. In 1994 it was said that "photonics is the foremost emerging technology that will affect manufacturing industries over the next few decades, surpassing the combined impact of the transistor and the microprocessor."[1]

The laser is essentially an optical amplifier. The word *laser* is an acronym that stands for **L**ight **A**mplification by the **S**timulated **E**mission of **R**adiation. The key words here are *amplification* and *stimulated emission*. The theoretical background of laser action as the basis for an optical amplifier was made possible by Albert Einstein as early as 1916, when he first predicted the existence of a new radiative process called stimulated emission. His theoretical work, however, remained largely unexploited until 1954, when C. H. Townes and coworkers developed a **M**icrowave **A**mplifier based on **S**timulated **E**mission of **R**adiation. It was

[1]See Angelo DePalma (1994).

called a *maser*. Shortly thereafter, in 1958, A. Schawlow and C. H. Townes adapted the principle of masers to light in the visible region, and in 1960, T. H. Maiman built the first laser device. Maiman's laser incorporated a ruby crystal for the laser amplifying medium and a Fabry-Perot optical cavity as the resonator. Within months of the arrival of Maiman's ruby laser, which emitted deep red light at a wavelength of 694.3 mm, A. Javan and associates developed the first gas laser, the helium-neon laser, which emitted light in both the infrared and visible spectrums, at 1.15 μm and 632.8 nm, respectively.

Following the birth of the ruby and helium-neon (He-Ne) lasers, other laser devices followed in rapid succession, each with a different laser medium and a different wavelength emission. For the early part of the 1960s, the laser was viewed by the world of industry and technology as a scientific curiosity. It was referred to in jest as "a solution in search of a problem." However, by the 1970s, all that had changed. The laser came into its own as a unique source of intense, coherent light. Today, fueled by many needs, new laser applications are continually being identified. Together with the fiber-optic cable and semiconductor optoelectronic devices, the laser and photonics have revolutionized optics and the optics industry.

In the first four sections of this chapter we concentrate on what might be called "simple laser physics." In turn, we look at the basic physics of laser action, the essential elements of a laser, the simplified operation of a laser, and the characteristics of laser light that make it so unique among light sources. In the final sections we review the different types of lasers available on the market today and list some of their many applications. We then close with a closer look at how lasers are used as surgical and diagnostic tools in ophthalmology.

12-1 BASIC PHYSICS OF LASER ACTION

In 1916, while studying the fundamental processes involved in the interaction of electromagnetic radiation with matter, Einstein showed that the existence of equilibrium between matter and radiation required a previously undiscovered radiation process called *stimulated emission*. According to Einstein, the interaction of radiation with matter could be explained in terms of three basic processes: stimulated absorption, spontaneous emission, and stimulated emission. The three processes are illustrated in Figure 12-1.

Stimulated absorption, or simply, absorption (Figure 12-1a) occurs whenever radiation containing photons of energy $hf = E_1 - E_0$ is incident on matter (atoms) having ground-state energy E_0 and a particular excited state energy E_1. The resonant photon energy hf raises the atom from energy state E_0 to E_1. In the process, the photon is absorbed and disappears.

Spontaneous emission, Figure 12-1b, takes place when atoms are in an excited state. No external radiation is needed to initiate the emission. In this process, when an atom in an excited energy state E_1 spontaneously gives up its energy and falls back to the ground state E_0, a photon of radiant energy hf equal to the energy difference $E_1 - E_0$ is released. The photon is emitted in a random direction. Even if an external radiation beam is present, the spontaneous photon is given off in a direction that is completely uncorrelated with the direction of the external beam.

Quite by contrast, *stimulated emission*, Figure 12-1c, requires the presence of external radiation. When an incident photon of resonant energy $hf = E_1 - E_0$

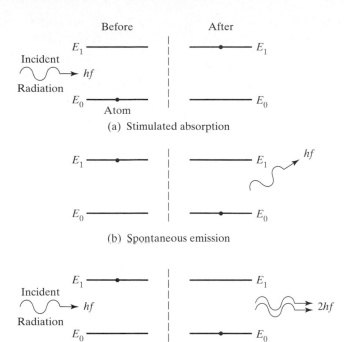

(a) Stimulated absorption

(b) Spontaneous emission

(c) Stimulated emission

Figure 12-1 Three basic processes that affect the passage of radiation through matter. Note that $hf = E_1 - E_0$.

passes by an atom in excited state E_1, it "stimulates" the atom to drop to the lower energy state E_0. In this process, not recognized before Einstein's insight, the atom releases a photon of the *same* energy, direction, phase, and polarization as that of the photon passing by. The net effect, then, is two identical photons in the place of one, or an *increase* in the intensity of the incident "beam." It is precisely this process of stimulated emission that makes possible the *amplification* of light in lasers.

Einstein *A* and *B* Coefficients. Einstein's proof for the existence of stimulated emission grew out of his desire to understand the basic mechanisms involved in the interaction between electromagnetic radiation and matter. A review of his study is both interesting and informative. As a model for this study, we shall assume that matter (a collection of atoms) is in thermodynamic equilibrium (temperature steady) with a blackbody radiation field of photon density $\rho(f)$, where we express ρ as a function of f, the frequency of the radiation. The atoms and the resonant radiation are contained in an enclosure at some steady temperature T and interact with one another. Figure 12-2 shows a simplified picture of two-level atoms and radiation (photons) bound inside an arbitrary unit volume. Since thermodynamic equilibrium is assumed, it follows that the number of atoms N_2 at energy level E_2, the number of atoms N_1 at energy level E_1, and the number of photons in the radiation field all remain constant. Emission and absorption processes that add and remove photons from the radiation field occur at a constant rate, leaving the total photon number (ρ times volume) unchanged. At the same time, for every N_2 atom moving to E_1 during an emission process, there will be an N_1 atom moving to E_2 during an absorption process, so that N_1 and N_2 do not change. This condition of balance is depicted in Figure 12-3, in terms of atoms moving from level E_1 to E_2 and from E_2 to E_1. The two emission rates are given by $A_{21}N_2$ and $B_{21}N_2\rho(f)$. The single absorption rate is given

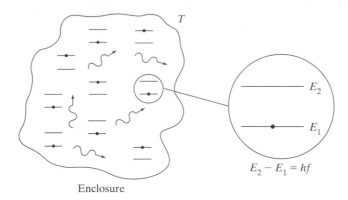

$$E_2 - E_1 = hf$$

Enclosure

Figure 12-2 A blackbody at temperature T emits radiation that interacts with atoms in the blackbody.

by $B_{12}N_1\rho(f)$. Each of the rates involves one of the important A and B coefficients defined by Einstein, namely A_{21}, B_{21}, and B_{12}.

Without going into the details of the elegant theory that enabled Einstein to give precise meaning to the three coefficients, we can describe more completely the emission and absorption rates that incorporate them and how a manipulation of atoms in levels E_2 and E_1 can result in the amplification of light.

The spontaneous emission term $A_{21}N_2$, in atoms per second, describes how fast atoms in level E_2 are dropping down to level E_1, thereby causing an emptying of level E_2. At the same time, for every drop, a photon of energy $hf = E_2 - E_1$ is added to the radiation field $\rho(f)$. As Einstein showed, the coefficient A_{21}, characteristic of the atom in question, is related to the atomic decay rate. Note that the spontaneous emission rate depends only on the type of atom involved and the number of atoms in level E_2. It does not depend on the existence or strength of the photon radiation field $\rho(f)$ surrounding the atom.

On the other hand, the stimulated emission rate $B_{21}N_2\rho(f)$, in atoms per second, depends directly on the strength of the radiation field $\rho(f)$ in the neighborhood of the atoms, as well as on the number of atoms N_2 in level E_2 and the coefficient B_{21}. The term $B_{21}N_2\rho(f)$ then causes an emptying of level E_2, and a filling of level E_1, with the simultaneous addition of a photon to the radiation field $\rho(f)$. The added photon is identical to the one that "stimulated" or caused the emission in the first place.

The stimulated absorption rate, $B_{12}N_1\rho(f)$, in atoms per second, is a process that depends also on the existence and strength of the radiation field $\rho(f)$. This term depletes the number of atoms N_1 in level E_1 and adds them to the number N_2 in level E_2, while at the same time removing a photon from the radiation field $\rho(f)$. Compared with the stimulated emission rate $B_{21}N_2\rho(f)$, it appears as an "inverse process."

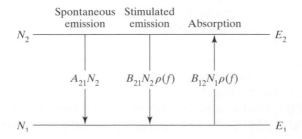

Figure 12-3 Radiative processes that affect the number of atoms of energy E_1 and E_2. The two emission processes remove atoms from level E_2 and add them to level E_1. The absorption process moves atoms from E_1 to E_2.

Based on Einstein's condition for equilibrium, in terms of atoms and photons involved in the three processes shown in Figure 12-3 and the analysis that he carried out, we can cite several important conclusions:

1. The Einstein coefficients A_{21}, B_{21}, and B_{12} are all interrelated. If one is known, by measurement or calculation, all are known. According to Einstein's work, A_{21} and B_{21} are related thus:

$$\frac{A_{21}}{B_{21}} = \frac{8\pi h f^3}{c^3} \tag{12-1}$$

where h is Planck's constant, f is the frequency of the radiation in the photon field $\rho(f)$, and c is the speed of light in a vacuum.

2. The stimulated emission coefficient B_{21} and the (stimulated) absorption coefficient B_{12} are equal, at least for the case of nondegenerate energy states,[2] so that

$$B_{12} = B_{21} \tag{12-2}$$

Note carefully that the terms $N_2 B_{21} \rho(f)$ and $N_1 B_{12} \rho(f)$ differ—even with $B_{12} = B_{21}$—since they depend on the population densities N_2 and N_1. If N_2 is greater than N_1 and a radiation field $\rho(f)$ interacts with the atoms, stimulated emission—$N_2 B_{21} \rho(f)$—exceeds absorption and photons are added to the field. If, however, N_1 is greater than N_2, absorption—$N_1 B_{12} \rho(f)$—exceeds stimulated emission and photons are removed from the field. The first case ($N_2 > N_1$) leads to an increase in $\rho(f)$, an amplification. The second case ($N_1 > N_2$) leads to a decrease in $\rho(f)$, an attenuation. For the laser to operate, it is necessary that N_2 be greater than N_1. This is the condition referred to as *population inversion*. Without a population inversion—a requirement that runs contrary to the energy level population densities generally found in atoms—laser action cannot occur.

3. The ratio B_{21}/A_{21} is found to be inversely proportional to the cube of the frequency f, as indicated in Eq. (12-1). Thus, the higher the frequency (the shorter the wavelength), the smaller B_{21} becomes in comparison with A_{21}. Since B_{21} is related to stimulated emission (which leads to photon amplification) and A_{21} is related to spontaneous emission (which contributes little, if any, to photon amplification), it would seem that lasers of short wavelength radiation (ultraviolet or X-ray, for example) would be more difficult to build and operate. Such has indeed been the case, even though lasers of shorter wavelength are being developed rather extensively.

4. Although the important relations between A_{21}, B_{21}, and B_{12} were derived by Einstein based on the condition of thermodynmic equilibrium, they are valid and hold under any condition. The laser, while operating, is hardly an enclosure in thermodynamic equilibrium. Yet the A and B coefficient relationships, because they are characteristic of the atom, are equally valid whether the atom is in the intense radiation field of a laser cavity or in a hot furnace that can be considered as a blackbody in thermodynamic equilibrium.

Two important ideas for the successful operation of a laser emerge from a review of Einstein's study of the interaction of electromagnetic radiation with

---

[2]A nondegenerate energy state is one for which there is one and only one corresponding energy level. Degenerate states exist when two or more states share the same energy level.

matter. The first is that there is a process, stimulated emission, that leads to light amplification. The second is that a population inversion of atoms in energy levels must be achieved if the stimulated emission process producing coherent photons is to out-rival the absorption process removing photons. We use these ideas in describing how the laser operates, but first let us examine the laser as a device and consider the individual parts essential to its operation.

12-2 ESSENTIAL ELEMENTS OF A LASER

The laser device is an optical oscillator that emits an intense, highly collimated beam of coherent radiation. The device consists basically of three elements: an external energy source or *pump*, an *amplifying medium*, and an optical cavity or *resonator*. These three elements are shown schematically in Figure 12-4: as a unit in Figure 12-4a and separately in Figure 12-4b, c, and d.

The Pump.   The pump is an external energy source that produces a population inversion in the laser medium. As explained in Section 12-1, amplification of a light wave or photon radiation field occurs only in a laser medium that exhibits a population inversion between two energy levels. Otherwise the light wave passing through the laser medium is simply attenuated.

Pumps can be optical, electrical, chemical, or thermal in nature, so long as they provide energy that can be coupled into the laser medium to excite the atoms from a lower energy state to a preferred, higher energy state and thereby create the required population inversion. For gas lasers, such as the He-Ne (helium-neon) laser, the most commonly used pump is an electrical discharge. The important parameters governing this type of pumping are the electronic cross sections for moving atoms to higher energy states and the lifetimes of the various energy levels in the He-Ne laser medium. In some gas lasers, the free electrons generated in the discharge process collide with and excite the laser atoms, ions, or molecules directly. In others, excitation occurs by means of inelastic

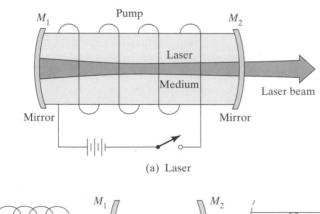

(a) Laser

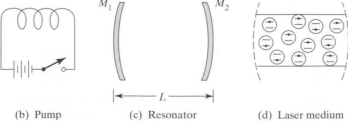

(b) Pump

(c) Resonator

(d) Laser medium

Figure 12-4   Basic elements of a laser. (a) Integral laser device with output laser beam. (b) External energy source, or *pump*. The pump can be an optical, electrical, chemical, or thermal energy source. The battery and helix pictured are only symbolic. (c) Empty optical cavity, or *resonator*, bounded by two mirrors. (d) Active cavity, or *laser medium*.

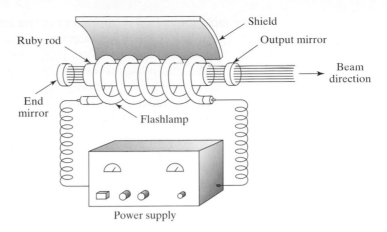

Figure 12-5 Components of a ruby laser system. The shield helps to reflect light from the flashlamp back into the ruby rod.

atom-atom (or molecule-molecule) collisions. In this latter approach, a mixture of two gases is used such that the two different species of atoms, say A and B, have excited states $A*$ and $B*$ that coincide. Energy may be transferred from one excited species to the other species in a process whose net effect can be symbolized by the relation $A* + B \rightarrow A + B*$, where atom A originally received its excitation energy from a free electron or by some other excitation process. A notable example is the He-Ne laser, where the laser-active neon atoms are excited by resonant transfer of energy from helium atoms in a metastable state. The helium atoms receive their energy from free electrons via collisions.

Although there are numerous other pumps or excitation processes, we cite only one more process of historical significance. The first laser, developed by T. Maiman at the Hughes Research Laboratories in 1960, was a pulsed ruby laser, which operated at the visible red wavelength of 694.3 nm. Figure 12-5 shows a drawing of the ruby laser device. To excite the triply ionized Cr^{+3} impurity atoms in the ruby rod, Maiman used a helical flashlamp filled with xenon gas. This particular method of exciting the laser medium is known as *optical pumping*. It is a very practical method for pumping liquid or solid media.

The Laser Medium. The amplifying medium or laser medium is the heart of a laser device. Many lasers are named after the type of laser medium used, for example, helium-neon (He-Ne), carbon dioxide (CO_2), and neodymium:yttrium aluminum garnet (Nd:YAG). The laser medium, which may be a gas, liquid, or solid, determines the wavelength of the laser radiation. Because of the large selection of laser media, the range of available laser wavelengths extends from the ultraviolet well into the infrared region, sometimes to wavelengths that are a sizable fraction of a millimeter. By the early 1990s, laser action had been observed in over half of the known elements, with more than 1000 laser transitions in gases alone. Two widely used transitions in gases are the 632.8-nm visible radiation from neon and the 10.6-μm infrared radiation from the CO_2 molecule. Other commonly used laser media and their corresponding radiations are listed in Table 12-4 near the end of this chapter.

In some lasers the amplifying medium consists of two parts, the laser host medium and the laser atoms. For example, the host of the Nd:YAG laser is a crystal of yttrium aluminum garnet (commonly called YAG), whereas the laser atoms are the trivalent neodymium ions. In gas lasers consisting of mixtures of gases, the distinction between host and laser atoms is generally not made.

The most important requirement of the amplifying medium is its ability to support a population inversion between two selected energy levels among the many existing excited energy states. This is accomplished by exciting (or pumping) more atoms into a higher energy level than exist in a lower level. As mentioned earlier, in the absence of pumping, any given energy level E for a laser material will have a smaller atom population than any energy level below it and generally a much, much smaller population than found in the ground state E_0. Pumping, sometimes vigorous pumping, is required to produce the "unnatural" condition of having more atoms in a higher energy state than in some state below it, thereby producing a population inversion. As it turns out though, due to the widely different spontaneous decay lifetimes of available atomic energy levels, only certain pairs of selected energy levels can be "inverted," even with vigorous pumping.

The Resonator. Given a suitable pump and a laser medium that can be inverted, the third basic element is a resonator, an optical "feedback" device that directs photons back and forth through the laser amplifying medium, thereby creating the radiation field $\rho(f)$. The resonator, or optical cavity, in its most basic form, consists of a pair of carefully aligned plane or curved mirrors centered along the optical axis of the laser system, as shown in Figure 12-4. One of the mirrors is chosen with a reflectivity as close to 100% as possible. The other is selected with a reflectivity somewhat less than 100% to allow part of the internally reflecting beam to escape and become the useful output beam of the laser device.

The geometry of the mirrors and their separation determine the structure of the electromagnetic field within the laser cavity. The exact distribution of the electric field pattern across the wavefront of the emerging laser beam, and thus the transverse irradiance of the beam, depends on the construction of the resonator cavity and mirror surfaces. Many transverse irradiance patterns, called *TEM modes*, are usually present in the output laser beam. By suppressing the gain of the higher-order modes—those with intense electric fields near the edges of the beam—the laser can be made to operate in a single fundamental mode, the TEM_{00} mode. The transverse variation in the irradiance of this mode is *Gaussian* in shape, with peak irradiance at the center and exponentially decreasing irradiance toward the edges.

12-3 SIMPLIFIED DESCRIPTION OF LASER OPERATION

We have described briefly the three basic elements that compose the laser device. How do these elements—pump, medium, and resonator—work together? We know basically that photons of a certain resonant energy must be created in the laser cavity, must interact with atoms, and must be amplified via stimulated emission, all the while bouncing back and forth between the mirrors of the resonator. We can gain a reasonably accurate, though qualitative, understanding of laser operation by studying Figures 12-6 and 12-7. Figure 12-6a shows, in four steps, what happens to a typical atom in the laser medium during the creation of a laser photon. Figure 12-6b shows the actual energy level diagram for a helium-neon laser, with the four steps described in Figure 12-6a clearly identified. Figure 12-7 then shows the same four-step process while focusing on the behavior of the atoms in the laser medium and the photon population in the laser cavity. Let us now examine these figures in turn.

In step 1 of Figure 12-6a, energy from an appropriate pump is coupled into the laser medium. The energy is sufficiently high to excite a large number of

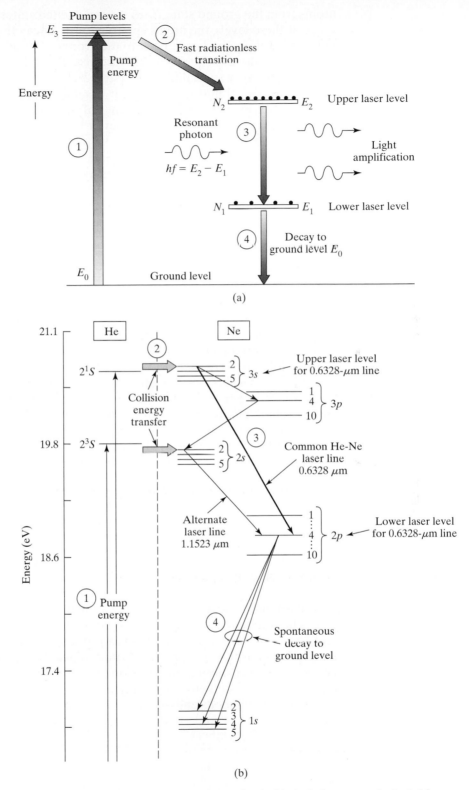

Figure 12-6 Four-step energy cycle associated with the lasing process for both (a) a general, four-level laser, and (b) a particular laser, the helium-neon laser. In (a) is shown a four-step energy cycle for a laser atom involved in the creation of laser photons. In (b) is shown the energy level diagram for the He-Ne laser and the production of 0.6328 μm laser line in terms of the four steps (circled numbers) outlined in (a).

atoms from the ground state E_0 to several excited states, collectively labeled E_3. Once at these levels, the atoms spontaneously decay through various chains involving lower energy levels, back to the ground state E_0. Many, however, preferentially start the trip back by a very fast (usually radiationless) decay from pump levels E_3 to a very special level, E_2 (step 2). Level E_2 is labeled the "upper laser level." It is special in the sense that it has a long decay lifetime. Whereas many excited levels in an atom decay in times of the order of 10^{-8} s, level E_2 is *metastable*, with a typical lifetime of the order of 10^{-3} s, 100,000 times longer than other levels. Thus as atoms funnel rapidly from pump levels E_3 to E_2, they begin to pile up at the metastable level, which acts like a bottleneck. In the process, N_2 grows to a large value. When level E_2 does decay, say by spontaneous emission, it does so to level E_1, labeled the "lower laser level." Level E_1 is an ordinary level that decays to ground state quite rapidly, so that the population N_1 cannot build to a large value and offset the buildup in level E_2. The net effect is the population inversion ($N_2 > N_1$) required for light amplification via stimulated emission.

Once the population inversion has been established and a photon of resonant energy $hf = E_2 - E_1$ passes by any one of the N_2 atoms in the upper laser level (step 3), stimulated emission can occur. When it does, laser amplification begins. Note carefully that a photon of resonant energy $E_2 - E_1$ can also be absorbed, disappearing in the process. Since N_2 is greater than N_1, however, and $B_{21} = B_{12}$ as shown earlier, the rate for stimulated emission, $B_{21}N_2\rho(f)$, exceeds that for absorption, $B_{12}N_1\rho(f)$. Then light amplification occurs. In that event there is a steady increase in the incident resonant photon population and lasing continues. This is shown schematically in step 3, where the incident resonant photon approaching from the left leaves the vicinity of an N_2 atom in duplicate. In step 4, one of the inverted N_2 atoms, which dropped to level E_1 during the stimulated emission process, now decays rapidly to ground state E_0. If the pump is still operating, this atom is ready to repeat the cycle, thereby ensuring a steady population inversion and a constant laser beam output.

In Figure 12-6b for the He-Ne laser, the pump energy (step 1) is supplied by an electrical discharge in the low-pressure gas mixture, thereby elevating ground state helium atoms to higher energy states, one of which is represented by an energy level labeled 2^1S, in accordance with established spectroscopic notation. Then by resonant collisional transfer—made possible because the 2^1S level of helium is nearly equal to the $3s_2$ level of neon—step 2 is achieved as excited helium atoms pass their energy over to ground state neon atoms, raising them to the neon $3s_2$ level. This process produces the population inversion required for effective amplification via stimulated emission of radiation.

The stimulated emission process (step 3) occurs between the neon levels $3s_2$ and $2p_4$, the transition with the highest probability[3] from $3s_2$ to any of the ten $2p$ states. This transition gives rise to photons of wavelength 0.6328 μm, photons that are amplified via stimulated emission and form the common red beam characteristic of helium-neon lasers. Finally, in step 4, the neon atom in energy state $2p_4$ decays by spontaneous emission to the $1s$ ground level. Once back in the ground state, it is again available to undergo collision with an excited helium atom and repeat the cycle. While Figure 12-6b specifically relates the four steps to the emission of the He-Ne 632.8-nm laser line, other transitions from the $3s$ to the $2s$ and $2p$ levels have also been made to lase. One such transition, leading to the 1.1523 μm line, is also indicated in the figure.

---

[3]A readable, comprehensive discussion of the helium-neon laser, with energy level diagrams and transition probabilities, is given in Thompson (1980).

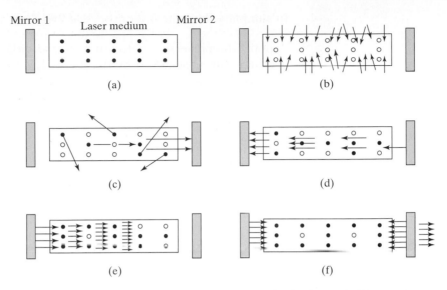

Figure 12-7 Step-by-step development of laser oscillation in a typical laser cavity.
(a) Quiescent laser. (b) Laser pumping. (c) Spontaneous and stimulated emission.
(d) Light amplification begins. (e) Light amplification continues. (f) Established laser
operation.

Now let us look at Figure 12-7. It shows essentially the same action but does so in terms of the behavior of the atoms in the laser medium and the photon population in the cavity. In (a) the laser medium is shown situated between the mirrors of the optical resonator. Mirror 1 is essentially 100% reflecting, while mirror 2 is partially reflecting and partially transmitting. Most of the atoms in the laser medium are in the ground state, indicated by the black dots. In (b), external energy (light from a flashlamp or from an electrical discharge) is pumped into the medium, raising most of the atoms to the excited levels (E_3 in Figure 12-6). Excited states are indicated by the open circles. During the pumping process, the population inversion is established. The light amplification process is initiated in (c), when excited atoms (some of the atoms at level E_2 in Figure 12-6) spontaneously decay to level E_1. Since this is spontaneous emission, the photons given off in the process radiate out randomly in all directions. Many, therefore, leave through the sides of the laser cavity and are lost. Nevertheless, there will generally be several photons—let us call them "seed" photons—directed along the optical axis of the laser. These are the horizontal arrows shown in (c) of Figure 12-7, directed perpendicularly to the mirrors. With the seed photons of correct (resonant) energy accurately directed between the mirrors and many N_2 atoms still in the inverted state E_2, the stage for stimulated emission is set. As the seed photons pass by the inverted N_2 atoms, stimulated emission adds identical photons in the same direction, providing an ever-increasing population of coherent photons that bounce back and forth between the mirrors. This buildup process, shown in (d) and (e), continues as long as there are inverted atoms and resonant energy photons in the cavity. Since output mirror 2 is partially transparent, a fraction of the photons incident on the mirror pass out through the mirror. These photons constitute the external laser beam, as shown in (f). Those that do not leave through the output mirror are reflected, recycling back and forth through the cavity-gain medium.

In summary then, the laser process depends on the following:

1. A population inversion between two appropriate energy levels in the laser medium. This is achieved by the pumping process and the existence of a metastable state.
2. Seed photons of proper energy and direction, coming from the ever-present spontaneous emission process between the two laser energy levels. These initiate the stimulated emission process.
3. An optical cavity that confines and directs the growing number of resonant energy photons back and forth through the laser medium, continually exploiting the population inversion to create more and more stimulated emissions, thereby creating more and more photons directed back and forth between the mirrors.
4. Coupling a certain fraction of the laser light wave (the cavity photon population) out of the cavity through the output coupler mirror to form the external laser beam.

12-4 CHARACTERISTICS OF LASER LIGHT

The laser light beam that flashed onto the laboratory scene in 1960 was truly a remarkable package of light energy. Its unique characteristics—never before seen in electromagnetic emissions from light sources—include its purity of color or wavelength, its high degree of "in-stepness" or phase correlation in space and time between its wavefronts, its pencil-like directionality, its intense brightness, and its ability to be almost point-focused. To gain a better appreciation of these truly singular characteristics, we examine them now one at a time, in more detail.

Monochromaticity. The light emitted by a laser is almost pure in color, almost of a single wavelength or frequency. Although we know that no light can be truly monochromatic (literally of one color), with unlimited sharpness in wavelength definition, laser light comes far closer than any other available source toward meeting this ideal limit.

The monochromaticity of light is determined by the fundamental emission process wherein atoms in excited states decay to lower energy states and emit light. In blackbody radiation, the emission process involves billions of atoms and many sets of energy-level pairs within each atom. The resultant radiation is hardly monochromatic, as we know. If we could select an identical set of atoms from this blackbody and isolate the emission determined by a single pair of energy levels, the resultant radiation, although weaker, would be decidedly more monochromatic. When such radiation is produced by nonthermal excitation, the radiation is often called *fluorescence*. Figure 12-8 depicts such an emission process. The fluorescence comes from the radiative decay of atoms between two energy levels E_2 and E_1, each having some width, as shown in the figure. The nature of the fluorescence, analyzed by a spectrophotometer, is shown in the *line shape* plot, a graph of spectral radiant exitance versus wavelength. Note carefully that the emitted light is not perfectly monochromatic but has a wavelength spread $\Delta\lambda$ about a center wavelength λ_0, where $\lambda_0 = c/f_0$ and $f_0 = (E_2 - E_1)/h$, for the transition from midlevel E_2 to midlevel E_1. While most of the light may be emitted at a wavelength λ_0, it is an experimental fact that some light is also emitted at wavelengths above and below λ_0, with different relative exitance, as shown by the line shape plot. Thus the emission is not monochromatic; it has a wave-

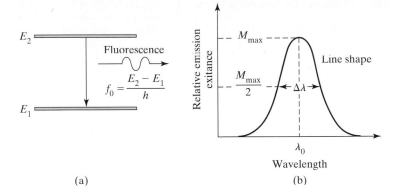

$$f_0 = \frac{E_2 - E_1}{h}$$

(a)

(b)

Figure 12-8 Fluorescence and its spectral content for a radioactive decay process between two energy levels in an atom. (a) Spontaneous decay process between well-defined energy levels. (b) Spectral content of fluorescence in (a), showing line shape and line width.

length spread given by $\lambda_0 \pm \Delta\lambda/2$, where $\Delta\lambda$ is often referred to as the *line width*. When the line width is measured at the half-maximum level of the line shape plot, it is called the FWHM line width, that is, *Full Width at Half Maximum*.

In the laser process, the line width $\Delta\lambda$ shown in Figure 12-8 is narrowed considerably, leading to light of a much higher degree of monochromaticity. Basically this occurs because the process of stimulated emission selectively amplifies photons of precisely the same energy, effectively narrowing the band of wavelengths emitted. This narrowing of the line width is shown qualitatively in Figure 12-9. To gain a quantitative appreciation for the monochromaticity of laser light, consider the data in Table 12-1, in which the line width of a high-quality He-Ne laser is compared to the line width of the spectral output of a typical sodium discharge lamp and to the very narrow cadmium red line found in the spectral emission of a low-pressure lamp. The conversion from $\Delta\lambda$ to Δf is made by using the approximate relationship, $\Delta f = c\Delta\lambda/\lambda_0^2$.

The data of Table 12-1 show that the He-Ne laser is 10 million times more monochromatic than the ordinary discharge lamp and about 100,000 times more so than the cadmium red line. Without significant filtering and consequent loss of brightness, no ordinary light source can approach the degree of monochromaticity present in the output beam of typical lasers

Coherence.   As we observed in the previous chapter, the optical property of light that most distinguishes the laser from other light sources is *coherence*. The laser is regarded, quite correctly, as the first truly coherent source. Other light sources, whether they be the sun, incandescent bulbs, or gas discharge lamps, are at best only partially coherent. To achieve some degree of acceptable coherence with such nonlaser sources, they must first be "stopped down" to limit

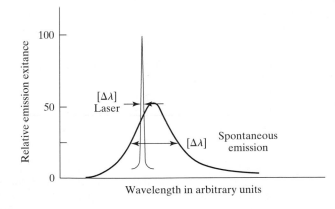

Figure 12-9   Qualitative comparison of line widths for laser emission and spontaneous emission involving the same pair of energy levels in an atom. The broad peak is the line shape of spontaneously emitted light between levels E_2 and E_1 *before* lasing begins. The sharp peak is the line shape of laser light between levels E_2 and E_1 *after* lasing begins.

TABLE 12-1   COMPARISON OF LINEWIDTHS

Light source	Center wavelength λ_0 (Å)	FWHM linewidth $\Delta\lambda$ (Å)	FWHM linewidth Δf (Hz)
Ordinarly discharge lamp	5896	$\cong 1$	9×10^{10}
Cadmium low-pressure lamp	6438	$\cong 0.013$	9.4×10^{8}
Helium-neon laser	6328	$\cong 10^{-7}$	7.5×10^{3}

the spatial extent of their emitting surfaces and then significantly filtered to decrease the line width of the polychromatic light. A consequence of these two modifications, however, is a serious reduction of emitted light intensity.

In contrast, a laser source—by the very nature of the process that produces amplification via stimulated emission—ensures both a narrow-band output and a high degree of phase correlation. Recall that in the process of stimulated emission, each photon added to the stimulating radiation has a phase, polarization, energy, and direction *identical* to that of the amplified light wave in the laser cavity. The laser light thus created and emitted is both *temporally* and *spatially* coherent. In fact, one can picture a real laser device as a powerful, fictitious point source, located at a distance from the target, giving off monochromatic light in a narrow cone angle. Figure 12-10 summarizes the basic ideas of coherence for nonlaser and laser light sources.

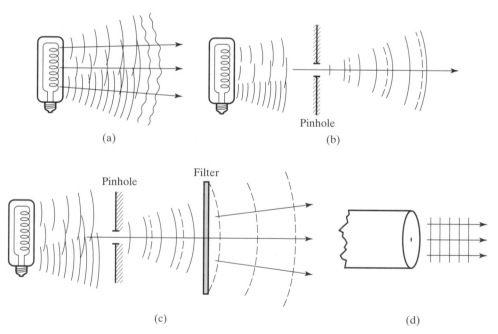

Figure 12-10   A tungsten lamp requires a pinhole and filter to produce partially coherent light. The light from a laser is naturally coherent. (a) Tungsten lamp. The tungsten lamp is an extended source that emits many wavelengths. The emission lacks both temporal and spatial coherence. The wave fronts are irregular and change shape in a haphazard manner. (b) Tungsten lamp with pinhole. An ideal pinhole limits the extent of the tungsten source and improves the spatial coherence of the light. However, the light still lacks temporal coherence since all wavelengths are present. Power in the beam has been severely reduced. (c) Tungsten lamp with pinhole and filter. Adding a good, narrow-band filter further reduces the power but improves the temporal coherence. Now the light is "coherent," but the available power is far below that initially radiated by the lamp. (d) Laser. Light coming from a laser has a high degree of spatial and temporal coherence. In addition, the output power can be very high.

For typical lasers, both the spatial and temporal coherence of the emitted light are far superior to that for light from other sources. The transverse spatial coherence of a single mode laser beam extends across the full width of the beam, whatever that may be. The temporal coherence, also called *longitudinal spatial coherence*, is many orders of magnitude above that of any ordinary light source. As noted earlier in Chapter 11, the *coherence time* t_c of a laser is a measure of the average time interval over which one can continue to predict accurately the phase of the laser light at a given point in space further along the propagation direction. The *coherence length* L_c, the distance of "perfect" phase correlation along the propagation direction, is related to the coherence time t_c by the equation $L_c = ct_c$, where c is the speed of light. For the He-Ne laser described in Table 12-1 the coherence time t_c is of the order of milliseconds—100 million times longer than the coherence time of 10^{-11} s for a typical sodium discharge lamp. Consequently, the coherence length for the He-Ne laser is thousands of kilometers, compared with millimeters for the sodium lamp.

Directionality.   When one sees the thin, pencil-like beam of a He-Ne laser for the first time, one is struck immediately by the high degree of beam directionality. No other light source, with or without the help of lenses or mirrors, generates a beam of such precise definition and minimum angular spread.

The astonishing degree of directionality of a laser beam is due to the geometrical design of the laser cavity. Figure 12-11 shows a specific cavity design and an external laser beam with an angular spread signified by the angle ϕ. The cavity mirrors shown are shaped with surfaces concave toward the cavity, thereby "focusing" the reflecting light back into the cavity and forming a *beam waist* at some position in the cavity. The nature of the beam inside the laser cavity and its characteristics outside the cavity are determined by solving the rather complicated boundary-value problem of standing electromagnetic waves in an open cavity. Although the details of this analysis are beyond the scope of this discussion, several results are worth examining. It turns out that the beam-spread angle ϕ is given by the relationship

$$\phi = \frac{1.27\lambda}{D} \tag{12-3}$$

where λ is the wavelength of the collimated, monochromatic light and D is the diameter of the laser beam at its beam waist. One cannot help but observe that Eq. (12-3) is quite similar to that obtained when treating the angular spread in

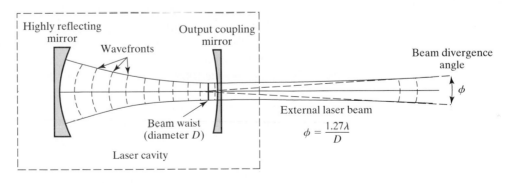

Figure 12-11   External and internal laser beam for a given cavity. Diffraction or beam spread, measured by the beam divergence angle ϕ, appears to be caused by an effective aperture of diameter D, located at the beam waist.

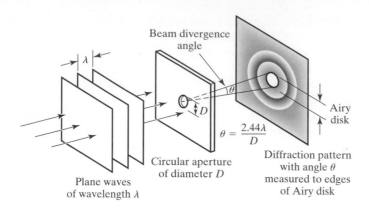

Beam divergence angle

λ

θ

D

$\theta = \dfrac{2.44\lambda}{D}$

Plane waves of wavelength λ

Circular aperture of diameter D

Diffraction pattern with angle θ measured to edges of Airy disk

Airy disk

Figure 12-12 Fraunhofer diffraction of plane waves through a circular aperture. Beam divergence angle θ is set by the edge of the Airy disk.

light generated by the diffraction of plane waves passing through a circular aperture (Chapter 16). The pattern consists of a central, bright circular spot, the *Airy disk*, surrounded by a series of bright rings. The essence of this phenomenon is shown in Figure 12-12. The diffraction angle θ, tracking the outside edge of the Airy disk, is given by

$$\theta = \frac{2.44\lambda}{D} \tag{12-4}$$

where λ is the wavelength of the collimated, monochromatic light and D is the diameter of the circular aperture. Both Eqs. (12-3) and (12-4) depend on the ratio of a wavelength to a diameter; they differ only by a constant coefficient. It is tempting, then, to think of the angular spread φ inherent in laser beams and given in Eq. (12-3) in terms of diffraction. If we treat the beam waist as an effective circular aperture located inside the laser cavity, then by controlling the size of the beam waist we control the diffraction or beam spread of the laser. The beam waist, in practice, is determined by the design of the laser cavity and depends on the radii of curvature of the two mirrors and the distance between the mirrors. Therefore, one ought to be able to build lasers with a given beam waist and, consequently, a given beam divergence or beam spread in the *far field*, that is, at sufficiently great distance L from the diffracting aperture such that

$$L \gg \frac{\text{area of aperture}}{\lambda}$$

This is indeed the case.

Several comments may be helpful in conjunction with Figures 12-11 and 12-12. Plane waves of uniform irradiance pass through the circular aperture in Figure 12-12; that is, the strength of the electric field is the same at all points along the wave front. In Figure 12-11, the wave fronts that pass "through" the effective aperture or beam waist are also plane waves, but the irradiance of the laser light is not uniform across the plane. For the lowest-order transverse mode (a particularly important one), the so-called TEM$_{00}$ mode or *Gaussian beam* mode, the irradiance of the beam decreases exponentially toward the edges of the beam in accordance with the Gaussian form e^{-8y^2/D^2}, where y measures the transverse beam direction and D is the beam width at a given position along the beam (see Figure 12-13). The circular aperture in Figure 12-12 is a true physical aperture; the beam waist in Figure 12-11 is not. It is interpreted as an *effective aperture* when beam spread from a laser is compared with light diffraction through a real aperture.

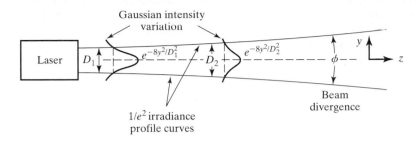

Figure 12-13 A Gaussian TEM$_{00}$ beam. The diameter (transverse width) of the beam is measured at the "$1/e^2$ profile" curves. These curves are the locus of points where the irradiance of the beam, on either side of the optical axis, has decreased to $1/e^2 = 0.135$ of its vlaue at the center of the beam. The Gaussian irradiance variation is shown at two positions along the beam, where the beam diameters are D_1 and D_2, respectively. The diameter D_2 is greater than D_1 because the beam is spreading according to the beam divergence angle $\phi = 1.27\lambda/D$.

With the help of Eq. (12-3), one can now develop a feel for the low beam spread, or high degree of directionality, of laser beams. Consider the following problem.

❏ **Example**

(a) He-Ne lasers (632.8 nm) have an internal beam waist of diameter near 0.5 mm. Determine the beam divergence of the output beam.

(b) We can control the beam waist D by laser cavity design and "select" the wavelength by choosing different laser media. What lower limit then might we expect for the beam divergence? How directional can lasers be? To develop answers to these questions, suppose we design a laser with a beam waist of 0.5-cm diameter and a wavelength in the ultraviolet of 200 nm. By what factor is the beam divergence improved?

Solution

(a) Using Eq. (12-3) for the beam divergence, we obtain

$$\phi = \frac{1.27\lambda}{D} = \frac{(1.27)(632.8 \times 10^{-9}) \text{ m}}{5 \times 10^{-4} \text{ m}} = 1.6 \times 10^{-3} \text{ rad}$$

This is a typical laser-beam divergence, indicating that the beam width increases about 1.6 cm for every 1000 cm of travel.

(b) Again, using Eq. 12-3, with the new values for λ and D, we get

$$\phi = \frac{1.27\lambda}{D} = \frac{(1.27)(200 \times 10^{-9}) \text{ m}}{5 \times 10^{-3} \text{ m}} = 5.1 \times 10^{-5} \text{ rad}$$

Thus the beam divergence angle becomes about 5×10^{-5} rad, roughly a 30-fold decrease in beam spread over the He-Ne laser described in part (a). This beam would spread about 1.6 cm for every 320,000 cm. Clearly, if beam waist size is at our command and lasers can be built with wavelengths below the ultraviolet, there is no limit to how parallel and directional the laser beam can be made. ∎

The high degree of directionality of the laser, or of any other light source, depends on the monochromaticity and coherence of the light generated. Ordinary sources are neither monochromatic nor coherent. Lasers, on the other hand, are superior on both counts, and as a consequence generate highly directional, quasi-collimated light beams.

Laser Source Intensity. It has been said that a 1-mW He-Ne laser is hundreds of times "brighter" than the sun. As difficult as this may be to imagine, calculations for luminance or visual brightness of a typical laser, compared to the sun, substantiate these claims. To develop an appreciation for the enormous difference between the radiance of lasers and thermal sources, we consider next a comparison of their photon output rates (photons per second).

❏ **Example**

    (a) Small gas lasers typically have power outputs of 1 mW. By contrast, neodymium-glass lasers, such as those designed for the production of laser-induced fusion, boast of power outputs near 10^{14} W! Using these two extremes and an average energy of 10^{-19} J per visible photon ($E = hf$), determine the approximate range of photon output from lasers.

    (b) For comparison, consider a broadband thermal source with a radiating surface equal to that of the beam waist of a 1-mW He-Ne laser with diameter of 0.5 mm (or area of $A = 2 \times 10^{-7}$ m^2). Let the surface emit radiation at a wavelength of 633 nm, with a line width $\Delta\lambda = 100$ nm and temperature $T = 1000$ K. Determine the photon output rate of this thermal source.

Solution

    (a) The rate of photon output is given by P/hf. Thus for a small gas laser,

$$\frac{P}{hf} = \frac{10^{-3} \text{ J/s}}{10^{-19} \text{ J/photon}} = 10^{16} \text{ photons/s}$$

and for the high power laser,

$$\frac{P}{hf} = \frac{10^{14} \text{ J/s}}{10^{-19} \text{ J/photon}} = 10^{33} \text{ photons/s}$$

    (b) The photon output rate for the broadband source can be calculated from the equation[4]

$$\text{thermal protons/s} \cong \frac{1}{\lambda^2} \frac{1}{e^{hf/kT} - 1} \Delta A \, \Delta f \tag{12-5}$$

The frequency $f = c/\lambda = 4.74 \times 10^{14}$ Hz, taking c as 3×10^8 m/s. The frequency bandwidth $\Delta f = (c/\lambda^2)\Delta\lambda = 7.5 \times 10^{13}$ Hz. The argument of the exponent is

$$\frac{hf}{kT} = \frac{(6.63 \times 10^{-34} \text{ J-s})(4.74 \times 10^{14} \text{ s}^{-1})}{(1.38 \times 10^{-23} \text{ J/K})(1000 \text{ K})} = 22.77$$

---

[4]One can obtain Eq. (12-5) from an expression for the spectral energy density of blackbody radiation given as $\rho(f) = (8\pi hf^3/c^3)(1/e^{hf/kT} - 1)$, by changing from $\rho(f)$ in units of energy per unit volume per unit frequency range to units that give thermal photons per second. In this process, one must change geometry accordingly, from an isotropic radiating volume ΔV for the blackbody radiation equation above to that of a radiating surface element ΔA in Eq. (12-5).

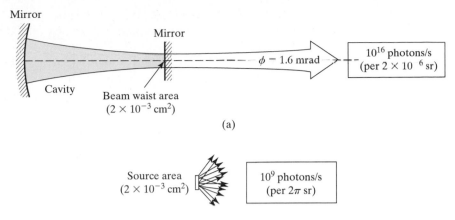

Figure 12-14 Comparison of photon output rates between a lower power He-Ne gas laser and a hot thermal source of the same radiating surface area. (a) 1-mW He-Ne laser ($\lambda = 633$ nm). (b) Broadband thermal source (Lambertian), with approximate values of $T = 1000$ K, $\Delta A = 2 \times 10^{-3}$ cm^2, $\lambda_0 = 633$ nm, and $\Delta\lambda = 100$ nm. Note that the laser emits all the photons in a small solid angle ($\approx 2 \times 10^{-6}$ sr) compared with the 2π solid angle of the thermal source.

where we have used the values for Planck's constant h and Boltzmann's constant k. Substituting all values into Eq. (12-5), the thermal photon output per second is

$$\left[\frac{1}{(633 \times 10^{-9})^2 \text{ m}^2}\right]\left[\frac{1}{7.74 \times 10^9}\right](2 \times 10^{-7} \text{ m}^2)(7.5 \times 10^{13} \text{ s}^{-1}) \cong 5 \times 10^9 \quad \blacksquare$$

We find a rate of only about 10^9 photons / s! This value is 7 orders of magnitude smaller than the photon output rate of a low-power 1-mW He-Ne laser and 24 orders of magnitude smaller than a powerful neodymium-glass laser. The comparison is summarized in Figure 12-14.

We see also from Figure 12-14 that the He-Ne laser emits 10^{16} photons/s into a very small solid angle of about 2×10^{-6} sr, whereas the thermal emitter, acting as a Lambertian source, radiates 10^9 photons/s into a forward, hemispherical solid angle of 2π sr. If we were to ask how many thermal photons/s are emitted by the thermal source into a solid angle equal to that of the laser, we would find the answer to be a mere 320 photons/s:

$$(10^9 \text{ photons/s})\left(\frac{2 \times 10^{-6} \text{ sr}}{6.28 \text{ sr}}\right) = 320 \text{ photons/s}$$

The contrast between 10^{16} photons/s for the laser source and 320 photons/s for the thermal source is now even more dramatic.

Carrying our comparison of source intensity between laser and nonlaser sources one step further, we can determine how the radiance [W/(cm^2-sr)] of a high-power, nonlaser source compares with that of a low-power laser.

❏ **Example**

We choose for the nonlaser source a super, high-pressure mercury lamp, capable of a source radiance L_e equal to 250 W/(cm^2-sr). Such lamps were about the best high-radiance sources available before the advent of the laser. For the

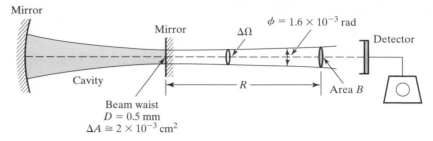

Figure 12-15 Radiance of a low-power, He-Ne laser. The radiance of the laser (beam waist) "seen" by the detector is about 10^6 W/cm²-sr.

laser source we choose a 4-mW He-Ne laser with beam-waist diameter of 0.5 mm and beam divergence angle of 1.6 mrad, emitting at a wavelength of 633 nm. The geometry is shown in Figure 12-15. It should be clear from the figure that the laser can be considered a radiating surface of area equal to that of the beam waist. The radiant power of the surface is equal to that of the laser, and the surface radiates only into the solid angle dictated by the geometry of the laser cavity.

Solution Using the definition for radiance L_e in terms of laser power Φ_e, source or beam-waist area ΔA, and solid angle $\Delta\Omega$,

$$L_e = \frac{\Phi_e}{\Delta A \, \Delta\Omega} \tag{12-6}$$

the radiance of the laser is found to be 10^6 W/cm²-sr, as follows: Referring to Figure 12-15, the solid angle $\Delta\Omega$ = area B/R^2, where, with the small angle approximation $\tan\theta \cong \theta$, we get for $\Delta\Omega$,

$$\Delta\Omega = \frac{\text{area } B}{R^2} = \frac{\pi}{R^2}\left[R\tan\left(\frac{\phi}{2}\right)\right]^2 \cong \frac{\pi\phi^2}{4}$$

Then substituting ΔA and $\Delta\Omega$ into Eq. (12-6), the radiance is

$$L_e = \frac{4\Phi_e}{\pi\phi^2\Delta A} = \frac{4(4\times10^{-3}\,\text{W})}{\pi(1.6\times10^{-3})^2(2\times10^{-3}\,\text{cm}^2)} = 1\times10^6\,\text{W/cm}^2\text{-sr}$$

Again, the comparison between laser source and high-intensity mercury lamp is dramatic. We conclude that where brightness and radiance are important in the selection of light sources, the laser stands alone. ∎

Focusability. Focusing light to a tiny, diffraction-limited spot is a challenge. It is customary in geometrical optics to show a positive lens focusing a beam of perfectly collimated light to a "point-image" (see Figure 12-16a). We know that a point image is an idealization, unattainable even in the limit of geometrical optics ($\lambda \to 0$), because perfect, aberration-free lenses do not exist. Nevertheless, the ideal of focusing light to a diffraction-limited point image has long been a goal. The laser, with its coherent, nearly collimated beam, has made that ideal attainable. Figure 12-16b shows the difficulty involved in focusing ordinary light to a tiny spot. First, the light emitted is incoherent. Second, the physical source of light cannot be too small, that is, approaching a point source, for then either the light generated would be insufficient or the source would melt. The

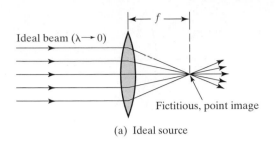

Ideal beam ($\lambda \rightarrow 0$)

Fictitious, point image

(a) Ideal source

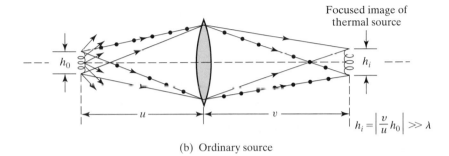

Focused image of
thermal source

h_0

h_i

u

v

$$h_i = \left| \frac{v}{u} h_0 \right| \gg \lambda$$

(b) Ordinary source

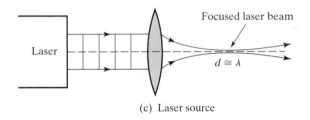

Focused laser beam

Laser

$d \cong \lambda$

(c) Laser source

Figure 12-16 Focused beams from various sources. (a) Ideal, collimated beam is focused to a fictitious "point" in accordance with geometrical optics. (b) Incoherent radiation from a thermal source is focused to a demagnified image of size $h_i \gg \lambda$. (c) Coherent laser beam is focused to a diffraction-limited spot of diameter $d \cong \lambda$.

combination of a nonpoint source and incoherent light leads to fairly large image sizes, fixed more or less by the laws of magnification in geometrical optics.

The laser, on the other hand, radiates intense, coherent light that appears to come from a distant "point source." By its unique properties then, it overcomes the precise limitations that frustrate one's attempts to focus thermal radiation to a tiny spot. Figure 12-16c shows a laser beam focused by a positive lens to a diffraction-limited spot of incredibly small diameter, approximately equal to the wavelength of the focused light. It can be shown that a laser beam with beam divergence ϕ, incident on a lens of focal length f, whose diameter is several times larger than the width of the incident beam, is focused to a diffraction-limited spot of diameter approximately equal to

$$d \cong f\phi \tag{12-7}$$

as shown in Figure 12-17. The beam divergence angle ϕ is equal to $1.27\lambda/D$, as given previously in Eq. (12-3). Note carefully that D in Eq. (12-3) refers to the diameter of the beam waist in the laser that "determined" the beam divergence, and d in Eq. (12-7) refers to the diameter of the laser beam focused by the positive lens. The location of the focused laser spot is essentially at the focal plane of the lens ($s' = f$), although a careful analysis shows that this is, in fact, an approximation, even if a good one.

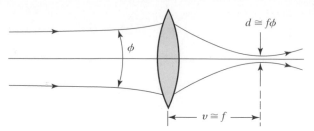

Figure 12-17 TEM$_{00}$ laser beam of beam spread angle ϕ.

With the help of Eq. (12-7), we can make several rough calculations to predict the spot size of focused laser beams. With a lens of focal length $f = 0.1$ m and incident laser light of beam divergence $\phi = 10^{-3}$ to 10^{-4} rad, spot sizes of the order of 10^{-4} to 10^{-5} m (or 100 μm to 10 μm) in diameter can be easily obtained. If we compare these diameters with the wavelength of the carbon dioxide laser ($\lambda = 10.6$ μm), we see at once that CO_2 laser light—indeed all laser light—can be focused to spot sizes of the order of a wavelength.

Equation (12-7) indicates that focusing laser light down to small spots can be achieved by lenses with short focal lengths and laser beams with small beam divergences. As long as aberration-free lenses of high quality are available, the focal length can be chosen to be as short as is practical. The beam divergence of a laser, usually determined at the time the laser is designed, can still be reduced with the additional optics found in *beam expanders*. In Figure 12-18, a collimated laser beam of width W_i and beam divergence ϕ_i is focused by the first lens of the beam expander (focal length f_1) to a spot of diameter $d = f_1 \phi_i$, in accordance with Eq. (12-7). The second lens, a distance f_2 from the focused spot, with $f_2 > f_1$, collects the light expanding from the focused spot and essentially recollimates it. The beam divergence of the expanded, recollimated beam is equal to

$$\phi_f = \left(\frac{W_i}{W_f}\right)\phi_i = \left(\frac{f_1}{f_2}\right)\phi_i \tag{12-8}$$

where $f_2/f_1 = W_f/W_i$ is the *beam expansion ratio*. The validity of Eq. (12-8) is not difficult to show. The incident beam, focused by the first lens, has a spot diameter $d_1 = f_1 \phi_i$. By the principle of reversibility of light rays, if the expanded beam

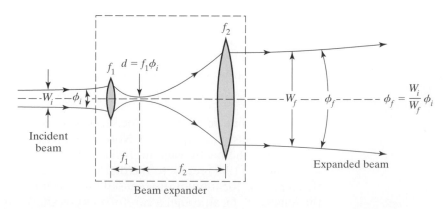

Figure 12-18 Beam expansion as a method of reducing divergence of a laser beam.

were to be redirected to the left and focused by the second lens, it would form the identical spot at the same location, so that $d_2 = f_2\phi_f$. Since $d_1 = d_2$ necessarily,

$$f_1\phi_i = f_2\phi_f$$

$$\phi_f = \frac{f_1}{f_2}\phi_i$$

If the beam expansion ratio is $f_2/f_1 = 5$, the beam divergence of the expanded laser beam is $\frac{1}{5}$ that of the incident beam. Evidently, by expanding the beam width by a factor of 5, one achieves a reduction in beam divergence by the same factor.

What has been gained? Suppose a laser beam is expanded 10-fold by a suitable beam expander. The outgoing beam then has had its beam divergence decreased by a factor of 10. If the expanded beam is focused to a tiny spot with a lens of arbitrary focal length f, the diameter of the spot will be $\frac{1}{10}$ that achievable with the unexpanded beam and the same lens. This 10-fold reduction in diameter leads to a 100-fold reduction in spot size area and thus for any given laser beam power, to a 100-fold increase in focused spot irradiance.

Laser energy focused onto small target areas makes it possible to drill tiny holes in hard, dense material, make tiny cuts or welds, make high-density recordings, and generally carry out industrial or medical procedures in target areas only a wavelength or two in size. In ophthalmology, for example, where Nd:YAG lasers are used in ocular surgery, target irradiances of 10^9 to 10^{12} W/cm^2 are required. Such irradiance levels are readily developed with the help of beam expanders and suitable focusing optics, as we will see in Section 12-6 and Problems 12-12 and 12-17.

12-5 LASER TYPES AND APPLICATIONS

To this point we have examined the basic physics of laser action, identified the essential parts that make up a laser, described in a general way how a laser operates, and studied the characteristics that make lasers such a unique source of light. Now we turn our attention first to the identification of some of the common lasers in existence today and then to the many ways lasers are used in almost every facet of our lives.

Laser types. Lasers are classified in many ways. Sometimes they are grouped according to the state of matter represented by the laser medium: gas, liquid, or solid. Sometimes they are classified according to how they are pumped: flashlamp, electrical discharge, chemical actions, and so on. Still other classifications divide them according to the nature of their output—pulsed or continuous wave (cw)—and according to their spectral region of emission (infrared, visible, or ultraviolet).

The lasers listed in Table 12-2 are, in a way, a cross section of the 50 or so types of lasers on the market today. A careful examination of Table 12-2 serves as an introduction to the state of laser technology. For each laser listed, the entries include data on emission wavelength, output power (or in some cases, energy per pulse), nature of output, beam diameter, and beam divergence. Table 12-2 includes examples of gas lasers, solid state lasers, ion lasers, liquid or dye lasers, excimer lasers, and diode lasers. Both pulsed and continuously operating (cw) lasers are represented. Taken as a whole, Table 12-2 includes lasers

TABLE 12-2 LASER PARAMETERS FOR SEVERAL COMMON LASERS*

Type	Wavelength	Power/energy	Type of output	Beam diameter (mm)	Beam divergence (mrad)
Helium-neon (gas)	632.8 nm	0.1–750 mW	cw	0.5–3	0.5–2.5
Helium-cadmium (gas)	325/442 nm	5–100 mW	cw	0.2–1.8	0.5–2.0
Carbon dioxide (gas)	10.6 μm	3–100 W	cw	1.5–4	1–10
Nitrogen (gas)	337 nm	1–300 mW	pulsed	3×6–6×32 (rectangular)	1.3 × 7
Argon ion (gas)	488/514.5 nm	5 mW–20 W	cw	0.7–2	0.4–1.5
Nd:YAG (solid)	1.064 μm	0.04–600 W	cw	0.75–6	2–18
Alexandrite (solid)	700–800 nm	0.1–2 W (TEM_{00})	pulsed	1–6	1–9
	700–800 nm	20–120 W (medical)	pulsed	1–6	< 1 (medical)
Erbium:YAG (solid)	2.94 μm	0.5–4J; 150–300 μs pulses	pulsed	5–7.5	3–15
Holmium:YAG (solid)	2.1 μm	0.85–4J; 50–1000 μs pulses	pulsed	4–6	3–15
Ti:sapphire (solid)	0.7–1.08 μm	0.3–5 W	cw	0.6–1.4	1–2
Dye (liquid)	380–1000 nm (tunable)	0.8–1.0 W	cw (pumped by Ar ion laser)	0.5–0.7	1.2–1.4
Argon fluoride (excimer)	193 nm	Up to 10 W av. pulse	pulsed	6×23–20×32 (rectangular)	2–6
Gallium arsenide (semiconductor diode)	780–900 nm	1–40 mW cw or av. pulse	cw/pulsed	N/A (diverges too rapidly)	200 × 600 (oval shape)

*In many instances, the values given above are representative of a wide selection of lasers available under the generic type described. A comprehensive listing for over 4000 choices of lasers, with varying specifications, is given each year in the *Laser Focus World Buyer's Guide*, Pennwell Publishing Co., Nashua, N.H., 03062.

whose wavelengths vary from 193 nm (deep ultraviolet) to 10.6 μm (far infrared); whose cw power outputs vary from 0.1 mW to 600 W; and whose beam divergences vary from 0.4 mrad or less (circular cross section) to 200 × 600 mrad (oval cross section). Had we also listed typical "wall-plug" efficiencies (laser beam power out over wall-plug electrical power in) we would have observed that efficiencies are typically very low, in the 0.1% to 2% range, only rarely achieving values in the 10 or 20% range. Despite these low efficiencies, lasers remain the world's most unique light source.

Laser Applications.   The variety of ways in which lasers can be applied seems to be limited only by human ingenuity. In some 30 years since its invention, the laser has found widespread use in industry, agriculture, construction, hospitals, entertainment, communications, and weaponry. In short, the laser has invaded the world of science and technology with remarkable vigor, and it continues to show great promise as a catalyst for exciting new developments in the explosive fields of photonics and lightwave technology.

It is, of course, the unique properties of the laser we have examined in this chapter—monochromaticity, directionality, coherency, and brightness—that account for its wide acceptance and usefulness. In each application, one or more of these properties is exploited to achieve a desired goal. Table 12-3 identifies some of the areas where lasers have made an impact, where their unique properties have been put to use. A quick glance at Table 12-3 certainly adds irony to the observation made in 1960, to which we referred at the beginning of the chapter, that the laser was "an invention looking for an application." Not only are the applications in Table 12-3 extensive and global in reach, but each one can be divided further to show more specialized applications. For example, in the medical field

TABLE 12-3 LASER APPLICATIONS IN THE 1990s

Materials processing	Communication
cutting	Fiber energy delivery
welding	Information processing
drilling	Optical sensing
heat treatment	Optical ranging
Medical	Laser planing
Marking and scribing	Laser printing and copying
Microfabrication	Metrology
Laser spectroscopy	Optical storage
Laser-initiated fusion	Holography
Tracking and targeting	Alignment
Weaponry	Bar code scanning
Heavy construction	Entertainment and display
Surveying	Subatomic manipulation

alone, the uses of the laser for diagnostic and therapeutic procedures—involving cutting, vaporizing, rupturing, welding, sealing, cauterizing, coagulating, and unblocking—are growing in hospitals across the country. Table 12-4 lists some of the current uses in the medical field, and the types of lasers, with wavelength and power levels, that are involved.

As the list indicates, the five laser types most actively used in medical fields are the excimer laser, the argon-ion laser, the tunable dye laser, the Nd:YAG laser, and the CO_2 laser. The excimer lasers, argon fluoride (ArF) and krypton fluoride (KrF), emit invisible laser energy in the deep ultraviolet. The Nd:YAG and CO_2 lasers emit laser energy at the other end of the optical spectrum, in the

TABLE 12-4 USE OF LASERS IN MEDICINE

Laser type	Wavelength	Power levels	Uses

Excimer
  Argon fluoride
  Krypton fluoride
	193 nm (ultraviolet; invisible)
	248 nm (ultraviolet; invisible)
		Pulsed, $\cong 1$ J/cm^2 with pulse width of about 10 ns ($\cong 10^8$ W/cm^2)	Cutting, incising, ablative photo-decomposition (ophthalmology, cardiology, and arthroscopy)

Argon-ion	488 nm (blue-green; visible)	Continuous; up to 10 W	Photocoagulation, welding, vaporization, (dermatology, ophthalmology, general surgery)
Tunable dye (pumped with an argon-ion laser)	631 nm (red; visible)	Continuous; 3–4 W	Photoactivation (treatment of tumors)
Nd:YAG	1.06 μm (infrared; invisible)	Continuous; up to maximum of 60–100 W	Photocoagulation, vaporization, perforation (ophthalmology, gastroenterology, dermatology, urology, and tumors)
Carbon dioxide (CO_2)	10.6 μm (infrared; invisible)	Continuous; up to 80 W	Tissue vaporization, incision, excision (dermatology, gynecology, gastroenterology, and neurosurgery)

near and far infrared, again invisible. The argon-ion laser emits a visible (blue-green) beam that can be seen and, therefore, more easily handled and aligned. The tunable dye laser, although generally quite low in output power, emits a beam that contains selectable wavelengths extending from the violet end of the visible region (400 nm) to the near infrared (900 nm). For the "invisible" lasers, spotter beams—generally He-Ne lasers—are used collaboratively to facilitate handling and focusing.

The CO_2 laser, one of the first to be used extensively by medical professionals, emits infrared (ir) energy at 10.6 μm that is almost totally absorbed by water. Since body cells and tissues are 70% to 90% water, a CO_2 laser can make a clean incision or, if focused more sharply, can vaporize malignant tissue. Because a laser cauterizes and seals capillaries as it cuts, surgery is essentially bloodless. One drawback of CO_2 lasers, to date, is their inability to be used readily with fiber optics: Thus, transmission of the beam from the laser to the target area is often accomplished with cumbersome, articulated optical arms.

Unlike the CO_2 laser, the Nd:YAG laser beam at 1.06 μm and the argon-ion laser beam at 488 nm mate well with fiber optics. The laser energy is thus transferred readily from the laser source to the operating table, without significant power loss. Using flexible systems that emerge from the marriage of laser beams and fiberoptic scopes (endoscopes), surgeons are able to perform major abdominal surgery, for example, through an incision as small as 1 inch.

The blue-green argon-ion laser beam is unique in that it is selectively absorbed by red and brown substances, such as red blood cells and melanin pigment spots. It can therefore stop retinal bleeding (photocoagulation) and weld detached retinas. It can fade port-wine birthmarks and tatoos, penetrating the nonpigmented upper layer of skin without harm.

The Nd:YAG laser, with fiberoptic scope, has also been used effectively in photocoagulation applications, controlling or stopping the bleeding of ulcers and large tumors or blood vessels deep in the body. The Nd:YAG laser has been used with some success to seal tiny bleeding vessels of the retina, thereby ameliorating a disease (retinopathy) that often leads to blindness. To accomplish this, the Nd:YAG laser is optically focused to a diameter of 1/1000 inch.

In one of its more recent, dramatic applications, a Nd:YAG laser beam has been focused to a 30-μm spot within the eye. By concentrating powers of over 1 billion W/cm^2, plasma breakdown occurs in the vitreous humor. Subsequent acoustic shock waves rupture unwanted, opacified membranes along the vision axis of the eye. This unique example of noninvasive, pain-free surgery (posterior capsulotomy) is treated in more depth in the section that follows.

Quite recently, both the Nd:YAG and the excimer lasers have been tested in an important assault on arteriosclerosis (hardening of the arteries) and critically blocked blood vessels of the heart. Again, with delicate fiberoptic scopes threaded carefully and slowly through major blood vessels, unwanted fatty deposits and plaque are literally blasted away.

To be added soon to the five workhorses listed in Table 12-4 are the following new candidates incorporating as dopants, holmium (Ho), and erbium (Er): Ho:YAG at 2.1 μm, Er:YAG at 2.94 μm, and Er^{3+}-fluoride fibers at 2.7 μm. Each is seen as a precise surgical tool for procedures involving vaporization, coagulation, incision, excision, and ablation.

In Table 12-4, we see that ophthalmology is a medical field that makes extensive use of lasers as a diagnostic and surgical tool. In the next section, we take a closer look at how lasers are used to treat ocular defects and diseases.

The choice of a laser system for treatment of particular disorders of the eye begins with answers to the following considerations:

- Is the target tissue to be seared, vaporized, lysed, or perforated? (The desired action on target determines the power level, beam divergence, spot size on target, and whether the laser mode of operation is to be pulsed or continuous.)
- What are the absorption characteristics of the ocular media along the laser beam path and particularly of the target area to be treated? (The answer here determines the type and wavelength of the laser to be used.)
- How is the laser energy to be delivered on target? (Delivery determines the nature of associated fiber optic beam-steering and beam-focusing elements required.)

Information on laser types and characteristics such as that found in Table 12-2, enables one to select a laser based on a consideration of the first question. Information on laser absorption in different media, as a function of wavelength, enables one to narrow the selection based on considerations of the second and third questions. Let us now consider absorption.

Absorption of Laser Energy. Since most ocular tissue is principally water, particular regions of laser transmission or laser absorption in the eye depend critically on the degree of absorption of the chosen wavelength in water. Figure 12-19 shows the relative absorption of laser energy, at wavelengths from 0.4 μm to 10.6 μm, in 1 mm of water and 1 mm of hemoglobin.

An examination of Figure 12-19 indicates that CO_2 laser energy (10.6 μm) is almost totally absorbed in tissue of high water content, while laser energy of argon ion (0.488 μm), frequency-doubled Nd:YAG (0.532 μm), krypton ion (0.647 μm), and ordinary Nd:YAG (1.06 μm) is almost totally transmitted. One can conclude that the cornea, mostly water, would be transparent to the argon, krypton, and Nd:YAG laser beam, while highly absorptive of the CO_2 beam. By contrast, the exact opposite is true for these same lasers in the important blood component, hemoglobin. Accordingly one would use CO_2 lasers to "cut" corneal tissue, but not for retinal procedures, and one would use one of the other four for photocoagulative treatments of the retinal surface, but not generally for interactions with water-laden tissue.

Fiberoptic-delivered laser energy has become increasingly attractive in laser surgery and medicine, and specifically in ophthalmology. The first fiber-delivered, high-powered lasers, appropriate for surgery, appeared in the early 1980s, a marriage between quartz fibers and Nd:YAG laser energy at 1.06 μm, providing tens of watts of power. Since then, improved fiber technology and increased choice of fiber media, along with additional laser sources—Ho:YAG at 2.1 μm; Er:YAG at 2.94 μm; Er^{3+}-doped fluoride at 2.7 μm—have broadened the acceptance of fiber-delivered laser energy in operating rooms.

A critical consideration for the pairing of laser energy with a fiberoptic delivery system revolves around the question of absorption of laser energy in the fiber medium. Figure 12-20 shows the attenuation of laser energy, in decibels per kilometer of fiber cable, for wavelengths from 800 nm to 1600 nm, for glass (silica), the preferred fiber material.

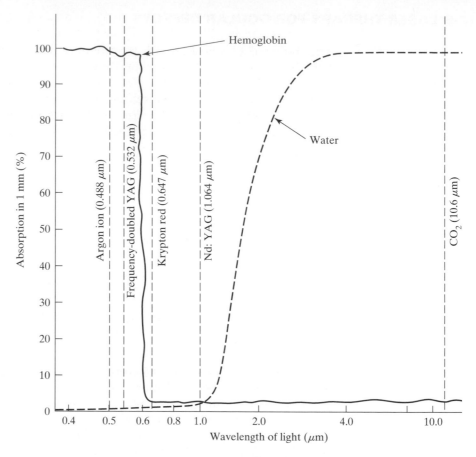

Figure 12-19 Absorption of laser light in ocular tissue for several important lasers. The percentage absorption is for 1 mm of water (dashed curve) and 1 mm of hemoglobin (solid curve). Note that radiation at 10.6 μm is almost totally absorbed in the cornea, mostly water, whereas that at 1.06 μm is almost totally transmitted.

One can see readily in Figure 12-20 that the attenuation of laser energy in glass reaches a minimum around 1500–1600 nm (1.55 μm). This corresponds to the same wavelength region wherein an increasing number of rare-earth lasers (holmium and erbium-doped) give peak emissions of laser energy. As a result, low-loss fibers coupled with laser energy at wavelengths suitable for cutting or vaporizing are being used increasingly in noninvasive surgery of ocular tissue.

Lasers in Ophthalmology. The laser candidates for ocular surgery currently include the CO_2 laser (10.6 μm) in the far infrared; Ho:YAG and

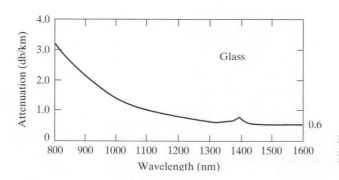

Figure 12-20 Spectral attenuation for all-glass multimode fibers. (Courtesy Corning Glass Works.)

Er:YAG lasers in the 1.5–3.0 μm range; semiconductor diode, alexandrite, titanium:sapphire and Nd:YAG in the far red to near infrared region (0.7 to 1.1 μm); argon or krypton-ion and excimer lasers in the deep blue and near-ultraviolet regions; and tunable dye lasers from the near ultraviolet to the near infrared.

Laser beams, both pulsed and continuous, selected from those listed above, are currently used for treating glaucoma, retinal bleeding, macular degeneration, retinal detachments, opacified intraocular membranes, iridectomies, and corneal sculpting. *Angle blockage glaucoma*, a disease of the eye characterized by increased fluid pressure within the eye and leading progressively to blindness, is treated with argon, pulsed Er:YAG, and Nd:YAG lasers. The treatment consists of repairing structural flaws in the eye that cause increased pressure, using the laser beam to open up blocked ducts or to create new canals for better drainage between the chambers of the eye. These lasers are also used effectively to treat *diabetic retinopathy*, the leading cause of blindness. To neutralize such organic disorders, the thermal energy in the laser beam is used to place thousands of tiny burns or welds at the back of the retina, thereby preventing the harmful growth or rupture of new, unwanted blood vessels (*neovascularization*). The thermal content of a laser beam (present in Nd:YAG, Er:YAG, and Ho:YAG lasers), so effective where retinal coagulation is required, is also used to tack or weld detached retinas that fall away from the choroid at peripheral positions. This procedure, in fact, was the first successful use of lasers in a clinical environment.

Lasers with wavelength emission in the deep ultraviolet, principally the excimer lasers, offer great promise for reshaping the eyeball to correct for myopic vision (*refractive surgery*) and for removing *cataracts*, an organic disorder marked by pockets of cloudy or opaque discoloration in the lens tissue of the eye, resulting in progressively degraded vision. One of the most popular uses of lasers in ocular therapy involves the Nd:YAG laser in *posterior capsulotomy*, a procedure that ruptures internal ocular membranes. The procedures we have described under the headings of corneal sculpting (refractive surgery), radial keratotomy, and posterior capsulotomy are treated in greater detail in the sections that follow.

Corneal Sculpting.   Currently there are several procedures available for changing the shape of the cornea to correct for myopia, astigmatism, or hyperopia. These include *radial keratotomy* (RK) for myopia, *astigmatic keratotomy* (AK) for astigmatism, automated lamellar keratoplasty (ALK) for hyperopia, and photorefractive keratectomy (PRK) for myopia.

Radial keratotomy (RK) involves a series of precise, radial, spokelike cuts, made with a sharp diamond blade near the center of the cornea. The procedure leads to flattening desired for correction of myopia. Astigmatic keratotomy (AK) reduces astigmatism by rounding out the corneal curvature—making incisions on the steeper parts of the cornea to change the oblong shape into a more spherical one. Automated lamellar keratoplasty (ALK) is carried out with a precise instrument, called a microkeratone, which first lifts the corneal cap, then removes a pre-determined amount of underlying corneal tissue, and then replaces the corneal cap. With variations in this procedure, the cornea can be either flattened or steepened, thereby correcting for myopia or moderate hyperopia. Photorefractive keratectomy (PRK) uses a tightly focused, pulsed excimer laser to achieve the radial cuts obtained in radial keratotomy with the diamond blade.

Because radial keratotomy was the first procedure performed to alter the curvature of the cornea, and of some historical significance, we now take a closer look at this ground-breaking procedure.

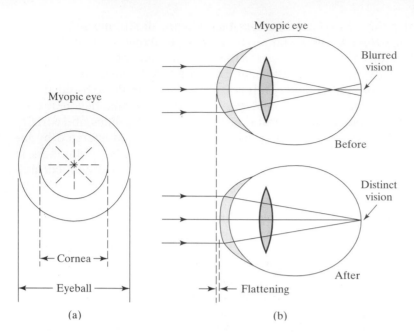

Myopic eye

Cornea

Eyeball

(a)

Myopic eye

Blurred vision

Before

Distinct vision

After

Flattening

(b)

Figure 12-21 Removing myopia by reshaping the cornea with radial keratotomy. (a) Frontal view showing radial cuts. (b) Side view before and after cuts.

Radial Keratotomy. As we have noted, radial keratotomy introduces radial cuts in the cornea of the elongated, myopic eyeball (see Figure 12-21). After the cuts have healed, the cornea flattens, thereby reducing the axial length of the eye. As a result, normal, or near-normal, vision is restored. This radical procedure, done mostly with a surgical blade in the hands of a skilled ophthalmologist, had its beginning in the Soviet Union in 1972. As the story is told, a rather myopic Soviet lad, Boris Petrov, was engaged in a schoolyard fight in Moscow. In the usual exchange, a hard punch struck one of the cold, thick lenses he was wearing, shattering it into multiple fragments. Some of the shards of glass, in one of those freak accidents of nature, embedded themselves in the lad's cornea. A Soviet ophthalmologist, Svyatoslav N. Fyodorov, who treated the youngster, did not hold out much hope for restored vision in the eye, even though the cuts were superficial. Astonishingly, though, as the corneal surface healed with all its scars, the cornea flattened out and most of the myopia disappeared. The lad saw better than ever before! Recognizing the significance of what he had witnessed, Fyodorov decided to replicate, under controlled conditions, what nature and a fist had accomplished so haphazardly. This he did, many times, in the procedure that has come to be known as radial keratotomy.

The laser was seen early on as a possible substitute for the diamond blade. In the beginning, the CO_2 laser, because of the high absorption of its 10.6 μm laser energy in water-laden corneal tissue, was considered a candidate for making the radial cuts on the cornea. However, due to the relatively long 10.6 μm wavelength of the CO_2 laser, compared with that of a pulsed excimer laser wavelength of nearly 0.2 μm, and the smaller focused spot achieved at lower wavelengths, the excimer laser is now the instrument of choice for photorefractive keratectomy. The surgical blade used in radial keratotomy is around 50 to 100 μm wide. CO_2 lasers at 10.6 μm can be focused down to less than 50 μm in width, and the excimer lasers to around 5 μm or less. Thus the excimer laser can become a very narrow "blade," making sharp, well-defined cuts in the cornea. The down side of the excimer laser, due to its low absorption in corneal tissue, is compensated for by using short pulses of high power density (power per unit area) to cut

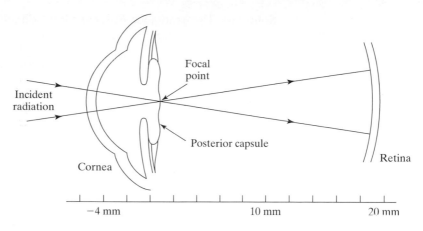

Figure 12-22 Side view of Nd:YAG laser beam focused on posterior capsule. (Reproduced with permission of Slack, Inc.)

the corneal tissue. The growing interest in photorefractive keratectomy is ample evidence that this laser procedure has arrived.

Posterior Capsulotomy. In a somewhat different procedure, identified as a posterior capsulotomy, a Nd:YAG laser is used to rupture unwanted, opacified membranes along the optical axis of the eye. The pulsed Nd:YAG laser emits a 1.06-μm laser beam that passes through the cornea, aqueous, and lens with relatively low absorption. The high-intensity laser beam penetrates the front part of the eye and comes to a sharp focus just beyond the implanted lens, some 4 mm from the front corneal surface (see Figure 12-22). Located there is the opacified membrane, the posterior segment of a weblike pillow (capsule) that previously contained the cataractous lens and now encases the lens implant.

Prior to laser surgery, the opacified membrane was removed invariably with invasive intervention by surgical instruments. In addition to the trauma involved with the operation, invasive surgery of the eye is always attended by possible introduction of foreign bodies and increased risk of infection. By contrast, the Nd:YAG laser surgery, performed on an outpatient basis in a matter of minutes, is neither traumatic nor infectious.[5] Posterior capsulotomy, carried out both in the United States and abroad as an outpatient procedure, focuses a short pulse of 1.06-μm laser radiation at the target. The typical pulse delivered by a high power Nd:YAG laser has an energy of 1 to 4 mJ and a pulse length of several nanoseconds. With a mode-locked laser, the pulse energy is about the same, but the pulse length is much shorter, only tens of picoseconds in duration. With such short times, the laser power, even for pulse energies as small as millijoules, reaches the megawatt range and higher.

❏ **Example**

Determine the irradiance (power per unit area) delivered by a Nd:YAG laser pulse of energy 4 mJ and pulse length 1 ns when the pulse is focused on target to a tiny spot of 30 μm diameter.

---

[5]Noninvasive laser intervention grew largely from the successful procedures developed by Daniele Aron-Rosa and coworkers at Trousseau Hospital in Paris and by Franz Fankhauser and colleagues at the University Eye Clinic in Bern, Switzerland.

Solution　The irradiance E_e is calculated directly as follows:

$$E_e = \frac{P}{A}, \quad \text{where } P = \frac{4 \times 10^{-3}\,\text{J}}{1 \times 10^{-9}\,\text{s}} = 4 \times 10^6\,\text{W}$$

and where

$$A = \frac{\pi D^2}{4} = \frac{(3.14)(30 \times 10^{-6})^2}{4} = 7.065 \times 10^{-10}\,\text{m}^2$$

Thus, the irradiance, or power density on target is

$$E = \frac{4 \times 10^6\,\text{W}}{7.065 \times 10^{-10}\,\text{m}^2} = 5.7 \times 10^{15}\,\text{W/m}^2 \quad \text{or} \quad 5.7 \times 10^{11}\,\text{W/cm}^2 \quad ■$$

As the example shows, when such high powers are focused sharply at the target in tiny spots ranging from 5 to 50 μm in diameter, irradiances, or power densities, of 10^{12} W/cm^2 are developed. Power densities of this magnitude are accompanied by very high electric fields that cause, first, a dielectric breakdown of optical tissue and, second, the formation of a plasma. The explosive growth of the plasma gives rise to a strong shock wave that travels radially outward, mechanically rupturing the nearby taut, opacified membrane. Neither the diverging laser radiation, which moves on toward the retina, nor the expanding shock wave, cause damage elsewhere in the eye.

An expanded view of these details is shown in Figure 12-23. The laser beam incident on the focusing lens has been beam-expanded by prior optics so that its beam divergence is very small (the beam is highly collimated). Because of this and the ability to focus highly collimated, coherent laser beams onto very small target areas, the beam converges to a "point" near the opacified membrane, producing the dielectric breakdown that ultimately leads to mechanical rupture of the membrane. The focusing optics brings the laser beam to a tight focus at the

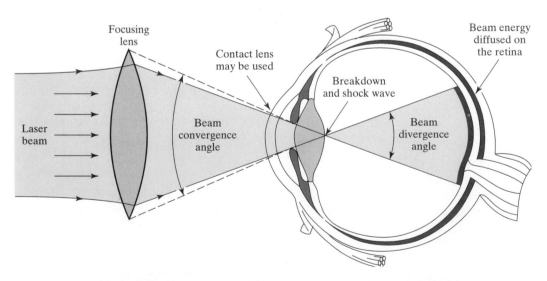

Figure 12-23　Focusing laser light to rupture membrane in posterior capsulotomy. The larger the beam convergence angle (about 17°)—limited by the pupil aperture—the larger the beam divergence angle and the less the effect of laser radiation on the retina.

target, and at the same time causes the laser beam to enter the eye with a relatively large beam convergence angle. By symmetry of angles, then, the beam expands rapidly past the target point, reaching the retina with values of harmless power density.

PROBLEMS

12-1. Why should one expect lasing at ultraviolet wavelengths to be more difficult to attain than lasing at infrared wavelengths? Develop your answer based on the ratio A_{21}/B_{21} and the meaning of the A_{21}, B_{21} coefficients. (*Hint:* See Eq. (12-1).)

12-2. Calculate the ratio of stimulated to spontaneous transitions for green light at 0.5 μm within a dense blackbody plasma at a temperature of 5000 K. That is, determine the value of the ratio

$$\frac{\text{Stimulated emission rate}}{\text{Spontaneous emission rate}} = \frac{B_{21}N_2\rho(f)}{A_{21}N_2}, \quad \text{where} \quad \frac{B_{21}}{A_{21}} = \frac{\lambda^3}{8\pi h}$$

What does the numerical value of this ratio imply?

12-3. (a) Given the center wavelength λ_0 and linewidth $\Delta\lambda$ (FWHM) for the three entries in Table 12-1 (ordinary discharge lamp, Cd low-pressure lamp, and He-Ne laser), verify that the FWHM frequency linewidth Δf (Hz) is as given in the table.

(b) How much "sharper" is a He-Ne laser line emission, for example, than a sodium line from a sodium discharge lamp?

12-4. Suppose that the coherence time of a light beam is roughly equal to the reciprocal of its frequency linewidth (FWHM). What then is the coherence time and coherence length of the He-Ne laser in Table 12-1?

12-5. A He-Ne laser has a beam waist (diameter) equal to about 1 mm. What is its beamspread angle in the far field?

12-6. Consider a broadband thermal source with a circular radiating surface of diameter 0.5 mm (roughly the size of beam waists in He-Ne lasers). Let the surface at a temperature of 1000 K emit light at 633 nm with a linewidth (FWHM) of 100 nm. Use Eq. (12-5) to show that the thermal photon output rate is about 5×10^9 photons/s.

12-7. Consider a 1-mW He-Ne laser emitting at 632.8 nm with a FWHM frequency linewidth of $\Delta f = 10^4$ Hz. Assume that the full-angle beam divergence ϕ in the far field is $\phi = 1.27\lambda/D$, where D is the diameter of the diffracting aperture (beam waist). See Figure 12-15.

(a) Show that the spectral radiance $\Delta L_e/\Delta f = \phi_e/(\Delta A \, \Delta\Omega \, \Delta f)$—where ϕ_e is the laser power, ΔA is the area of a radiating surface, $\Delta\Omega$ is the solid angle the laser radiates into, and Δf is the spectral bandwidth (FWHM) of the laser emission—is independent of the value of the diameter D of the diffracting aperture.

(b) Obtain a numerical value for the spectral radiance (in W/m²-sr-Hz) for the laser described above.

12-8. With reference to the beam expander shown in Figure 12-18, suppose $f_2/f_1 = 10$ and the beam divergence of the incident beam is 1 mrad.

(a) What is the beam divergence of the expanded beam?

(b) If the expanded beam is focused by a subsequent lens of power 10 diopters, what is the beam waist at the focal spot?

(c) If the power in the incident beam is 1 mW, what is the irradiance (W/m²) at the focal spot?

12-9. For a Nd:YAG laser, there are four pump levels located at 1.53 eV, 1.653 eV, 2.119 eV, and 2.361 eV above the ground state energy level.

(a) What is the wavelength associated with the photon energy required to populate each of the pump levels?

(b) Knowing that a Nd:YAG laser emits photons of wavelength 1.064 μm, determine the *quantum efficiency* associated with each of the four pump levels. (Note: The quantum efficiency is defined as the ratio of the energy of the laser photon to the energy of the pump photon.)

12-10. To operate a Nd:YAG laser, 2500 W of "wall-plug" power are required for a power supply that drives the arc lamps. The arc lamps provide pump energy to create the population inversion. The overall laser system, from power in (to the power supply) to power out (laser output beam), is characterized by the following component efficiencies:

80%—power supply operation
30%—arc lamps for pump light energy
70%—optical reflectors for concentrating pump light on laser rod
15%—for spectral match of pump light to Nd:YAG pump levels
50%—due to internal cavity/rod losses

    (a) Taking the efficiencies into account sequentially as they "occur," how much of the initial 2500 W is available for power in the output beam?
    (b) What is the overall operational efficiency (wall-plug efficiency) for this laser? (Note: *Wall-plug* efficiency is defined as the ratio of wall plug power in to laser power out.)

12-11. From an examination of Figure 12-19, determine which lasers would be suitable for **(a)** photocoagulation of bleeding vessels on the retina; **(b)** thermal cutting of corneal layers.

12-12. The CO_2 laser used to make corneal incisions has an average power of 5 W and a beam divergence of 2.2 mrad. After emerging from the laser, it is sent through a 5× beam expander and then focused onto the cornea by a 3.3-cm focal length germanium lens.
    (a) Why is germanium (and not glass) used as a lens material?
    (b) In Section 12-4, where *laser focusability* is discussed, it is shown that the angular beam spread (divergence) of an expanded laser beam is equal to the divergence of the incident beam divided by the beam expansion factor. What then is the beam divergence of this CO_2 laser beam after passing through the 5× beam expander?
    (c) Using the approximate formula, $D = f\phi$, where f is the focal length and ϕ is the beam divergence of the expanded beam, determine the diameter D of the focal spot on the cornea. (Refer to Section 12-4 for a discussion of this equation.)
    (d) What is the power density (irradiance) of the focused CO_2 laser beam on the cornea?

12-13. For the posterior capsulotomy surgery described in this chapter, the following data are typical:

Laser: Nd:YAG
Wavelength: 1.06 μm
Pulsewidth: 10 ns
Energy per pulse: 10 mJ
Beam divergence at focusing lens: 0.1 mrad
Power of focusing lens: 20 D

    (a) What is the average power per pulse?
    (b) What is the spot size of the focused Nd:YAG laser beam at the opaque membrane in the interior of the eye? (See Problem 12-12c.)
    (c) Assuming that none of the incident power is lost, what is the beam irradiance on target?

12-14. When working with lasers, one is admonished to "never stare into the beam." With low-power helium-neon lasers (milliwatt level), one might be tempted to consider this caution somewhat loosely. Even though the power in a laser beam may be fairly small, the focusing action of the eye can concentrate the laser power onto the retinal surface to create a large irradiance level. To illustrate this effect, consider a 4-mW He-Ne laser that emits a nearly collimated beam of 7 mm diameter at 632.8 nm, with a beam divergence of 1.5 mrad.
    (a) Calculate the irradiance of the beam in W/cm² just outside the laser.

(b) Assume that the entire beam enters a dark-adapted eye (pupil diameter of 7 mm) that is staring at the beam and focused at infinity. Treating the eye as a simple thin lens of 17 mm focal length, determine the irradiance of the focused spot on the retinal surface. (Refer to Problem 12-12c for help in calculating the size of the focused spot.)

(c) By approximately what factor does the focusing action of the eye increase the irradiance? What do you conclude about the seriousness of the admonition?

12-15. Laser protective eyewear filters are used by personnel who operate laser equipment. The optical density (OD) for a specific filter is given by the formula OD = $\log_{10}(E_p/\text{MPE})$, where E_p is the energy per unit area (J/cm^2) in the laser pulse and MPE is the "maximum permissible exposure" to the eye, also in units of J/cm^2. Suppose you are to determine an appropriate filter for use with a Nd:YAG laser (1.06 μm) that emits single laser pulses of 80 mJ. For safety reasons, this laser is rated as a Class IV laser with an MPE of 5.0×10^{-6} J/cm^2.

(a) Determine the energy per unit area in J/cm^2 of the 80-mJ laser emitting a 7-mm diameter laser pulse.

(b) What should be the OD of the protective filter that absorbs 1.06 μm radiation?

12-16. Which of the lasers listed in Table 12-2, based on wavelength considerations alone, would be reasonable candidates for use in focusing intense laser radiation into the vitreous chamber of the eye or onto the retina, if minimum absorption of this radiation is to occur in the cornea, aqueous region, and lens as the beam is directed centrally through the eye?

12-17. A pulsed erbium-YAG solid state laser delivers 200 μs pulses with an average power of 2 J/s at a wavelength of 2.94 μm and an average beam divergence of 8 mrad.

(a) What is the average power per pulse delivered by this laser?

(b) If a 5-cm focal length lens is available to focus this laser beam on a certain target where an irradiance of about 800×10^6 W/cm^2 is desired, what should be the beam expansion factor of a suitable beam expander to achieve this?

(c) Concerning the optical materials of the focusing lens and beam expander, explain an important factor to be considered in achieving optimum transmission of the laser energy.

12-18. Figure 12-20 gives the attenuation, in db/km, of EM radiation propagated through glass fibers for wavelengths from 0.80 to 1.60 μm.

(a) Based on laser parameters given in Table 12-2, identify those lasers that would have attenuation values of about 1.5 db/km or less.

(b) Using the power loss formula $P(z) = P_0 \times 10^{-\alpha_{db}z/10}$, where $P(z)$ is the power level (watts) at position z (km), α_{db} is the attenuation in db/km, and P_0 is the initial power, determine the fractional power transmission $P(z)/P_0$ after propagating 1500 m.

(c) Figure 12-20 indicates that the "ideal" laser wavelength for low-loss transmission through glass fibers is 1.55 μm with $\alpha_{db} \cong 0.6$ db/km. What percent of initial power remains after 1500 m of propagation?

REFERENCES

DePalma, Angelo. 1994. "Photonics Technology Will Surpass the Transistor," *Optics and Photonics News*, May.

Heavens, O. S., and R. W. Ditchburn. 1991. *Insight into Optics*. New York: John Wiley & Sons.

Hecht, E., and A. Zajac. 1979. *Optics*. Reading, Mass.: Addison-Wesley Publishing Company. Ch. 14.

Hitz, C. Breck. 1985. *Understanding Laser Technology*. Tulsa, Okla.: PennWell Publishing Company.

Iga, K. 1994. *Fundamentals of Laser Optics*. New York: Plenum Press. Ch 2, 3, 4.

Laufer, Gabriel. 1996. *Introduction to Optics and Lasers in Engineering*. New York: Cambridge University Press.

LENGYEL, B. A. 1971. *Lasers*, 2d ed. New York: John Wiley and Sons.

MÖLLER, K. D. 1988. *Optics*. Mill Valley, Calif.: University Science Books. Ch 15, 17.

O'SHEA, D. C., W. R. CALLEN, and W. T. RHODES. 1978. *Introduction to Lasers and Their Applications*. Reading, Mass.: Addison-Wesley Publishing Co.

SALEH, B. E. A., and M. C. TEICH. 1991. *Fundamentals of Photonics*. New York: John Wiley and Sons.

SIEGMAN, A. E. 1986. *Lasers*. Mill Valley, Calif.: University Science Books.

SLINEY, D. H., and M. L. WOLBARSHT. 1980. *Safety with Lasers and Other Optical Sources: A Comprehensive Handbook*. New York: Plenum Press.

THOMPSON, G. H. B. 1980. *Physics of Semiconductor Laser Devices*. New York: Wiley-Interscience.

UNGER, H. G. 1970. *Introduction to Quantum Electronics*. New York: Pergamon Press.

VERDEYEN, J. T. 1981. *Laser Electronics*. Englewood Cliffs, N.J.: Prentice-Hall.

WALDMAN, GARY. 1983. *Introduction to Light*. Englewood Cliffs, N.J.: Prentice Hall.

WELFORD, W. T. 1981. *Optics*, 2d ed. Oxford University Press.

The following are trade journals that serve the fields of optics and photonics. They are useful to those who wish to consult monthly updates on the latest in technology, devices, and advances in the field.

Photonics Spectra, Laurin Publishing Company, Pittsfield, Mass.

Laser Focus World, PennWell Publishing Company, Nashua, N.H.

13

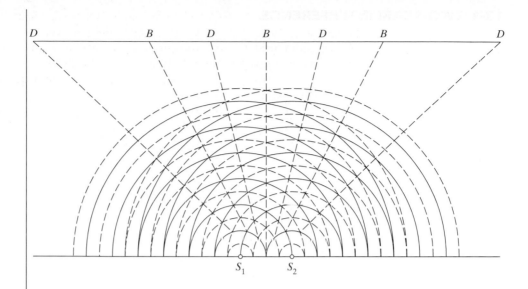

Interference Phenomena

INTRODUCTION

Like standing waves, treated in Chapter 11, the phenomenon of interference depends on the superposition of two or more individual waves under rather strict conditions that will soon be clarified. When interest lies primarily in the effects of increasing or decreasing the amplitude of the resultant light waves, due precisely to their superposition, these effects are usually said to be due to the interference of light. When increases in amplitude (*constructive interference*) and decreases in amplitude (*destructive interference*) occur alternately in a spatial display, the interference is said to produce a pattern of *fringes*, as in the double-slit interference pattern described later in this chapter. The same conditions may lead to the enhancement of one visible wavelength interval or color at the expense of the others, in which case interference colors are produced, as in oil slicks and soap films. The simplest explanation of these phenomena can be successfully undertaken by treating light as a wave motion. In this and following chapters, several such applications, considered under the general heading of interference, are presented.

13-1 TWO-BEAM INTERFERENCE

We consider first the interference of two waves, represented by their electric fields E_1 and E_2. We assume that the electric fields oscillate in the same, or nearly the same direction, so that we can treat them as scalar quantities. In cases of interference, the waves typically originate from a single source S (see Figure 13-1), separate, and reunite after traveling along different paths. At some general point P, reached by the two waves E_1 and E_2 after traveling total distances of x_1 and x_2, respectively, the waves interact to produce a disturbance whose electric field E_P is given by the principle of superposition. Thus,

$$E_P = E_1 + E_2$$

where

$$E_1 = E_{01} \cos(kx_1 - \omega t + \varepsilon_1) \tag{13-1}$$

$$E_2 = E_{02} \cos(kx_2 - \omega t + \varepsilon_2) \tag{13-2}$$

allowing for an initial phase difference of $(\varepsilon_1 - \varepsilon_2)$. The waves have the same frequency and wavelength and, thus, the same k.

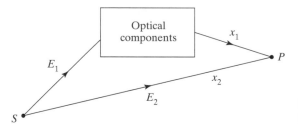

Figure 13-1 General scheme for observing interference.

Now E_1 and E_2 are rapidly varying functions with optical frequencies of the order of 10^{14} to 10^{15} Hz for visible light. Thus both E_1 and E_2 average to zero over very short time intervals. For example, E_1 oscillates between values of $+E_{01}$ and $-E_{01}$, with a time average of zero. However, notice that the *square* of E_1 varies between E_{01}^2 and zero, with a non-zero time average that is a measure of the energy of the wave. The radiant power density, or *irradiance*, E_e (in watts per m^2) is the quantity that measures the time average of the square of the wave amplitude. Measurement of the waves by their effect on the eye or some other light detector depends on the energy, and thus the irradiance, of the light beam. Unfortunately, the standard symbol for irradiance, except for subscript, is the same as that for the electric field. To avoid confusion, we use, temporarily, the symbol I for irradiance. Thus, arguing from Eqs. (11-17) with $\langle E^2 \rangle = \frac{1}{2}E_0^2$,

$$I = \varepsilon_0 c \langle E^2 \rangle \tag{13-3}$$

The notation $\langle E^2 \rangle$ stands for the time average of E^2, c is the speed of light in free space, and ε_0 is the permittivity of free space. In this case of 2-wave interference, the irradiance at P is given by

$$I_p = \varepsilon_0 c \langle E_P^2 \rangle = \varepsilon_0 c \langle (E_1 + E_2)^2 \rangle$$

or

$$I_p = \varepsilon_0 c \langle E_1^2 \rangle + \varepsilon_0 c \langle E_2^2 \rangle + 2\varepsilon_0 c \langle E_1 E_2 \rangle \tag{13-4}$$

By Eq. (13-3), the time averages of the first two terms of Eq. (13-4) correspond to the irradiances of the individual waves, I_1 and I_2. The last term depends on an interaction of the waves and is called the *interference term*, I_{12}. We may then write

$$I_p = I_1 + I_2 + I_{12} \tag{13-5}$$

If light behaved without interference, like classical particles, we would then expect simply, $I_p = I_1 + I_2$. The presence of the third term, I_{12} is indicative of the wave nature of light, which can produce increases or decreases of the irradiance through interference.

Consider now the interference term,

$$I_{12} = 2\varepsilon_0 c \langle E_1 E_2 \rangle \tag{13-6}$$

where E_1 and E_2 are given by Eqs. (13-1) and (13-2). Their product,

$$E_1 E_2 = E_{01} E_{02}[\cos (kx_1 - \omega t + \varepsilon_1)][\cos (kx_2 - \omega t + \varepsilon_2)]$$

can be simplified by first expanding the cosine factors, interpreted as the difference of two angles. For the sake of compactness, let us define

$$\alpha \equiv kx_1 + \varepsilon_1 \quad \text{and} \quad \beta \equiv kx_2 + \varepsilon_2$$

so that

$$E_1 E_2 = E_{01} E_{02} \cos (\alpha - \omega t) \cos (\beta - \omega t)$$

Expanding and multiplying the cosine factors, we arrive at

$$\langle E_1 E_2 \rangle = E_{01} E_{02}[\cos \alpha \cos \beta \langle \cos^2 \omega t \rangle + \sin \alpha \sin \beta \langle \sin^2 \omega t \rangle$$
$$+ (\cos \alpha \sin \beta + \sin \alpha \cos \beta)\langle \sin \omega t \cos \omega t \rangle]$$

where time averages are indicated for each time-dependent factor. Over any number of complete cycles, one can easily show that

$$\langle \cos^2 \omega t \rangle = \tfrac{1}{2}, \quad \langle \sin^2 \omega t \rangle = \tfrac{1}{2}, \quad \text{and} \quad \langle \sin \omega t \cos \omega t \rangle = 0$$

Thus,

$$\langle E_1 E_2 \rangle = \tfrac{1}{2} E_{01} E_{02}(\cos \alpha \cos \beta + \sin \alpha \sin \beta) = \tfrac{1}{2} E_{01} E_{02} \cos (\alpha - \beta)$$

or, writing out $(\alpha - \beta)$,

$$\langle E_1 E_2 \rangle = \tfrac{1}{2} E_{01} E_{02} \cos [k(x_1 - x_2) + \varepsilon_1 - \varepsilon_2)] \tag{13-7}$$

where the expression in square brackets is the total phase difference between E_1 and E_2, as given in Eqs. (13-1) and (13-2):

$$\delta = k(x_1 - x_2) + (\varepsilon_1 - \varepsilon_2) \tag{13-8}$$

Combining Eqs. (13-6), (13-7) and (13-8),

$$I_{12} = \varepsilon_0 c E_{01} E_{02} \cos \delta \tag{13-9}$$

Similarly, the irradiance terms I_1 and I_2 of Eq. (13-5) can be shown to yield

$$I_1 = \varepsilon_0 c \langle E_1^2 \rangle = \tfrac{1}{2} \varepsilon_0 c E_{01}^2 \tag{13-10}$$

$$I_2 = \varepsilon_0 c \langle E_2^2 \rangle = \tfrac{1}{2} \varepsilon_0 c E_{02}^2 \qquad (13\text{-}11)$$

so that we may write, finally,

$$I_p = I_1 + I_2 + 2\sqrt{I_1 I_2} \cos \delta \qquad (13\text{-}12)$$

Depending on whether $\cos \delta > 0$ or $\cos \delta < 0$ in Eq. (13-12), the interference term either augments or diminishes the sum of the individual irradiances I_1 and I_2, leading to constructive or destructive interference, respectively. On the other hand, if the initial phase difference $(\varepsilon_1 - \varepsilon_2)$ in Eq. (13-8) varies randomly, the waves are said to be mutually *incoherent*, and $\cos \delta$ becomes a time-dependent factor whose average is zero. Even though interference is always occurring, no pattern can be sustained long enough to be detected. Thus some degree of *coherence*, that is, $\langle \cos \delta \rangle \neq 0$, is necessary to observe interference. In particular, if the two waves originate from independent sources, such as incandescent bulbs or gas-discharge lamps, the waves are mutually incoherent and their interference is not detected. (Laser sources, even though independent, can possess sufficient mutual coherence for interference to be observed over short periods of time.) The other term in the total phase angle δ is, from Eq. (13-8), $k(x_1 - x_2)$. As the point of observation P determined by the distances x_1 and x_2 in Figure 13-1 varies, $\cos \delta$ takes on alternating maximum and minimum values and interference fringes occur, with spatially separated regions of maximum and minimum irradiance.

To be more specific, when $\cos \delta = +1$, constructive interference yields the maximum irradiance

$$I_{max} = I_1 + I_2 + 2\sqrt{I_1 I_2} \qquad (13\text{-}13)$$

This condition occurs whenever the phase difference $\delta = 2m\pi$, where m is any integer or zero. On the other hand, when $\cos \delta = -1$, destructive interference yields the minimum, or background irradiance

$$I_{min} = I_1 + I_2 - 2\sqrt{I_1 I_2} \qquad (13\text{-}14)$$

a condition that occurs whenever $\delta = 0, \pi, 3\pi, 5\pi, \ldots = (2m + 1)\pi$. A plot of irradiance I versus phase δ, in Figure 13-2a, exhibits periodic fringes. Destructive interference is complete, that is, cancellation is complete, when $I_1 = I_2 = I_0$. Then Eqs. (13-13) and (13-14) give

$$I_{max} = 4I_0 \quad \text{and} \quad I_{min} = 0$$

Resulting fringes, shown in Figure 13-2b, now exhibit better contrast. A measure of *fringe contrast* (also called *visibility*) with values between 0 and 1 is given by the quantity

$$\text{fringe contrast} = \frac{I_{max} - I_{min}}{I_{max} + I_{min}} \qquad (13\text{-}15)$$

In the experimental utilization of fringe patterns it is therefore usually desirable to arrange that the interfering beams E_1 and E_2 have the same amplitudes.

Another useful form of Eq. (13-12), for the case of interfering beams of equal amplitude I_0, is found by writing

$$I = I_0 + I_0 + 2\sqrt{I_0^2} \cos \delta = 2I_0(1 + \cos \delta)$$

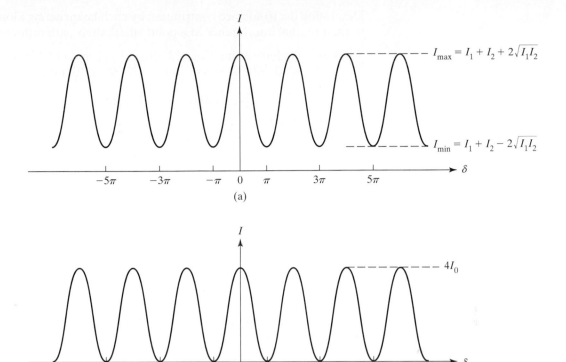

Figure 13-2 Irradiance of interference fringes as a function of phase. Fringe contrast is enhanced in (b), where the background irradiance $I_{min} = 0$ when $I_1 = I_2$.

and then making use of the trigonometric identity

$$1 + \cos \delta \equiv 2 \cos^2\left(\frac{\delta}{2}\right)$$

The irradiance for two equal interfering beams is then

$$I = 4I_0 \cos^2\left(\frac{\delta}{2}\right) \tag{13-16}$$

Notice that energy is not conserved at each point in the region of superposition, that is, $I \neq 2I_0$, but that over at least one spatial period of the fringe pattern, $I_{av} = 2I_0$. This situation is typical of interference and diffraction phenomena: If the power density falls below the average at some points, it rises above the average at other points in such a way that the total pattern satisfies the principle of energy conservation.

❑ **Example**

Two interfering beams with parallel electric fields are given by

$$E_1 = 2 \cos\left(kx_1 - \omega t + \frac{\pi}{3}\right) \qquad (\text{kV/m})$$

$$E_2 = 5 \cos\left(kx_2 - \omega t + \frac{\pi}{4}\right) \qquad (\text{kV/m})$$

Determine the irradiance contributed by each beam acting alone and that due to their mutual interference at a point where their path difference is zero.

Solution We have, using Eqs. (13-10), (13-11), and (13-12), with $\varepsilon_0 = 8.854 \times 10^{-12}\,(\text{C-s})^2/\text{kg-m}^3$ and $c = 2.998 \times 10^8$ m/s,

$$I_1 = \tfrac{1}{2}\varepsilon_0 c E_{01}^2 = \tfrac{1}{2}\varepsilon_0 c (2000)^2 = 5309\ \text{W/m}^2$$

$$I_2 = \tfrac{1}{2}\varepsilon_0 c E_{02}^2 = \tfrac{1}{2}\varepsilon_0 c (5000)^2 = 33{,}180\ \text{W/m}^2$$

$$I_{12} = 2\sqrt{I_1 I_2}\,\cos\delta = 2\sqrt{(5309 \times 33180)}\,\cos\!\left(\frac{\pi}{3} - \frac{\pi}{4}\right) = 25{,}640\ \text{W/m}^2$$

To find the fringe contrast near the region of superposition we must calculate

$$I_{\max} = I_1 + I_2 + 2\sqrt{I_1 I_2} = 5309 + 33180 + 2\sqrt{5309 \times 33180} = 65{,}034\ \text{W/m}^2$$

$$I_{\min} = I_1 + I_2 - 2\sqrt{I_1 I_2} = 5309 + 33180 - 2\sqrt{5309 \times 33180} = 11{,}945\ \text{W/m}^2$$

The contrast is then given by $(I_{\max} - I_{\min})/(I_{\max} + I_{\min})$, or

$$\text{fringe contrast} = \frac{65{,}034 - 11{,}945}{65{,}034 + 11{,}945} = 0.690$$

If the amplitudes of the two waves were equal, then $I_{\max} = 4I_0$, $I_{\min} = 0$, and the fringe contrast would be 1. ■

13-2 YOUNG'S DOUBLE-SLIT EXPERIMENT

The decisive experiment performed by Thomas Young in 1802 is shown schematically in Figure 13-3. Monochromatic light is first allowed to pass through a single small hole in an aperture in order to approximate a single point source S. The light spreads out in spherical waves from the source according to Huygens' prin-

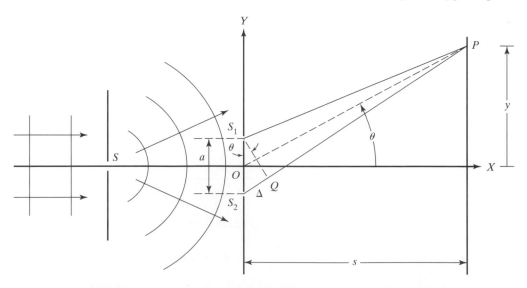

Figure 13-3 Schematic for Young's double-slit experiment. The holes S_1 and S_2 are usually slits, with the long dimension extending into the page.

ciple and is allowed to fall on the two closely spaced holes, S_1 and S_2, in an aperture. The holes thus become two coherent sources of light, whose interference can be observed on a screen some distance away. If the two holes are equal in size, light emanating from the holes have comparable amplitudes, and the irradiance at any point of superposition is given by Eq. (13-16), $I = 4I_0 \cos^2 (\delta/2)$. For observation points, such as P, located on the screen a distance s from the aperture, the phase difference δ between the two waves arriving at P must be determined to calculate the resultant irradiance there. Clearly, if

$$S_2P - S_1P = m\lambda \qquad m = 0, 1, 2, \ldots$$

the waves arrive in phase, and maximum irradiance or brightness results. If, on the other hand,

$$S_2P - S_1P = (m + \tfrac{1}{2}\lambda)$$

the requisite condition for destructive interference or darkness is met. Practically speaking, the hole separation a is much smaller than the screen distance s, allowing a simple expression for the path distance, $S_2P - S_1P$. Using P as a center, let an arc S_1Q be drawn of radius S_1P, so that it intersects the line S_2P at Q. Then $S_2P - S_1P$ is equal to the segment Δ, as shown. The first approximation is to regard arc S_1Q as a straight line segment that forms one leg of the right triangle, S_1S_2Q. If θ is the angle between the plane of the aperture and S_1Q, then $\Delta = a \sin \theta$. The second approximation identifies the angle θ with the angle between the optical axis OX and the line drawn from the point O midway between S_1 and S_2 to the point P at the screen. Observe that the corresponding sides of the two angles θ are related such that $OX \perp S_1S_2$ and OP is almost exactly perpendicular to S_1Q. The condition for constructive interference at a point P on the screen is then, to a very good approximation,

$$S_2P - S_1P = \Delta = m\lambda \cong a \sin \theta \qquad (13\text{-}17)$$

whereas for destructive interference,

$$\Delta = (m + \tfrac{1}{2})\lambda = a \sin \theta \qquad (13\text{-}18)$$

where m is zero or of integral value. The irradiance on the screen, at a point determined by the angle θ, is found using Eq. (13-16) and the relationship between path difference Δ and phase difference δ,

$$\delta = \left(\frac{2\pi}{\lambda}\right)\Delta$$

The result is

$$I = 4I_0 \cos^2\left(\frac{\pi\Delta}{\lambda}\right) = 4I_0 \cos^2\left(\frac{\pi a \sin \theta}{\lambda}\right)$$

For points P near the optical axis, where $y \ll s$, we may approximate further: $\sin \theta \cong \tan \theta \cong y/s$, so that I is given by the simplified form,

$$I = 4I_0 \cos^2\left(\frac{\pi a y}{\lambda s}\right) \qquad (13\text{-}19)$$

By allowing the cosine function in Eq. (13-19) to become alternately ± 1 and 0, the conditions expressed by Eqs. (13-17) and (13-18) for constructive and destructive interference are reproduced.

If we use $\sin \theta \cong y/s$ and Eq. (13-17), we see that bright fringes form at values of y along the screen given by

$$y_m = \frac{m\lambda s}{a}, \quad m = 0, 1, 2, \ldots \tag{13-20}$$

From Eq. (13-20) we find further that a constant separation Δy exists between irradiance maxima, corresponding to successive values of m given by

$$\Delta y = \frac{\lambda s}{a} \tag{13-21}$$

with minima situated midway between them. Thus fringe separation is proportional both to wavelength and screen distance and inversely proportional to the hole spacing. Reducing the hole spacing expands the fringe pattern formed by each color. Measurement of the fringe separation provides a means of determining the wavelength of the light. The single hole, used to secure a degree of spatial coherence, may be eliminated if laser light, both highly monochromatic and spatially coherent, is used to illuminate the double slit. In the observational arrangement just described, fringes are observed on a screen placed perpendicular to the optical axis at some distance from the aperture, as indicated in Figure 13-4. Fringe maxima coincide with integral orders of m, and fringe minima fall halfway between maxima.

❏ **Example**

Light from a narrow slit passes through two identical and parallel slits, 0.2 mm apart. Interference fringes are seen on a screen 1 m away, with a separation Δy of 3.29 mm. What is the wavelength of the light? How does the irradiance along the screen vary, if the contribution of either slit alone is I_0?

Solution   From Eq. (13-21),

$$\lambda = \frac{a\Delta y}{s} = \frac{(0.0002)(3.29 \times 10^{-3})}{1} = 658 \text{ nm}$$

According to Eq. (13-19), $I = 4I_0 \cos^2 (\pi a y/\lambda s)$. In this case,

$$I = 4I_0 \cos^2\left[\frac{\pi(0.0002)y}{(658 \times 10^{-9})(1)}\right] = 4I_0 \cos^2(955y) \qquad ■$$

An alternative way to view the formation of bright (B) positions of constructive interference and dark (D) positions of destructive interference is shown in Figure 13-5. The crests and valleys of spherical waves from S_1 and S_2 are shown approaching the screen. Along directions marked B, wave crests (or wave valleys) from both slits coincide, producing maximum irradiance. Along directions marked D, on the other hand, the waves are seen to be out of step by half a wavelength, and destructive interference results.

Obviously, fringes should be present in the entire half-space surrounding the slits, where light from the slits is allowed to interfere, though the irradiance

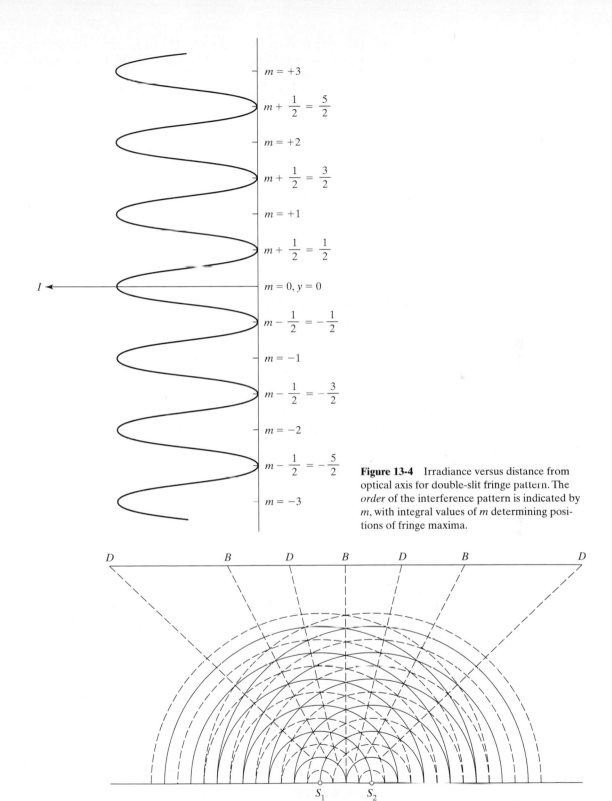

Figure 13-4 Irradiance versus distance from optical axis for double-slit fringe pattern. The *order* of the interference pattern is indicated by m, with integral values of m determining positions of fringe maxima.

$m = +3$

$m + \dfrac{1}{2} = \dfrac{5}{2}$

$m = +2$

$m + \dfrac{1}{2} = \dfrac{3}{2}$

$m = +1$

$m + \dfrac{1}{2} = \dfrac{1}{2}$

$m = 0, y = 0$

$m - \dfrac{1}{2} = -\dfrac{1}{2}$

$m = -1$

$m - \dfrac{1}{2} = -\dfrac{3}{2}$

$m = -2$

$m - \dfrac{1}{2} = -\dfrac{5}{2}$

$m = -3$

I

D B D B D B D

S_1 S_2

Figure 13-5 Alternating bright and dark interference fringes are produced by light from two coherent sources S_1 and S_2. Along directions where crests (solid circles) from S_1 intersect crests from S_2, brightness (B) results. Along directions where crests meet valleys (dashed circles), darkness (D) results.

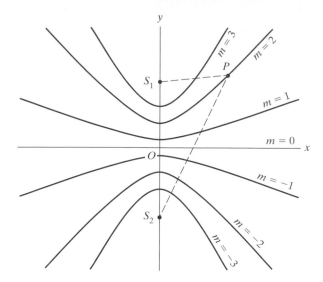

Figure 13-6 Bright fringe surfaces for two coherent point sources. The distances from S_1 and S_2 to any point P on a fringe differ by an integral number of wavelengths. The surfaces are generated by rotating the pattern about the y-axis.

is greatest in the forward direction. If we now imagine two coherent *point sources* S_1 and S_2 of light radiating in *all* directions, then the condition given by Eq. (13-17) for bright fringes,

$$S_2P - S_1P = m\lambda \tag{13-22}$$

defines a family of bright fringe surfaces in the space surrounding the holes. To visualize this set of surfaces, we may take advantage of the inherent symmetry in the arrangement. In Figure 13-6, the intersection of several bright fringe surfaces with a plane that includes the two sources is shown. Each surface corresponds to an integral value of order m. The surfaces are hyperbolic, since Eq. (13-22) is precisely the condition for a family of hyperbolic curves with parameter m. Inasmuch as the y-axis (or S_1S_2 axis) is an axis of symmetry, the corresponding bright fringe surfaces are generated by rotating the entire pattern about the y-axis. One should then be able to visualize the intercept of these three-dimensional surfaces with the plane of an observational screen placed anywhere in the vicinity. In particular, a screen placed perpendicular to the OX axis, as in Figure 13-3, intercepts hyperbolic arcs that appear as straight-line fringes near the axis, whereas a screen placed perpendicular to the OY axis shows concentric circular fringes centered on the axis. Because the fringe system extends throughout the space surrounding the two sources, the fringes are said to be *nonlocalized*.

The holes S, S_1, and S_2 of Figure 13-3 are usually replaced by parallel, narrow slits (oriented with their long sides perpendicular to the page in Figure 13-3) to illuminate more fully the interference pattern. The effect of the array of point sources along the slits, each set producing its own fringe system as just described, is simply to elongate the pattern parallel to the fringes, without changing their geometrical relationships. This is true even when two points along a source slit are not mutually coherent.

13-3 DOUBLE-SLIT INTERFERENCE WITH VIRTUAL SOURCES

Interference fringes may sometimes appear in arrangements when only one light source is present. It is possible, through reflection or refraction, to produce virtual images that, acting together or with the actual source, behave as two coherent

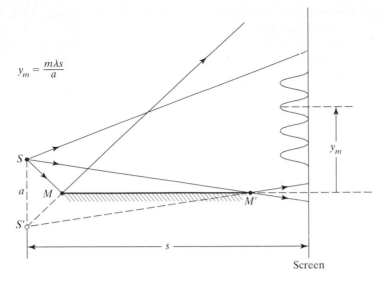

$$y_m = \frac{m\lambda s}{a}$$

Screen

Figure 13-7 Interference with Lloyd's mirror. Coherent sources are the point source S and its virtual image, S'.

sources that can produce an interference pattern. Figures 13-7, 13-8, and 13-9 illustrate three such examples. These examples are not only of some historic importance; they also serve to impress us with the variety of ways unexpected fringe patterns may appear in optical experiments, especially when the extremely coherent light of a laser is being used. In Figure 13-7, interference fringes are produced due to the superposition of light at the screen that originates at the actual source S and, by reflection, also originates effectively from its virtual source S' below the plane mirror MM'. Where the direct and reflected beams strike the screen, fringes appear. The position of bright fringes is given by the double slit Eq. (13-20), where a is equal to SS', that is, twice the distance of source S above the mirror plane. The arrangement is known as *Lloyd's mirror*. If the screen contacts the mirror at M', the fringe at their intersection is found to be dark. Since at this point the optical path difference between the two interfering beams vanishes, one should expect a bright fringe. The contrary experimental result is explained by requiring a phase shift of π radians for the air-glass reflection.

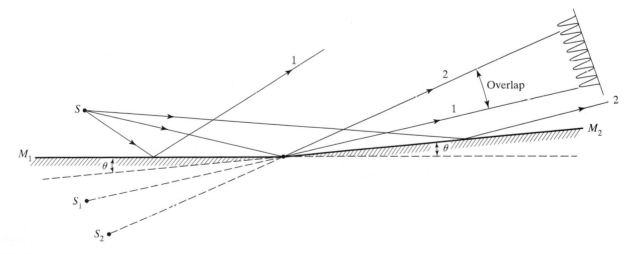

Figure 13-8 Interference with Fresnel's mirrors. Coherent sources are the two virtual images of point source S, formed in the two plane mirrors M_1 and M_2. Direct light from S is not allowed to reach the screen.

Another closely related arrangement is *Fresnel's mirrors*, Figure 13-8. Interference occurs between the light reflected from each of two mirrors M_1 and M_2, inclined at a small angle θ. Two representative rays, originating at the real point source S, are shown reflected from each mirror. The ray from S, directed toward the intersection of the mirrors, is shown reflecting as two rays from the two mirror surfaces on either side of the region of intersection. The pair of reflected rays labeled 1, if extended backward, intersect at the point S_1, their virtual source. Similarly, the pair of reflected rays labeled 2, if extended backward, intersect at the point S_2, their virtual source. Thus interference fringes appear in the region of overlap, as indicated. This is a case in which interference occurs between two coherent, virtual sources, as in a double slit configuration.

Figure 13-9 shows *Fresnel's biprism*, a thin prism of apex angle α that refracts light from a small source S in such a way that it appears to come from two coherent, virtual sources, S_1 and S_2, located to the rear of the biprism. The two rays labeled 1 are refracted by the prism such that they now appear to come from a virtual source S_1. Similarly, the two rays labeled 2 are refracted by the prism such that they appear to come from a virtual source S_2. Interference fringes due to the two virtual sources appear in the overlap region on the screen, as shown.

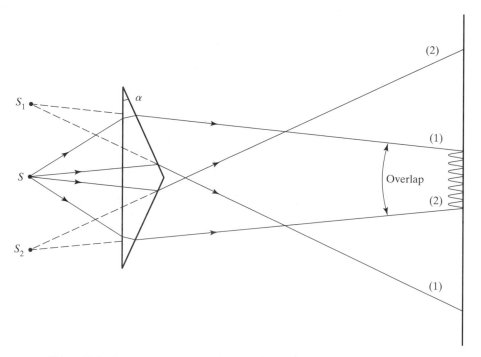

Figure 13-9 Interference with Fresnel's biprism. Coherent sources are the virtual images S_1 and S_2 of source S, formed by refraction in the two halves of the prism.

13-4 INTERFERENCE IN DIELECTRIC FILMS

The familiar appearance of colors on the surface of oily water and soap films and the beautiful iridescence often seen in mother-of-pearl, peacock feathers, and butterfly wings are associated with the interference of light in single or multiple thin surface layers of a transparent material. There exists a variety of situations in which such interference can take place, affecting the nature of the interference pattern and the conditions under which it can be observed. Variables in the situ-

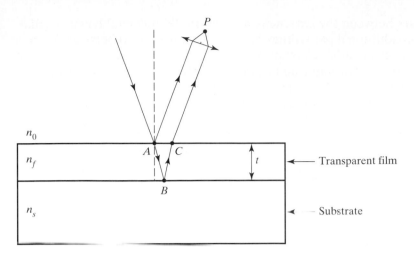

Figure 13-10 Double-beam interference from a film. Rays reflected from the top and bottom plane surfaces of the film are brought together at P by a lens.

ation include the size and spectral width of the source and the shape and reflectance of the film.

Consider the case of a film of transparent material bounded by parallel planes, such as might be formed by an oil slick, a metal oxide layer, or an evaporated coating on a flat, glass substrate (Figure 13-10). A beam of light incident on the film surface at A divides into reflected and refracted portions. This separation of the original light into two parts, preliminary to recombination and interference, is usually referred to as *amplitude division*, in contrast to a situation like Young's double slit in which separation is said to occur by *wave front division*. The refracted beam reflects again at the film-substrate interface at B and leaves the film at C, in the same direction as the beam reflected at A. Part of the beam may reflect internally again at C and continue to experience multiple reflections within the film layer until it has lost its intensity. There will thus exist multiple parallel beams emerging from the top surface, although with rapidly diminishing amplitudes. Unless the reflectance of the film is large, a good approximation to the more complex situation of multiple reflection is to consider only the first two emerging beams. The two parallel beams leaving the film at A and C can be brought together by a converging lens, the eye, for example. The two beams intersecting at P superpose and interfere. Since the two beams travel different paths from point A onward, a relative phase difference develops that can produce constructive or destructive interference at P. The optical path difference Δ, in the case of *normal incidence*, is the additional path length ABC traveled by the refracted ray times the refractive index of the film. Thus

$$\Delta = n(AB + BC) = n(2t) \qquad (13\text{-}23)$$

where t is the film thickness. For example, if $2nt = \lambda_0$, the wavelength of the light in vacuum, the two interfering beams on the basis of optical path difference alone would be in phase and produce constructive interference. However, an additional phase difference, due to the phenomenon of phase change on reflection, must be considered. Suppose that $n_f > n_0$ and $n_f > n_s$. In fact, often $n_0 = n_s$ because the media bounding the film are identical, as in the case of a water film (soap bubble) in air. Then the reflection at A occurs with light going from a lower index n_0 toward a higher index n_f, a condition usually called *external reflection*. The reflection at B, on the other hand, occurs for light going from a higher index n_f toward a lower index n_s, the condition of *internal reflection*. A relative phase shift of π occurs between the externally and internally reflected beams so that, equivalently, an additional path difference of $\lambda/2$ is introduced between the two beams. The net optical path difference between the beams is then $\lambda + \lambda/2$, which puts them precisely *out of phase* and destructive interference results at P. If, instead, both reflections are external ($n_0 < n_f < n_s$) or if both reflections are internal ($n_0 > n_f > n_s$), no relative phase difference due to reflection needs to be taken into account. In that case, constructive interference occurs at P.

A frequent use of such single-layer films is in the production of *antireflecting coatings* on optical surfaces, often found in lenses for cameras and binoculars. In most cases, the light enters the film from air so that $n_0 = 1$. Furthermore, if $n_s > n_f$, no relative phase shift between the two reflected beams occurs, and the optical path difference alone determines the type of interference to be expected. If the film thickness is $\lambda_f/4$, where λ_f is the wavelength of the light in the film, then $2t = \lambda_f/2$ and the optical path difference $2n_f t = \lambda_0/2$, since $\lambda_0 = n_f \lambda_f$. Destructive interference occurs at this wavelength and to some extent at neighboring wavelengths, which means that the light reflected from such a film is the light in the incident spectrum minus the light in the wavelength region around λ_0. If the incident light is white and λ_0 is in the visible region, the reflected light is colored. Extinction of a region of the spectrum by nonreflecting films of $\lambda/4$ thickness is, of course, more effective if the amplitudes of the two reflected beams are equal. In general, one can say that for constructive interference the two amplitudes add (being in phase), and for destructive interference the amplitudes subtract (being exactly out of phase). For the difference to be zero, that is, for destructive interference to be complete, the amplitudes must be equal. It can be shown that in the case of normal incidence, the *reflection coefficient r*—the ratio of reflected to incident electric field amplitudes—is given by

$$r = \frac{1 - n}{1 + n} \tag{13-24}$$

Here n is the *relative index* n_2/n_1. The amplitudes of the electric field reflected internally and externally from the film of Figure 13-10 are equal, assuming a nonabsorbing film, if the relative indices are equivalent for these cases, that is, if

$$\frac{n_f}{n_0} = \frac{n_s}{n_f} \quad \text{or} \quad n_f = \sqrt{n_0 n_s} \tag{13-25}$$

Since usually $n_0 = 1$, the requirement that reflected beams be of equal amplitude is met by choosing a film whose refractive index is the square root of the substrate's refractive index. A suitable film material for the application may or may not exist and some compromise is made. For example, to reduce the reflectance of lenses employed in optical instruments designed to handle white light, the film thickness of $\lambda_f/4$ is determined with a λ_f in the center of the visible spectrum—or wherever the detection system is most sensitive. In the case of the eye, this is the yellow-green portion near 550 nm. Assuming $n = 1.50$ for the glass lens, ideally

$$n_f = \sqrt{(1)(1.50)} = 1.22$$

The nearest practical film material with a matching index is MgF_2, with $n = 1.38$. The desired loss of reflected light near the middle of the spectrum results in a predominance of the blue and red ends of the spectrum, so that the coatings appear purple in reflected light.

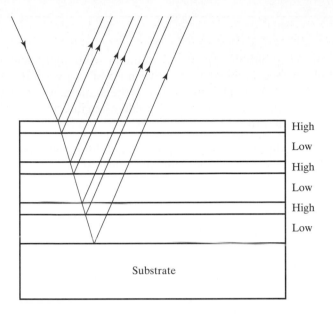

Figure 13-11 Multilayer dielectric mirror of alternating high and low index. Each film is $\lambda_f/4$ in optical thickness.

High
Low
High
Low
High
Low

Substrate

As another example, consider a multilayer stack of alternating high-low index dielectric films (Figure 13-11). If each film has an optical thickness of $\lambda_f/4$, a little analysis shows that in this case all emerging beams are in phase. Multiple reflections in the region of λ_0 increase the total reflected intensity and the quarter-wave stack performs as an efficient mirror. Such multilayer stacks can be designed to satisfy extinction or enhancement of reflected light over a greater portion of the spectrum than the single layer film.

Returning now to the single-layer film, we want first to generalize the condition for constructive and destructive interference by allowing for the case of incident rays that make an arbitrary angle with the normal to the film surface. When the incident rays are not normal to the film, the path difference between light reflected from the top and bottom of the film is no longer simply $2n_f t$. A combination of some geometry together with Snell's law shows that in this case,

$$\Delta = 2n_f t \cos \theta_t \qquad (13\text{-}26)$$

where the angle θ_t is the angle of refraction into the film. Of course this angle is related to the angle of incidence θ_i through Snell's law. Notice that for normal incidence, $\theta_i = \theta_t = 0$ and Eq. (13-26) reduces to $\Delta = 2n_f t$, as expected. The corresponding phase difference is $\delta = k\Delta = (2\pi/\lambda_0)\Delta$. The net phase difference must also take into account possible phase differences arising on reflection at interfaces separating media of different refractive indices, as discussed previously. Nevertheless, if we call Δ_p the optical path difference given by Eq. (13-26) and Δ_r the equivalent path difference arising from phase change on reflection, we can state quite generally the conditions for constructive interference:

$$\Delta_p + \Delta_r = m\lambda \qquad (13\text{-}27)$$

and destructive interference:

$$\Delta_p + \Delta_r = (m + \tfrac{1}{2})\lambda \qquad (13\text{-}28)$$

where $m = 0, 1, 2, \ldots$.

❏ **Example**

A portion of a soap bubble appears green under white light incident normally on the bubble surface. Taking the refractive index of soapy water to be near that of pure water ($n = 1.33$), and using 540 nm as the nominal wavelength for green light, estimate the thickness of the soap bubble when interference occurs in first order and in second order.

Solution The light reflected from the soap bubble undergoes a π-phase change at the first surface of the water film but not at the second. Thus the equivalent path difference of the reflected beams due to reflection alone is $\Delta_r = \lambda/2$. If green light is seen on reflection, this is a case of constructive interference at $\lambda = 540$ nm. According to Eq. (13-27),

$$\Delta_p + \Delta_r = m\lambda$$

where $\Delta_p = 2n_f t \cos \theta$ and $\Delta_r = \lambda/2$, we have

$$2n_f t \cos \theta_t + \frac{\lambda}{2} = m\lambda$$

or

$$2n_f t \cos \theta_t = (m - \tfrac{1}{2})\lambda$$

For normal incidence, $\theta_t = 0$. Assuming first order interference ($m = 1$),

$$2(1.33)(t)(1) = (1 - \tfrac{1}{2})(540)$$

with length units in nanometers. Thus, $t = 270/2.66 = 102$ nm. For second order ($m = 2$) interference, $t = 810/2.66 = 305$ nm. ■

13-5 FRINGES OF EQUAL THICKNESS

If the film is of varying thickness t, the optical path difference $\Delta = 2n_f t \cos \theta_t$ varies even without variation in the angle of incidence. Thus, if the direction of the incident light is fixed, say at normal incidence, a bright or dark fringe will be associated with a particular thickness for which Δ satisfies the condition for constructive or destructive interference, respectively. For this reason, fringes produced by a variable-thickness film are called *fringes of equal thickness*. A typical arrangement for viewing these fringes is shown in Figure 13-12a. An extended source is used in conjunction with a beam splitter set at an angle of 45° to the incident light. The beam splitter in this position enables light to strike the film at normal incidence, while at the same time providing for the transmission of part of the reflected light into the detector (eye). Fringes, often called *Fizeau fringes*, are seen by the eye as if localized at the film, from which the interfering rays diverge. At normal incidence, $\cos \theta_t = 1$ and $\Delta = 2n_f t$. Thus the condition for bright and dark fringes, Eqs (13-27) and (13-28), is

$$2n_f t + \Delta_r = \begin{cases} m\lambda, & \text{bright} \\ (m + 1/2)\lambda, & \text{dark} \end{cases} \tag{13-29}$$

where Δ_r is either $\lambda/2$ or zero, depending on whether there is or is not a relative phase shift of π between the rays reflected from the top and bottom surfaces of

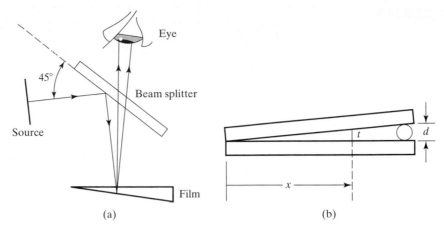

Figure 13-12 Interference from a wedge-shaped film, producing localized fringes of equal thickness. (a) Viewing assembly. (b) Air wedge formed with two microscope slides.

the film. One way of forming a suitable wedge for experimentation is to use two clean, glass microscope slides, wedged apart at one end by a thin spacer, perhaps a hair, as in Figure 13-12b. The resulting air layer between the slides produces Fizeau fringes when the slides are illuminated by monochromatic light. For this film, the two reflections are from glass to air (internal reflection) and from air to glass (external reflection), so that Δ_r in Eq. (13-29) is $\lambda/2$. As t increases in a linear fashion along the length of the slides from $t = 0$ to $t = d$, Eqs. (13-29) are satisfied for consecutive orders of m, and a series of equally spaced, alternating bright and dark fringes is seen at the air layer by reflected light. These fringes are virtual, localized fringes and cannot be projected on a screen.

If the extended source of Figure 13-12a is the sky and white light is incident at some angle on a film of variable thickness, as in Figure 13-13, the film may appear in a variety of colors, like the colors on an oil slick after a rain. Suppose that in a small region of the film the thickness is such as to produce constructive interference for wavelengths in the red portion of the spectrum at some order m. If the wavelengths at which constructive interference occurs again for orders $m + 1$ and $m - 1$ are outside the visible spectrum, the reflected light appears red. This can occur readily for low orders and therefore for thin films.

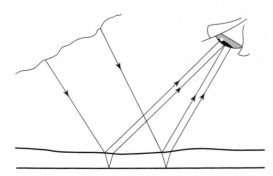

Figure 13-13 Interference by an irregular film illuminated by an extended source. Variations in film thickness, as well as angle of incidence, determine the wavelength region reinforced by interference.

PROBLEMS

13-1. Two beams having parallel electric fields are described by

$$E_1 = 3 \sin\left(kr_1 - \omega t + \frac{\pi}{5}\right)$$

$$E_2 = 4 \sin\left(kr_2 - \omega t + \frac{\pi}{6}\right)$$

with amplitudes in kV/m. The beams interfere at a point where the phase difference due to path is $\pi/3$ (the first beam having the longer path). At the point of superposition, calculate **(a)** the irradiances I_1 and I_2 of the individual beams; **(b)** the irradiance I_{12} due to their interference; **(c)** the net irradiance; **(d)** the fringe visibility.

13-2. Two harmonic light waves with amplitudes of 1.6 and 2.8 interfere at some point on a screen. What fringe contrast or visibility results there if the two electric field vectors are parallel and if they are perpendicular?

13-3. The ratio of the amplitudes of two beams forming an interference fringe pattern is 2/1. What is the fringe contrast? What ratio of amplitudes produces a fringe contrast of 0.5?

13-4. (a) Show that if one beam of a two-beam interference setup has an irradiance of N times that of the other beam, the fringe visibility is given by

$$V = \frac{2\sqrt{N}}{N + 1}$$

(b) Determine the beam irradiance ratios for visibilities of 0.96, 0.9, 0.8, and 0.5.

13-5. A mercury source of light is positioned behind a glass filter, which allows transmission of the 546.1-nm green light from the source. The light is allowed to pass through a narrow, horizontal slit positioned 1 mm above a flat mirror surface. Describe both qualitatively and quantitatively what appears on a screen 1 m away from the slit.

13-6. Two slits are illuminated by light that consists of two wavelengths. One wavelength is known to be 436 nm. On a screen, the fourth minimum of the 436-nm light coincides with the third maximum of the other light. What is the wavelength of the unknown light?

13-7. In a Young's experiment, narrow double slits 0.2 mm apart diffract monochromatic light onto a screen 1.5 m away. The distance between the fifth minima on either side of the zeroth-order maximum is measured to be 34.73 mm. Determine the wavelength of the light.

13-8. A quasi-monochromatic beam of light illuminates Young's double-slit setup, generating a fringe pattern having 5.6-mm separation between consecutive dark bands. The distance between the plane containing the apertures and the plane of observation is 10 m, and the two slits are separated by 1.0 mm. Sketch the experimental arrangement. Why is an initial single slit necessary? What is the wavelength of the light?

13-9. In an interference experiment of the Young type, the distance between slits is 0.5 mm, and the wavelength of the light is 600 nm.
(a) If it is desired to have a fringe spacing of 1 mm at the screen, what is the proper screen distance?
(b) If a thin plate of glass ($n = 1.50$) of thickness 100 micrometers is placed over one of the slits, what is the lateral fringe displacement at the screen?
(c) What path difference corresponds to a shift in the fringe pattern from a peak maximum to the (same) peak half-maximum?

13-10. Light of continuously variable wavelength illuminates normally a thin oil (index of 1.30) film on a glass surface. Extinction of the reflected light is observed to occur at wavelengths of 525 and 675 nm in the visible spectrum. Determine the thickness of the oil film and the orders of the interference.

13-11. A thin film of MgF$_2$ ($n = 1.38$) is deposited on glass so that it is antireflecting at a wavelength of 580 nm under normal incidence. What wavelength is minimally reflected when the light is incident instead at 45°?

13-12. A nonreflecting, single layer of a lens coating is to be deposited on a lens of refractive index $n = 1.78$. Determine the refractive index of coating material and the thickness required to produce zero reflection for light of wavelength 550 nm.

13-13. Remember that the energy of a light beam is proportional to the square of its amplitude.

   (a) Determine the percentage of light energy reflected in air from a single surface separating a material of index 1.40 for light of $\lambda = 500$ nm.

   (b) When deposited on glass of index 1.60, how thick should a film of this material be in order to reduce the reflected energy by destructive interference?

   (c) What is then the effective percent reflection from the film layer?

13-14. A soap film is formed using a rectangular wire frame and held in a vertical plane. When illuminated normally by laser light at 632.8 nm, one sees a series of localized interference fringes that measure 15 per cm. Explain their formation.

13-15. A beam of white light (a continuous spectrum from 400 to 700 nm, let us say) is incident at an angle of 45° on two parallel glass plates separated by an air film 0.001 cm thick. The reflected light is admitted into a prism spectroscope. How many dark "lines" are seen across the entire spectrum?

13-16. Two microscope slides are placed together but held apart at one end by a thin piece of tin foil. Under sodium light (589 nm) normally incident on the air film formed between the slides, one observes exactly 40 bright fringes from the edges in contact to the edge of the tin foil. Determine the thickness of the foil.

13-17. Plane plates of glass are in contact along one side and held apart by a wire 0.05 mm in diameter, parallel to the edge in contact, and 20 cm distant. Using filtered green mercury light, directed normally on the air film between plates, interference fringes are seen. Calculate the separation of the dark fringes. How many dark fringes appear between the edge and the wire?

REFERENCES

BAUMEISTER, PHILIP, and GERALD PINCUS. 1970. "Optical Interference Coatings." *Scientific American* (Dec.): 58.

FEYNMAN, RICHARD P., ROBERT B. LEIGHTON, and MATTHEW SANDS. 1975. *The Feynman Lectures in Physics*, vol. 1. Reading, Mass.: Addison-Wesley Publishing Company. Ch. 28, 29.

FINCHAM, W. H. A., and M. H. FREEMAN. 1980. *Optics*, 9th ed. London: Butterworths. Ch. 14.

GHATAK, AJOY K. 1972. *An Introduction to Modern Optics*. New York: McGraw-Hill Book Company. Ch. 4.

HEAVENS, O. S., and R. W. DITCHBURN. 1991. *Optics*. New York: John Wiley & Sons.

HECHT, EUGENE, and ALFRED ZAJAC. 1974. *Optics*. Reading, Mass.: Addison-Wesley Publishing Company. Ch. 9.

KLEIN, MILES V. 1970. *Optics*. New York: John Wiley and Sons. Ch. 5.

LONGHURST, R. S. 1967. *Geometrical and Physical Optics*, 2d ed. New York: John Wiley and Sons. Ch. 7, 8.

WELFORD, W. T. 1981. *Optics*, 2d ed. Oxford University Press.

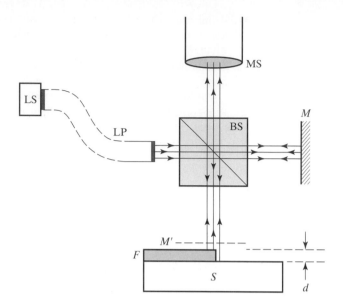

14

Interference Applications

INTRODUCTION

An instrument designed to exploit the interference of light and the fringe patterns that result from optical path differences in any of a variety of ways is called an *optical interferometer*. This general description of the instrument should reflect the wide variety of designs and uses of interferometers. Applications extend also to acoustic and radio waves, but here we are interested in the optical region. In this chapter we discuss first an apparatus designed to view *Newton's rings* and its application to the measurement of thin films. This is followed by a description of the prototype of interferometers, the *Michelson interferometer*, together with an indication of only a few of its many applications. Finally, we discuss the elements of *holographic* techniques, which are also dependent on the phenomenon of optical interference.

14-1 NEWTON'S RINGS

Fizeau fringes, as described in the previous chapter, are fringes of equal thickness associated with interference of a thin film. The fringe contours thus reveal directly any nonuniformities in the thickness of the film. Figure 14-1 shows an arrangement in which the Fizeau fringes produced are referred to as *Newton's rings*. The arrangement can be put to practical use to determine, for example, the

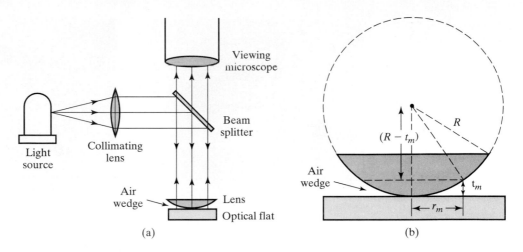

Figure 14-1 (a) Newton's rings apparatus. Interference fringes of equal thickness are produced by the air wedge between lens and optical flat. (b) Essential geometry for production of Newton's rings.

quality of the spherical surface of a lens. An air wedge, formed between the spherical surface of a lens and an optically flat surface, as shown in Figure 14-1a, is illuminated with normally incident monochromatic light. The light might be from a sodium lamp or a mercury lamp fitted with a filter to isolate one of its spectral lines. Interference between light reflected from the top and bottom of the air film produces the Fizeau fringes. For a perfectly spherical surface resting on a flat surface, the equal-thickness contours—and therefore the fringes viewed—are concentric circles around the point of contact with the optical flat. At that point, $t = 0$ and the path difference Δ_r between reflected rays is $\lambda/2$, as a result of the phase change on reflection from the top of the optical flat. The center of the fringe pattern thus appears dark, and Eq. (13-29), with $n_f = 1$ for air,

$$2t + \Delta_r = (m + \tfrac{1}{2})\lambda$$

gives $m = 0$ for the order of the destructive interference. Irregularities in the surface of the lens show up as distortions in the concentric ring pattern. This arrangement can also be used as an optical means of measuring the radius of curvature of the lens surface. A geometrical relation exists between the radius r_m of the mth-order dark fringe, the corresponding air-film thickness t_m, and the radius of curvature R of the air film or the lens surface. Referring to Figure 14-1b and making use of the Pythagorean theorem, we have

$$R^2 = r_m^2 + (R - t_m)^2$$

or

$$R = \frac{r_m^2 + t_m^2}{2t_m} \tag{14-1}$$

The radius r_m of the mth dark ring is measured and the corresponding thickness t_m of air is determined from the interference condition of Eq. (13-29). Thus R can be found from Eq. (14-1). A little thought should convince one that light transmitted through the optical flat will also show circular interference fringes. As shown in Figure 14-2, this transmitted pattern (Figure 14-2b) differs

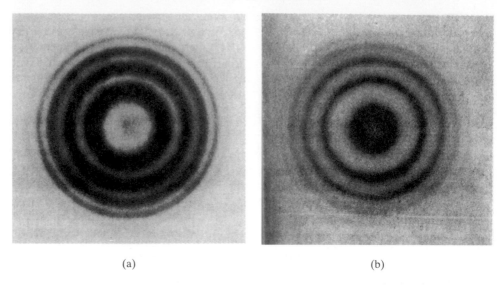

(a)   (b)

Figure 14-2 Newton's rings in (a) reflected light, and (b) transmitted light are complementary. (From M. Cagnet, M. Francon, and J. C. Thrierr, *Atlas of Optical Phenomenon*, Plate 9, Berlin: Springer-Verlag, © 1962. Used by permission.)

in two important respects from the Fizeau fringes formed by reflection (Figure 14-2a). First, the fringes show poor contrast, because the two transmitted beams with largest amplitudes have quite different values and result in incomplete cancellation. Second, the center of the fringe pattern is bright rather than dark, and the entire fringe system by transmission is complementary to the system by reflection.

❑ **Example**

A plano-convex lens ($n = 1.523$) of $\frac{1}{8}$ diopter power is placed, convex surface down, on an optically flat surface. Using a traveling microscope and sodium light ($\lambda = 589.3$ nm), interference fringes are observed. Determine the radii of the first and tenth dark rings.

Solution In this case, $\Delta_r = \lambda/2$, so that the equation

$$2t_m + \Delta_r = (m + \tfrac{1}{2})\lambda$$

leads to an air-film thickness at the mth dark ring given by $t_m = m\lambda/2$. The ring radii r_m are given by Eq. (14-1). On neglecting the very small term in t_m^2, this is $r_m^2 = 2Rt_m$. The radius of curvature of the convex surface of the lens is found from the lensmaker's equation:

$$\frac{1}{f} = (n - 1)\left(\frac{1}{R_1} - \frac{1}{R_2}\right)$$

With $f = 8$ m, $n = 1.523$, and $R_2 \to \infty$, this gives $R_1 = R = 4.184$ m. Then for the mth ring, in general, we have

$$r_m^2 = 2Rt_m = 2R\left(\frac{m\lambda}{2}\right) = mR\lambda$$

For the first ($m = 1$) dark ring,

$$r_1^2 = (1)(4.184)(589.3 \times 10^{-9}) = 2.466 \times 10^{-6} \text{ m}^2$$

and for the tenth ($m = 10$) dark ring,

$$r_{10}^2 = (10)(4.184)(589.3 \times 10^{-9}) = 24.66 \times 10^{-6} \text{ m}^2$$

or $r_1 = 1.57$ mm and $r_{10} = 4.97$ mm. ∎

It is ironic that the phenomenon we have been describing, involving so intimately the wave nature of light, should be known as Newton's rings after one who championed the corpuscular theory of light. Probably the first measurement of the wavelength of light was made by Newton, using this technique. Consistent with his corpuscular theory, however, Newton interpreted this quantity as a measurement of the distance between the "easy fits of reflection" of light corpuscles.

14-2 FILM-THICKNESS MEASUREMENT BY INTERFERENCE

Fringes of equal thickness also provide a sensitive optical means for measuring thin films. A sketch of one possible arrangement is shown in Figure 14-3. Suppose the film F to be measured has a thickness d. The film has been deposited on some substrate S. Monochromatic light is channeled from a light source LS through a fiber-optic light pipe LP to a right-angle beam-splitting prism BS, which transmits one beam to a flat mirror M and reflects the other to the film surface. After reflection, each beam is transmitted by the beam splitter into a microscope MS, where they are allowed to interfere. Equivalently, the beam reflected from the mirror M can be considered to arise from its virtual image M'. The virtual mirror M' is constructed by imaging M through the beam-splitter reflecting plane. This construction makes it clear that the interference pattern results from interference due to the air film between the reflecting plane at M' and the film F. In practice, mirror M can be moved toward or away from the beam splitter to equalize optical path lengths and can be tilted to make M' more or less parallel to the

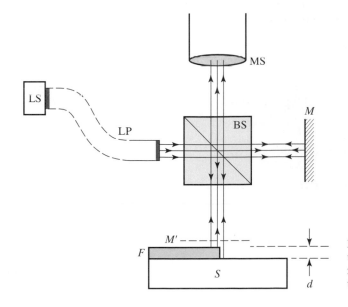

Figure 14-3 Film-thickness measurement. Interference fringes produced by light reflected from the film surface and substrate allow a determination of the film thickness d.

film surface. Furthermore, the beam splitter and mirror assembly form one unit that can be attached to the microscope in place of its objective lens. When M' and the film surface are not precisely parallel, the usual Fizeau fringes due to a wedge (Section 13-5) are seen through the microscope, which has been prefocused on the film. The light beam striking the film is allowed to cover the edge of the film F, so that two fringe systems are seen side by side, corresponding to air films that differ by the required film thickness at their juncture. Figure 14-4a shows a typical photograph of the fringe systems, made through a microscope. The translation of one fringe system relative to the other provides a means of determining d, as follows. For normal incidence, bright fringes satisfy Eq. (13-27),

$$\Delta_p + \Delta_r = 2nt + \Delta_r = m\lambda$$

where t represents the thickness of the air film at some point. If the air-film thickness t now changes by an amount $\Delta t = d$, the order of interference m changes accordingly. Since the change in the term $2nt$ on the left side of the equation must be matched by a change in the term $m\lambda$ on the right side, we have

$$2n\,\Delta t = 2d = (\Delta m)\lambda$$

where we have set $n = 1$ for an air film. Increasing the thickness t by $\lambda/2$, for example, changes the order of any fringe by $\Delta m = 1$, that is, the fringe pattern translates by one whole fringe. For a shift of fringes of magnitude Δx (Figure 14-4b) the change in m is given by $\Delta m = \Delta x/x$, resulting in

$$d = \left(\frac{\Delta x}{x}\right)\left(\frac{\lambda}{2}\right) \tag{14-2}$$

Since both fringe spacing x and fringe shift Δx can be measured with a stable microscope—or from a photograph like that of Figure 14-4—the film thickness

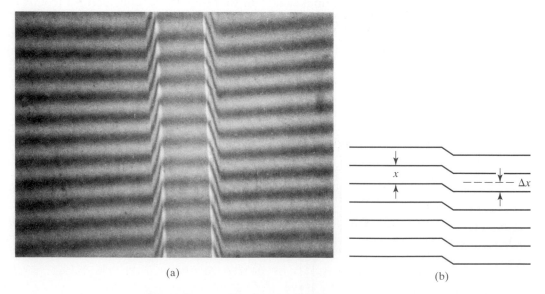

(a)　　　　　　　　　　　　　　　　　　(b)

Figure 14-4 (a) Photograph of interference fringes produced by the arrangement shown in Figure 14-3. The troughlike depression evident in the interference pattern was made by evaporating the film over a thin, straight wire. (b) Sketch of one side of the trough shown in the photo. The fringe pattern shifts by an amount Δx at the film edge. (Used by permission of J. Feldott.)

d is determined. When using monochromatic light, the net shift of fringe systems is ambiguous because a shift $\Delta x = 0.5x$, for example, will look exactly like a shift $\Delta x = 1.5x$. This ambiguity may be removed in one of two ways. If the shift is more than one fringe width, this situation is apparent when viewing white-light fringes formed in the same way. The superposition of colors that form the white-light fringes creates a pattern whose center at $m = 0$ is unique, serving as an unambiguous index of fringe location. The integral shift of fringe patterns is then easily seen and can be combined with the monochromatic measurement Δx described previously. A second method is to prepare the film so that its edge is not sharp but tails off gradually. In this case each fringe of one set can be followed down the film edge into the corresponding fringe of the second set, as in Figure 14-4. If the film cannot be provided with a gradually tailing edge, a thin film of silver, for example, can be evaporated over both the film and substrate. The step in the metal film will usually be somewhat sloped, but the total step will be the same as the thickness of the film to be measured. A one-to-one correspondence between individual fringes of each set can then be made visually.

14-3 THE MICHELSON INTERFEROMETER

The Michelson interferometer makes use of one of many designs to accomplish precise measurements based on the interference of light, in which the basic measuring stick is the wavelength of light. To achieve interference between two coherent beams of light, an interferometer divides an initial beam into two or more parts that travel diverse optical paths and then reunite to produce an interference pattern. One criterion for broadly classifying interferometers distinguishes the manner in which the initial beam is separated. *Wave front division interferometers* sample portions of the same wave front of a coherent beam of light, as in the case of Young's double slit, or adaptations like those using Lloyd's mirror or Fresnel's biprism. *Amplitude division interferometers* instead use some type of *beam splitter* that divides the initial beam into two parts. The Michelson interferometer is of this type. Usually the beam splitting is managed by a semireflecting metallic or dielectric film; it can be accomplished by other means as well, including double refraction and diffraction. Another means of classification distinguishes between those interferometers that make use of the interference of *two beams*, as in the case of the Michelson interferometer, and other interferometers that operate with *multiple beams*.

The Michelson interferometer, first introduced by Albert Michelson in 1881, has played a vital role in the development of modern physics. This simple and versatile instrument was used, for example, to establish experimental evidence for the validity of the special theory of relativity, to detect and measure hyperfine structure in line spectra, to measure the tidal effect of the Moon on Earth, and to provide a substitute standard for the meter in terms of wavelengths of light. Michelson himself pioneered much of this work. A schematic of the Michelson interferometer is shown in Figure 14-5a. From an extended source of light S, a beam of light (labeled 1) is split by a beam splitter (BS) by means of a thin, semitransparent front-surface metallic or dielectric film, deposited on glass. The interferometer is therefore of the amplitude-splitting type. Reflected beam 2 and transmitted beam 3, of roughly equal amplitudes, continue to fully-reflecting mirrors M2 and M1, respectively, where their directions are reversed. On returning to the beam splitter, beam 2 is now partly transmitted and beam 3 is partly reflected by the semitransparent film so that they come together again and

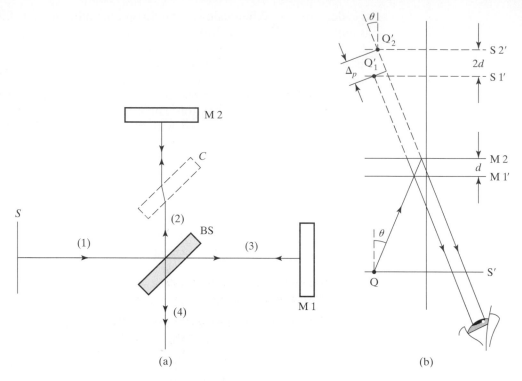

Figure 14-5 (a) The Michelson interferometer. (b) Equivalent optics for the Michelson interferometer.

leave the interferometer as beam 4. The useful aperture of this double-beam interferometer is such that all rays striking M1 and M2 will be normal or nearly so. Thus beam 4 includes rays that have traveled different optical paths and demonstrates interference. At least one of the mirrors is equipped with tilting adjustment screws that allow the surface of M1 to be made perpendicular to that of M2. One of the mirrors is also movable along the direction of the beam by means of an accurate track and micrometer screw. In this way the difference between the optical paths of beams 2 and 3 can be gradually varied. Notice that beam 3 traverses the beam splitter three times before leaving the beam splitter as part of beam 4, whereas beam 2 traverses it only once. In some applications, where white light is used, it is essential that the optical paths of the two beams be made precisely equal. Although this can be accomplished at one wavelength by appropriately increasing the distance of M2 from BS, the correction would not suffice at another wavelength because of the dispersion of the glass. To compensate for all wavelengths at once, a plate C, made of the same material and dimensions as BS, is inserted parallel to BS in the path of beam 2. Any small, remaining inequalities in optical paths can be removed by allowing the compensator to rotate, thus varying the optical path through the thickness of the glass plate.

The actual interferometer in Figure 14-5a possesses two optical axes at right angles to one another. A simpler but equivalent optical system, having a single optical axis, can be displayed by working with virtual images of source S and mirror M1 via reflection in the BS mirror. These positions are most simply found by regarding the assembly including S, M1, and beams 1 and 3 of Figure 14-5a as rotated counterclockwise by 90° about the point of intersection of the beams with the BS mirror. The resulting geometry is shown in Figure 14-5b. The new position of the source plane is S', and the new position of the mirror M1 is M1'. Light from a point Q on the source plane S' then effectively reflects from both mirrors M2 and M1', shown parallel and with an optical path difference of d. The two reflected beams appear to come from the two virtual images, Q_1' and Q_2', of object point Q. Since the images S1' and S2', formed by the reflection of the source plane S in the mirrors M1' and M2 respectively, must be separated by twice the mirror separation d, the distance between Q1' and Q2' is $2d$, and the optical path difference between the two beams emerging from the interferometer is

$$\Delta_p = 2d \cos \theta \tag{14-3}$$

where the angle θ measures the inclination of the beams relative to the optical axis. For a normal beam, $\theta = 0$ and $\Delta_p = 2d$. We expect this result since, if one mirror is farther from BS than the other by a distance d, the extra distance traversed by the beam taking the longer route includes distance d twice, once before and once after reflection. If, in addition, $\Delta = m\lambda$, so that the two beams interfere constructively, it follows that they will do so repeatedly for every $\lambda/2$ translation of one of the mirrors.

The optical system of Figure 14-5b is now equivalent to the case of interference due to a plane, parallel air film, illuminated by an extended source. Virtual fringes of equal inclination may be seen by looking into the beam splitter along ray 4, with the eye or a telescope focused at infinity. Assuming that the two interfering beams are of equal amplitude, the irradiance of the fringe system of circles concentric with the optical axis, is given, as in Eq. (13-16), by

$$I = 4I_0 \cos^2\left(\frac{\delta}{2}\right) \tag{14-4}$$

where the phase difference is

$$\delta = k\Delta = \left(\frac{2\pi}{\lambda}\right)\Delta \tag{14-5}$$

The net optical path difference is $\Delta = \Delta_p + \Delta_r$, as usual. A relative π phase shift (equivalent to a $\lambda/2$ path difference) between the two beams occurs because beam 2 experiences two external reflections but beam 3 experiences only one. For dark fringes, then,

$$\Delta_p + \Delta_r = 2d \cos \theta + \frac{\lambda}{2} = \left(m + \frac{1}{2}\right)\lambda$$

or, more simply,

$$2d \cos \theta = m\lambda, \quad m = 0, 1, 2, \ldots \quad \text{dark fringes} \tag{14-6}$$

If d is of such magnitude that the normal rays forming the center of the fringe system satisfy Eq. (14-6), that is, the center fringe is dark, then its order, given by

$$m_{\text{max}} = \frac{2d}{\lambda} \tag{14-7}$$

is a large integer. Neighboring dark fringes decrease in order outward from the center of the pattern, as $\cos\theta$ decreases from its maximum value of 1. This ordering of fringes may be inverted for convenience by associating another integer p with each fringe of order m, where

$$p = m_{max} - m = \frac{2d}{\lambda} - m \qquad (14\text{-}8)$$

Using Eq. (14-8) to replace m in Eq. (14-6), we arrive at

$$p\lambda = 2d(1 - \cos\theta), \quad p = 0, 1, 2, \ldots \quad \text{dark fringes} \qquad (14\text{-}9)$$

where now the central fringe is of order zero and the neighboring fringes increase in order, outward from the center. Figure 14-6 illustrates the relationship between orders m and p for the arbitrary case where $m_{max} = 100$. Eq. (14-6) or (14-9) indicates that, as d is varied, a particular point in the fringe pattern ($\theta = $ constant) corresponds to gradually changing values of order m or p. Integral values occur whenever the point coincides with a dark fringe. Equivalently, this means that as d is varied, fringes of the pattern appear to contract toward the center where they disappear, or else expand outward from the center where they seem to originate, depending on whether the optical path difference is decreasing or increasing.

At a particular point at or near the center of the pattern ($\theta = 0°$), a change in d produces a corresponding change in the order m. If, initially, d has the value d_1, say, we can write Eq. (14-6), with $\cos\theta = 1$,

$$2d_1 = m_1\lambda$$

After a translation of the mirror to a mirror separation of d_2, the equation satisfies

$$2d_2 = m_2\lambda$$

Subtracting these equations, we have

$$2(d_2 - d_1) = (m_2 - m_1)\lambda$$

or

$$m_2 - m_1 = \frac{2(d_2 - d_1)}{\lambda} \qquad (14\text{-}10)$$

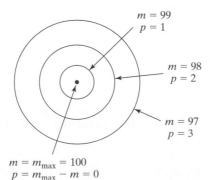

$m = 99$
$p = 1$

$m = 98$
$p = 2$

$m = 97$
$p = 3$

$m = m_{max} = 100$
$p = m_{max} - m = 0$

Figure 14-6 Alternate orderings of fringes.

Thus for a mirror translation $(d_2 - d_1)$, a number $(m_2 - m_1)$ of fringes is counted passing a fixed point at or near the center ($\cos \theta = 1$) of the fringe pattern, an exercise facilitated by the use of telescope cross hairs. Eq. (14-10) suggests an experimental way of either measuring λ when Δd is known, or calibrating the micrometer translation screw when λ is known.

❏ **Example**

Fringes are observed due to monochromatic light in a Michelson interferometer. When the movable mirror is translated by 0.073 mm, a shift of 300 fringes is observed. What is the wavelength of the light? What displacement of the fringe system takes place when a flake of glass of index $n_g = 1.51$ and 0.005 mm thickness t is placed in one arm of the interferometer? (Assume the light beam to be normal to the glass surface.)

Solution   Using Eq. (14-10),

$$\lambda = \frac{2(d_2 - d_1)}{(m_2 - m_1)} = \frac{(2)(0.073)}{300} = 4.87 \times 10^{-4} \text{ mm} = 487 \text{ nm}$$

With the glass inserted, one arm is effectively lengthened by a path difference of $d_2 - d_1 = n_g t - n_{air} t$, so that

$$m_2 - m_1 = \frac{2(d_2 - d_1)}{\lambda} = \frac{2(n_g - 1)t}{\lambda} = \frac{2(0.51)(0.005 \times 10^{-3})}{487 \times 10^{-9}} = 10.5$$

or the fringe system is displaced by 10.5 fringes.   ■

14-4 HOLOGRAPHY

Holography is one of the many flourishing fields that owes its success to the laser. Although the technique was invented in 1948 by the British scientist Dennis Gabor, before the advent of coherent laser light, the assurance of success was made possible by the laser. Emmett Leith and Juris Upatnieks at the University of Michigan first applied laser light to holography in 1962 and also introduced an important off-axis technique of illumination that we explain presently. The spectacular improvement in three-dimensional photography made possible by the hologram has aroused interest in nonscientific circles as well, so that the fast-multiplying applications of holography today also include its use in art and advertising. We are giving a brief description of holography in this chapter on interference applications because the *hologram* is, in fact, a complex system of interference patterns on a photographic plate.

Conventional versus Holographic Photography.   We are aware that a conventional photograph is a two-dimensional version of a three-dimensional scene, bringing into focus every part of the scene that falls within the depth of field of the lens. As a result the photograph lacks the perception of depth or the parallax with which we view a real-life scene. In contrast, the hologram provides a record of the scene that preserves these qualities. The hologram succeeds in effectively "freezing" and preserving for later observation the intricate wave front of light that carries all the visual information contained in the scene. When a hologram is viewed this wave front is reconstructed or released, and we view what we would have seen at the original scene when looking through the "window" defined by the hologram. The reconstructed wave front provides depth perception and parallax, allowing us to look to some extent around the edge of an object to see what is behind. It may be manipulated by a lens, for example, in the same way as the original wave front. Thus a *hologram*, as its etymology suggests, includes the "whole message."

The real-life qualities of the image provided by a hologram stem from the preservation of information relating to the phase of the wave front in addition to its amplitude or irradiance. Recording devices like ordinary photographic film and photomultipliers are sensitive only to the radiant energy received. In a developed photograph, for example, the *optical density* of the emulsion at each point is a function of the optical energy received there and the subsequent light-sensitive chemical reaction that reduces silver to its metallic form. When energy alone is recorded, the phase relationships of waves arriving from different directions and distances, and hence the visual life-likeness of the scene, are lost. To record these phase relationships as well, it is necessary to convert phase information into amplitude information. The interference of light waves provides the requisite means. Recall that when waves interfere to produce a large *amplitude*, they must be in *phase*, and when the amplitude is a minimum, the waves are out of phase, so that various contributions effectively cancel one another. If the wave front of light from a scene is made to interfere with a coherent *reference* wave front, then the resultant interference pattern encodes information about the phase relationships of each part of the wave front with the reference wave and, therefore, with every other part.

In conventional photography, a lens is used to focus the scene onto a film. All the light originating from a single point of the scene and collected by the lens is focused to a single conjugate point in the image. We can say that a one-to-one relationship exists between object and image points. By contrast, a hologram is made, as we shall see, without use of a lens or any other focusing device. The hologram is a complex interference pattern of microscopically spaced fringes, not an image of the scene. Each point of the hologram receives light from every point of the scene, or, to put it another way, every object point illuminates the entire hologram. There is no one-to-one correspondence between object points and points in the wave front before reconstruction occurs. The hologram is a record of this wave front.

Hologram of a Point Source. To see how the process is realized in practice, both making the hologram and using the hologram to reconstruct the original scene, we begin with a very basic example, the hologram of a point source. In Figure 14-7a, plane wave fronts of coherent, monochromatic radiation illuminate a photographic plate. In addition, spherical wave fronts reach the plate after scattering from object point O. The plate, when photographically developed, then shows a series of concentric interference rings about X as a center. Point P falls on such a ring, for example, if the optical path difference $OP - OX$ is an integral number of wavelengths, ensuring that the *reference beam* of plane wave front light arrives at P in step with the scattered *subject beam* of light. The developed plate is called a *Gabor zone plate*—or zone lens—with circular transmitting zones whose transmittance is a gradually varying function of radius. This plate is a *hologram* of point O. The hologram itself is a series of circular interference fringes that do not resemble the object, but the object may be reconstructed (as in Figure 14-7b) by placing the hologram back into the reference

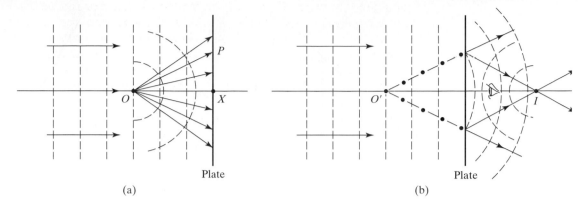

Figure 14-7 Hologram of a point source O is constructed in (a) and used in (b) to reconstruct the wave front. A virtual image O' and a real image I are formed in reconstruction. Plane and spherical wave fronts are represented by dashed lines, real light rays by solid lines, and virtual extensions of real rays by dot-dash lines.

beam without the presence of the object O. Just as light directed from O originally interferes with the plane-wave reference beam to produce the zone rings, so the same reference beam, in passing through the hologram, is reinforced along the same directions so that it appears to come from image point O'. The image at O' is a *virtual image* of the original object point O, seen on reconstruction by looking into the hologram (see the eye in Figure 14-7b). The hologram also directs light to a real image point I, as shown, but this is not the image ordinarily viewed. When object point O is replaced by an extended object or three-dimensional scene, each point of the scene produces its own Gabor zone pattern on the film. The hologram is now a highly complex montage of overlapping zones in which is encoded all the information of the wave front from the scene. On reconstruction, each set of zones produces its own real and virtual images, and the original scene is reproduced.

One usually views the virtual image by looking into the hologram. Figure 14-7b shows that when viewing the virtual image in this way, undesirable light forming the real image is also intercepted. Leith and Upatnieks (1965) introduced an off-axis technique, as mentioned previously, using one or more mirrors to bring in the reference beam from a different angle so that the directions of the reconstructed real and virtual wave fronts are separated. One of many holographic techniques for producing such an off-axis reference beam is shown in Figure 14-8. A combination of pinhole and lens is used to expand the beam from a laser. The expanded beam is then split by a semireflecting plate BS to produce two coherent beams. One beam, the *reference beam E_R*, is directed by two plane mirrors M_1 and M_2 onto the photographic plate, as shown. The other beam, E_S, reflects diffusely from the subject and some of this beam, which we call the *subject beam*, also strikes the film, where it interferes with the reference beam and produces the hologram. The reconstruction of the original wave front is accomplished, as in the case of the point object, by situating the hologram in the reference beam again, at the original angle between them. Of course, the original subject is now absent. When illuminated by the reference beam, the hologram bends rays of light in such a way as to reproduce both a virtual image (being viewed by the eye in Figure 14-9) and a real image. The hologram H shows the same essential features as the hologram of the point object. Photography by holography is a two-step process. Recall that in the making of a hologram, no lens is used, and

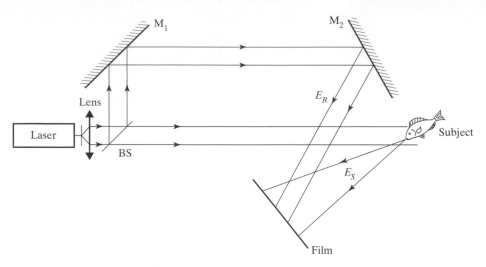

Figure 14-8   Off-axis holographic system.

the presence of the coherent reference beam is essential. Of course, the holographic system must be vibration-free to within a fraction of the wavelength of the light during the exposure, a condition that is easily satisfied when high-power laser pulses of very short duration are used to freeze undesirable motion.

Hologram Properties.   As stated earlier, the entire hologram receives light from each object point in the scene. As a result, any portion of the hologram contains information of the whole scene. If a hologram is cut up into small squares, each square is a hologram of the whole scene, although the reduction in aperture degrades the resolution and of course reduces the brightness of the image. The situation is much the same as when looking through a small, square aperture placed in front of a window. The same scene is viewed, though with slightly varying perspective, as the opening is moved to different parts of the window.

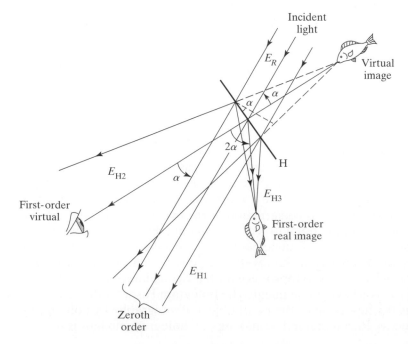

Figure 14-9   Reconstruction of hologram formed in Figure 14-8. Beam E_{H1} corresponds to zero-order diffraction of the incident beam by the hologram. Beam E_{H2} is observed looking into the hologram and appears to emanate from the virtual image. Beam E_{H3} focuses as a real image of the subject.

Each view is complete, exhibiting both depth and parallax. Another interesting property of a hologram is that a contact print of the hologram, which interchanges the optically dense and transparent regions, has the same properties in use. The "negative" of a hologram alters neither the fringe contrast nor their spacing, and hence does not modify the stored information. Furthermore, the hologram may contain a number of separate exposures, each taken with the film at a different angle relative to the reference beam and with different wavelengths of light. On reconstruction, each scene appears in its own light and without mutual interference when viewed along the directions of the original scene.

We have presented here a basic introduction to holography, omitting the subject of white-light holograms ("rainbow" holograms that produce images in their own original color) and omitting discussion of many fascinating applications of holography. We hope this introduction may serve as a motive for additional reading.

PROBLEMS

14-1. Newton's rings are formed between a spherical lens surface and an optical flat. If the tenth bright ring of green light (546.1 nm) is 7.89 mm in diameter, what is the radius of curvature of the lens surface?

14-2. The fifth dark ring of interference is measured to have a radius of 1.044 mm in a Newton's rings experiment. The plano-convex lens has a radius of curvature of 50.0 cm. Determine **(a)** the wavelength of the light; **(b)** the air-gap thickness at the location of the fifth dark ring.

14-3. Newton's rings are viewed both with the space between lens and optical flat empty and filled with a liquid. Show that the ratio of the radii observed for a particular order fringe is very nearly the square root of the liquid's refractive index.

14-4. A Newton's rings apparatus is illuminated by light with two wavelength components. One of the wavelengths is 546 nm. If the eleventh bright ring of the 546-nm fringe system coincides with the tenth ring of the other, what is the second wavelength? What is the radius at which overlap takes place and what is the thickness of the air film there? The spherical surface has a radius of 1 m.

14-5. A fringe pattern found using an interference microscope objective is observed to have a regular spacing of 1 mm. At a certain point in the pattern, the fringes are observed to shift laterally by 3.4 mm. If the llumination is green light of 546.1 nm, what is the dimension of the "step" in the film that caused the shift?

14-6. When one mirror of a Michelson interferometer is translated by 0.0114 cm, 523 fringes are observed to pass the crosshairs of the viewing telescope. Calculate the wavelength of the light.

14-7. When looking into a Michelson interferometer illuminated by the 546.1-nm light of mercury, one sees a series of straight-line images that number 12 per cm. Explain their occurrence.

14-8. A thin sheet of fluorite of index 1.434 is inserted normally into one beam of a Michelson interferometer. Using light of wavelength 589 nm, the fringe pattern is found to shift by 35 fringes. What is the thickness of the sheet?

14-9. Looking into a Michelson interferometer, one sees a dark central disk surrounded by concentric bright and dark rings. One arm of the device is 2 cm longer than the other, and the wavelength of the light is 500 nm. Determine **(a)** the order of the central disk, and **(b)** the order of the sixth dark ring from the center.

14-10. A Michelson interferometer is used to measure the refractive index of a gas. The gas is allowed to flow into an evacuated glass cell of length L placed in one arm of the interferometer. The wavelength is λ.

(a) If N fringes are counted as the pressure in the cell changes from vacuum to atmospheric pressure, what is the index of refraction n in terms of N, L, and λ?

(b) How many fringes are counted if the gas is carbon dioxide, for which $n = 1.00045$, using a 10-cm cell length and sodium light at 589 nm?

14-11. A Michelson interferometer is used with red light of wavelength 632.8 nm and is adjusted for a path difference of 20 μm. Determine the angular radius of the **(a)** first (smallest diameter) ring observed, and **(b)** tenth ring observed.

14-12. A polished surface is examined using a Michelson interferometer, with the polished surface replacing one of the mirrors. A fringe pattern characterizing the surface contour is observed using He-Ne light of wavelength 632.8 nm. Fringe distortion over the surface is found to be less than 1/4 the fringe separation at any point. What is the maximum depth of polishing defects on the surface?

14-13. Suppose a hologram is to be made of a moving object using a 1-ns laser pulse at a wavelength of 633 nm. What is the permissible speed of the motion such that the object does not move more than $\lambda/10$ during the exposure?

14-14. Let us suppose that as a theoretical limit, one bit of information can be stored in each λ^3 of hologram volume. At a wavelength of 492 nm and a refractive index of 1.30, determine the storage capacity of 1 mm^3 of hologram volume. (Note: A bit is a unit of information storage capacity, such as the binary digit 0 or 1.)

REFERENCES

BAUMEISTER, PHILIP, and GERALD PINCUS. 1970. "Optical Interference Coatings." *Scientific American* (Dec.): 58.

CAULFIELD, H. JOHN, ed. 1979. *Handbook of Optical Holography*. New York: Academic Press.

CAULFIELD, H. JOHN. 1984. "The Wonder of Holography." *National Geographic* 165 (March): 364.

CHAN, VINCENT W. S. 1995. "All-Optical Networks." *Scientific American* (Sept.): 72.

CONNES, PIERRE. 1968. "How Light is Analyzed." *Scientific American* (Sept.): 72.

FEYNMAN, RICHARD P., ROBERT B. LEIGHTON, and MATTHEW SANDS. 1975. *The Feynman Lectures in Physics*, vol. 1. Reading, Mass.: Addison-Wesley Publishing Company. Ch. 28, 29.

FINCHAM, W. H. A., and M. H. FREEMAN. 1980. *Optics*, 9th ed. London: Butterworths. Ch. 14.

FRANCON, M. 1974. *Holography*. New York: Academic Press.

FRANCON, MAURICE. 1966. *Optical Interferometry*. New York: Academic Press.

HARIHARAN, P. 1985. *Optical Interferometry*. Orlando, Fla.: Academic Press.

HECHT, EUGENE, and ALFRED ZAJAC. 1972. *Optics.* Reading, Mass.: Addison-Wesley Publishing Company. Ch. 9.

JEONG, TUNG H. 1991. *Holography*. Bellingham SPIE-International Society for Optical Engineering.

KLEIN, MILES V. 1970. *Optics*. New York: John Wiley and Sons. Ch. 5.

LEITH, EMMETT N., and JURIS UPATNIEKS. 1965. "Photography by Laser." *Scientific American* (June): 24.

PSALTIS, DEMETRI, and FAI MOK. 1994. "Holographic Memories." *Scientific American* (November): 70.

RASTOGI, PRAMOD K. 1996. *Holographic Interferometry Principles and Methods*. New York: Springer-Verlag New York, Inc.

ROBINSON, GLEN M., DAVID M. PERRY, and RICHARD W. PETERSON. 1991. "Optical Interferometry of Surfaces." *Scientific American* (July): 66.

SMITH, F. DOW. 1968. "How Images Are Formed." *Scientific American* (Sept.): 96.

SMITH, HOWARD MICHAEL. 1975. *Principles of Holography*. New York: John Wiley and Sons.

STROKE, GEORGE W. 1969. *An Introduction to Coherent Optics and Holography*, 2d ed. New York: Academic Press.

TOLANSKY, SAMUEL. 1973. *An Introduction to Interferometry*. New York: John Wiley and Sons.

VEST, C. M. 1979. *Holographic Interferometry*. New York: John Wiley and Sons.

WALDMAN, GARY. 1983. *Introduction to Light*. Englewood Cliffs, N. J.: Prentice Hall.

15

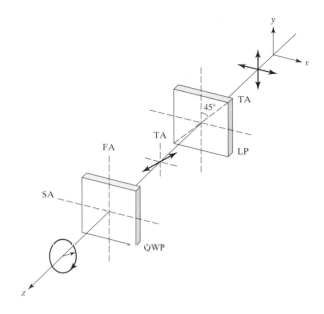

Polarized Light

INTRODUCTION

The representation of a plane wave of electromagnetic radiation by a drawing such as Figure 11-2 is not applicable to ordinary light. In a plane wave, unlike the case of ordinary light, the electric field $\boldsymbol{E}$-vector always oscillates parallel to a fixed direction in space. Light of such character is said to be *linearly polarized*. The same can be said of the magnetic field $\boldsymbol{B}$-vector, which in a plane wave always maintains an orientation perpendicular to the electric field vector such that the direction of wave propagation is at every point normal to both the electric and magnetic field vectors. Ordinary light, however, is produced by a number of independent atomic sources whose radiation is not synchronized. Consider a beam of ordinary light, such as that produced by a hot filament. The resultant $\boldsymbol{E}$-field vector, generated from a collection of radiating atoms, does not maintain a constant direction of oscillation, nor does it vary spatially in any other *regular* manner. Though perpendicular to the direction of the light beam, the electric field is otherwise equally divided among all directions. Such ordinary light is simply said to be *unpolarized*. Of course, a beam of light can consist of a mixture of unpolarized light with a certain amount of polarized light, in which case it is said to be *partially polarized*. In partially polarized light, the electric field vector favors one direction more than others, though all directions are present.

The possibility of producing polarized light is essentially related to its transverse character. If light were a longitudinal wave, the production of polarized light in the ways to be described would simply not be possible. Thus the polarization of light constitutes experimental proof of its transverse character.

The most important processes that produce polarized light are discussed in this chapter under the following topics: (1) *dichroism*, (2) *reflection*, (3) *scattering*, and (4) *birefringence*. *Optical activity* is described as a mechanism that *modifies* polarized light. Finally, *photoelasticity* is briefly discussed as a useful application.

15-1 MODES OF POLARIZATION

Consider a ray of light directed perpendicularly out of the page at the origin O of the axis system shown in Figure 15-1a. Let the E-field of the light at O be represented by the vector shown. Since the E-field varies continuously in magnitude and alters direction every half-period, the figure shows the magnitude and direction of E at a particular instant. Figures 15-1b, c, and d show the E vector at

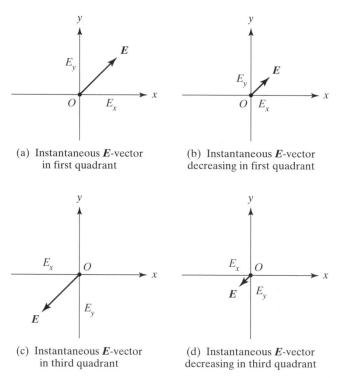

(a) Instantaneous E-vector in first quadrant

(b) Instantaneous E-vector decreasing in first quadrant

(c) Instantaneous E-vector in third quadrant

(d) Instantaneous E-vector decreasing in third quadrant

(e) Instantaneous E-vector oscillating between second and fourth quadrants

Figure 15-1 Representations of the instantaneous E-vector of a light ray traveling in the $+z$ direction (out of the page), shown at different instants of its oscillation (a, b, c, d). The maximum extension of the E-vector in either direction is the amplitude of the E-wave. The oscillations of the E-vector are equivalent to scillations of the two orthogonal components, E_x and E_y. In (e), the oscillation takes place in the second and fourth quadrants because E_x and E_y are of opposite sign. In all cases, since the E-vector remains along a line, the light is said to be *linearly polarized*.

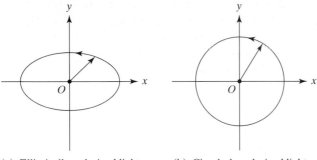

(a) Elliptically polarized light   (b) Circularly polarized light

Figure 15-2 (a) An ellipse is traced out by the E-vector because the E_x and E_y components are out of phase. In this case, the phase difference Φ is 90°. The light is said to be *elliptically polarized*. (b) A circle is traced out by the E-vector as a special case when the components E_x and E_y of the ellipse in (a) are equal in magnitude. The light is said to be *circularly polarized*.

other instants in its periodic oscillations as the light passes through the origin point at O. In these instances, the E_x and E_y components of E are either both positive or both negative, and the vector alternates between the first and third quadrants. If the E_x and E_y components of E have opposite signs, the alternations of E lie in the second and fourth quadrants, as shown in Figure 15-1e. In all these cases, the E-vector oscillates along a fixed direction (or line) in space and the light is said to be *linearly polarized*. There are situations in which the E-vector does not remain in the same direction, but in which the tip of the E-vector periodically traces out an ellipse. The light is then said to be *elliptically polarized*. As a special case, if the tip of the E-vector periodically traces out a circle, it is said to be *circularly polarized*. These instances are suggested in Figures 15-2a and b.

Various modes of polarization can be envisioned as resulting from independent oscillations of the x- and y-components of the E-vector when they have the same period or frequency. Depending on the amplitude and the *phase* of the component oscillations, it is possible to produce a resultant E-vector that describes linear, elliptical, or circular polarization. The phase of the oscillations describes the degree to which they are "in step" or "out of step" with one another.

The process we are describing here is actually a rather common one, the result of superposing two orthogonal oscillations. In general, the resulting pattern is called a *Lissajous figure*. Perhaps the simplest model is a pendulum bob set oscillating along perpendicular directions at the same time. Depending on the amplitude and synchronization of the two oscillations, the pendulum bob can be made to trace out an ellipse or, as special cases of an ellipse, a circle or a straight line. In Figure 15-3 are shown various Lissajous patterns resulting from orthogonal oscillations of various amplitudes and phase relationships. The angle Φ is used to express the *phase difference* in each case. A phase difference of zero means the two vibrations are in step, increasing (or decreasing) together and passing through zero at the same time. A phase difference of $\Phi = 90°$ means they are out of step by one-quarter period. A phase difference of $\Phi = 180°$ means they are out of step by one-half a period, and so on.[1]

---

[1]In all cases shown in Figure 15-3, the orthogonal oscillations have the same period or frequency. If the frequencies are also allowed to vary, more complicated Lissajous figures result. These are not needed in discussing polarized light since both components of the E- or B-fields necessarily have the same frequency, given by the ratio of the speed of light divided by its wavelength: $f = c/\lambda$.

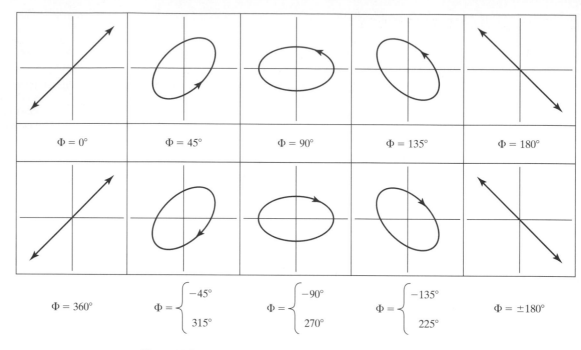

Figure 15-3 Lissajous figures as a function of the relative phase difference Φ for orthogonal vibrations of unequal amplitude. An angle lead greater than 180° may also be represented as an angle lag of less than 180°. For all figures we have adopted the phase lag convention $\Phi = \Phi_y - \Phi_x$.

❏ **Example**

Show that when a phase difference of 180° is introduced between the E_x and E_y components of linearly polarized light making an angle α with the x-axis, the result is linearly polarized light making an angle of $-\alpha$ with the x-axis; equivalently, the E-vector is rotated clockwise by 2α.

Solution   Refer to Figure 15-2a. The E-vector for linearly polarized light can be considered the resultant at each instant of its changing E_x and E_y components, which are always in phase. Thus when E_x is increasing from the origin, so is E_y. When E_x reaches its maximum magnitude, so does E_y, and so on. The angle α remains constant and depends on the simultaneous values of E_x and E_y such that $\tan \alpha = E_y/E_x$. Consider introducing a phase difference of $\Phi = 180°$, without changing maximum values of E_x and E_y, as might be accomplished by sending the light through a *half-wave plate* (described in following sections). This means, for example, that when E_x is increasing from zero, E_y is *decreasing* from zero, and when E_x reaches its maximum positive value, E_y reaches its maximum *negative* value. The resultant of the two components is now an E-vector that varies linearly like that of Figure 15-2e. Because the maximum values of E_x and E_y have not changed, the E-vector makes the same angle α with the x-axis in the second and fourth quadrants. Compared with its original orientation, the change is equivalent to rotating the original E-vector clockwise by an angle of 2α. This result is also evident by comparing the Lissajous figures in Figure 15-3 for $\Phi = 0°$ and $\Phi = 180°$.   ∎

Various devices can serve as optical elements that transmit light but modify the state of polarization. The physical mechanisms underlying their operation will be discussed in the following sections. Here we wish merely to categorize such polarizers in terms of their effects, which are basically three in number.

Linear Polarizer. The linear polarizer selectively transmits all or most of the E-vibrations, or components of E-vibrations, in a given direction, while removing most of the vibrations in the perpendicular direction. In most cases, the selectivity is not 100% efficient, so that the transmitted light is partially polarized. Figure 15-4 illustrates the operation schematically. Unpolarized light traveling in the $+z$ direction passes through a plane polarizer, whose preferential axis of transmission (or *transmission axis* TA) is vertical. The unpolarized light is represented by two perpendicular (x and y) vibrations, since any direction of vibration present can be resolved into components along these two directions. The light transmitted includes components along only the TA direction and is therefore linearly polarized in the vertical, or y-direction. The horizontal components of the original light have been removed by absorption. In the figure, the process is assumed to be 100% efficient.

Phase Retarder. The phase retarder does not remove either of the component orthogonal E-vibrations but introduces a phase difference between them. If light corresponding to each vibration travels with different speeds through such a *retardation plate*, there will be a cumulative phase difference Φ between the two orthogonal vibrations as they emerge.

Symbolically, Figure 15-5 shows the effect of a retardation plate on unpolarized light in a case where the vertical component travels through the plate faster than the horizontal component. This is suggested by the schematic separation of the two components on the optical axis. The fast axis (FA) and slow axis (SA) directions of the plate are also indicated. When the net phase difference $\Phi = 90°$, the retardation plate is called a *quarter-wave plate*; when it is $180°$, it is called a *half-wave plate*.

Rotator. The rotator has the effect of rotating the direction of linearly polarized light incident on it by some particular angle. Vertical linearly polarized

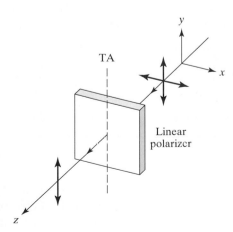

Figure 15-4 Operation of a linear polarizer.

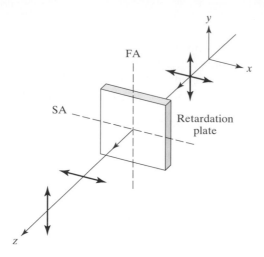

Figure 15-5 Operation of a phase retarder.

light is shown incident on a rotator in Figure 15-6. The effect of the rotator element is to transmit linearly polarized light whose direction of vibration has, in this case, been rotated counterclockwise by an angle θ.

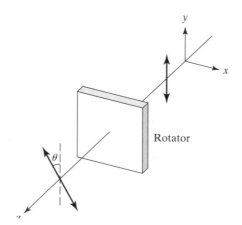

Figure 15-6 Operation of a rotator.

❏ **Example**

Show how the combination of a linear polarizer and a quarter-wave plate can be used to produce circularly polarized light.

Solution Refer to Figure 15-7. Let the linear polarizer (LP) produce light vibrating at an angle of 45°, as shown. This is accomplished by allowing unpolarized light to pass through the LP with its TA at 45°. The light is then allowed to pass through the quarter-wave plate (QWP) where it is divided equally between the vertical fast axis (FA) and horizontal slow axis (SA). On emerging, a phase difference of $\Phi = 90°$ results in circularly polarized light. This outcome can also be deduced from Figure 15-3, where the elliptical Lissajous figure for a phase difference of 90° reduces to a circle in the case of equal amplitudes. ∎

The three basic types of polarizing elements just discussed—the linear polarizer, the phase retarder, and the rotator—owe their special operation to various actual physical mechanisms that have a different effect on the two

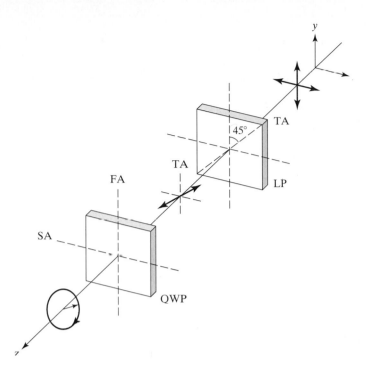

Figure 15-7 Production of circularly polarized light.

perpendicular, transverse $\boldsymbol{E}$-oscillations of unpolarized light. In the following articles, we describe several of these physical mechanisms.

15-3 DICHROISM: POLARIZATION BY SELECTIVE ABSORPTION

A *dichroic polarizer* selectively *absorbs* light with its transverse $\boldsymbol{E}$-vibrations along a unique direction that is characteristic of the dichroic material and easily *transmits* light with $\boldsymbol{E}$-vibrations perpendicular to that unique direction. The preferred direction for transmission is called the *transmission axis* of the polarizer. In the ideal polarizer, the transmitted light is linearly polarized in the same direction as the transmission axis. The state of polarization of the light can most easily be tested by a second dichroic polarizer, which then serves as an *analyzer*, shown in Figure 15-8. When the TA of the analyzer is oriented at 90° relative to the TA of the polarizer, the light is effectively extinguished. As the analyzer is rotated about the direction of the light, the light transmitted by the pair increases, reaching a maximum when their TAs are aligned. If I_0 represents the maximum transmitted intensity, then the *Law of Malus* states that the irradiance for any relative angle θ between the TAs is given by

$$I = I_0 \cos^2 \theta \tag{15-1}$$

Malus' law is easily understood with the help of Figure 15-9. Notice that the amplitude of the light emerging along the TA of the analyzer is $E_0 \cos \theta$. The irradiance I (in W/m^2) is then proportional to the square of this amplitude,

$$E_0^2 \cos^2 \theta = I_0 \cos^2 \theta$$

yielding Malus' law.

The impressive ability of dichroic materials to absorb light strongly with the electric field $\boldsymbol{E}$ along one direction and to transmit light easily with the field $\boldsymbol{E}$

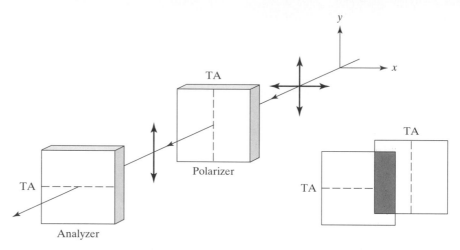

Figure 15-8 Crossed dichroic polarizers functioning as a polarizer-analyzer pair. No light is transmitted through the analyzer.

along a perpendicular direction can perhaps best be understood by reference to a standard experiment with microwaves, illustrated in Figure 15-10. Wavelengths of microwaves range roughly from 1 mm to 1 m. It is found that when a vertical wire grid, whose spacing is much smaller than the wavelength, intercepts microwaves with vertical linear polarization, little or no radiation is transmitted. Conversely, when the grid intercepts waves polarized in a direction perpendicular to the wires, there is efficient transmission of the waves. The TA of the wire grid is thus in the horizontal direction. The explanation of this behavior involves a consideration of the interaction of electromagnetic radiation with the metal wires that together behave as a dichroic polarizer. Within the vertical metal wires, the mobile free electrons are set in oscillatory motion by the oscillations of the electric field of the incident radiation. In motion, the electrons absorb the energy

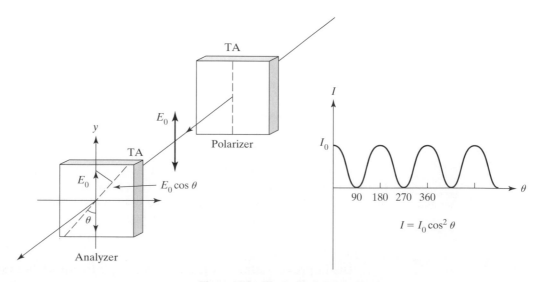

Figure 15-9 Illustration of Malus' law.

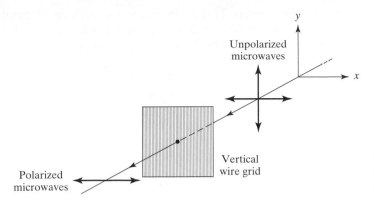

Figure 15-10 Action of a vertical wire grid on microwaves. Effective absorption of the vertical component of the radiation occurs when $\lambda \gg$ grid spacing.

of the radiation,[2] canceling its propagation in the forward direction. If the incident radiation is polarized instead in the horizontal direction, the free oscillation of the electrons is inhibited by the narrow width of the vertical wires. In this event, little absorption of the incident radiation can occur and there is forward propagation of the radiation. If now the grid, together with its TA, is rotated by 90°, vertical E-vibrations are transmitted through the horizontal wires and horizontal E-vibrations are canceled. The wire grid polarizes microwaves just as a dichroic absorber polarizes optical radiation.

To produce a similar effect using the much smaller optical wavelengths, the conduction paths analogous to the grid wires must be much closer together. The most common dichroic absorber for light is Polaroid H-sheet, invented in 1938 by Edwin H. Land. When a sheet of clear, polyvinyl alcohol is heated and stretched, its long, hydrocarbon molecules tend to align in the direction of stretching. The stretched material is then impregnated with iodine atoms, which become associated with the linear molecules and provide conduction electrons to complete the analogy to the wire grid. Some naturally occurring materials, such as the mineral tourmaline, also possess dichroic properties to some degree. All that is required in principle is that the electrons be much freer to respond to an incident electromagnetic wave in one direction than in an orthogonal direction. The ideal condition is one of total extinction when polarizers are crossed with their TAs at 90°. This ideal is not quite achieved in Polaroid H-sheet, which is less effective at the blue end of the spectrum. Consequently, when a Polaroid H-sheet is crossed with another such sheet acting as an analyzer, the combination contributes a dark blue tint to the transmitted light.

15-4 POLARIZATION BY REFLECTION FROM DIELECTRIC SURFACES

Light that is specularly reflected from dielectric surfaces is at least partially polarized, having a preferred E-direction. This is most easily confirmed by looking through a piece of polarizing filter (such as that found in Polaroid glasses) while rotating it about the direction of reflected light entering the eye. When the preferred E-vibrations of the reflected light are perpendicular to the TA of the filter, they are absorbed and the objects from which light is specularly reflected into

[2]Actually, the oscillating electrons act as tiny dipole antennas. Once set into motion by the incident electric field, they re-radiate electromagnetic waves. It can be shown, however, that in metals the re-radiated wave is 180° out of phase with the incident wave, so that cancellation occurs. Thus no wave is propagated in the forward direction.

the eye appear reduced in brightness. This is precisely the working principle of Polaroid sunglasses. Since the preferred E-vibration in light reflected from ground level into the eye turns out to be horizontal, the TA of the Polaroid lenses in a pair of sunglasses is fixed in the vertical direction.

To appreciate the physics that underlies this phenomenon, consider Figure 15-11, which shows a narrow beam of light incident at an arbitrary angle on a smooth, flat dielectric surface. The unpolarized incident beam is conveniently represented by two perpendicular E-vibrations. One E-vibration (Figure 15-11a) is transverse or perpendicular to the plane of incidence, and is represented as a dot. The other (Figure 15-11b) is parallel to the plane of incidence, that is, the plane of the page, which includes the incident ray and the normal drawn to the point of incidence. It is shown as a double-headed arrow. Standard notation refers to the perpendicular component (dot) as the TE or *transverse electric* mode and the parallel component (arrow) as the TM or *transverse magnetic* mode. The TM mode is so named because, when the electric field is parallel to the plane of incidence, the magnetic field is necessarily perpendicular to the plane of incidence.

Consider first the TE component (Figure 15-11a). The action of the electric field on the electrons in the surface of the dielectric is to stimulate oscillations along the same direction, perpendicular to the page. Thus, according to the laws of reflection and refraction, the original beam is partially reflected and partially transmitted, as shown, each partial beam having E-vibrations perpendicular to the page. Consider next the TM component of the incident beam (Figure 15-11b). Again, the beam is partially reflected and partially transmitted, each partial beam having E-vibrations along the direction of the E-field of the incident beam just inside the dielectric. Now notice that at a certain angle of incidence, the reflected ray would have to have all of its E-vibrations parallel to its own direction. It would no longer be a transverse electromagnetic wave! For this angle of incidence of a TM beam, therefore, the reflected ray disappears.[3] Thus, an unpolarized incident beam, having both TE and TM vibrations, produces a reflected beam that is pure TE. In this case, the unpolarized incident beam is rendered linearly polarized by reflection. This situation arises for only one angle of incidence, the *polarizing angle* θ_p, also called *Brewster's angle*. Examination of Figure 15-11c shows that the unique orientation leading to the missing TM reflected wave occurs when the reflected and refracted rays are perpendicular to one another, or

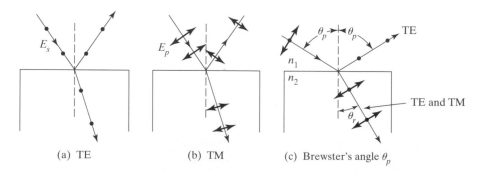

(a) TE   (b) TM   (c) Brewster's angle θ_p

Figure 15-11 Specular reflection of light at a dielectric surface: (a) TE mode, (b) TM mode, and (c) polarization at Brewster's angle.

[3]Another way of arriving at the same conclusion is to notice that the oscillating electrons in the surface of the dielectric, behaving as dipole antennas, cannot radiate along the direction of their own vibrations. When that direction matches the direction of the reflected wave, the TM incident wave produces no reflected beam.

$\theta_p + \theta_r = 90°$. We can find the polarizing angle by combining Snell's law with trigonometric relations as follows:

$$n_1 \sin \theta_p = n_2 \sin \theta_r = n_2 \sin (90 - \theta_p) = n_2 \cos \theta_p$$

$$\frac{\sin \theta_p}{\cos \theta_p} = \frac{n_2}{n_1}$$

$$\theta_p = \tan^{-1}\left(\frac{n_2}{n_1}\right) \qquad \text{Brewster's law} \qquad (15\text{-}2)$$

❏ **Example**

Find the angles of incidence and refraction that lead to linear polarization of a beam of unpolarized light when reflected from a plane glass surface whose refractive index is 1.50.

Solution   The polarizing angle is calculated as follows:

$$\theta_p = \tan^{-1}\left(\frac{n_2}{n_1}\right) = \tan^{-1}\left(\frac{1.50}{1}\right) = 56.3°$$

Since the polarizing angle and the angle of refraction must be complementary, the angle of refraction is $\theta_r = 90 - 56.3 = 33.7°$.

Of course, the angle of refraction could also be found using Snell's law:

$$\sin \theta_r = \left(\frac{n_1}{n_2}\right) \sin \theta_p = \left(\frac{1}{1.50}\right) \sin 56.3° = 0.5546$$

$$\theta_r = \sin^{-1} 0.5546 = 33.7° \qquad ■$$

Polarizing angles exist both for external reflection ($n_2 > n_1$) and internal reflection ($n_2 < n_1$). However they are clearly not the same. For example, by reversing the roles of n_1 and n_2 in the calculation above, one finds that for light reflected on its way out of the glass the polarizing angle is 33.7° rather than 56.3°. Note however that the polarizing angles for internal and external reflection are complementary, a result easily shown to be generally true.

While reflection at the polarizing angle from a dielectric surface can be used to produce linearly polarized light, the method is relatively inefficient. For reflection from air to glass, as in the example just given, only 15% of the TE component is found in the reflected beam. This deficiency can be remedied to a degree by stepwise intensification of the reflected beam as in a *pile-of-plates* polarizer (Figure 15-12). The use of repeated reflections at Brewster's angle by

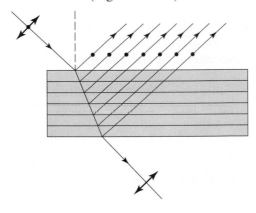

Figure 15-12   Pile-of-plates polarizer.

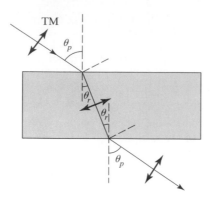

Figure 15-13 Brewster window. Brewster's law is satisfied for the TM mode at both surfaces.

multiple layers of the dielectric both increases the intensity of the TE component in the integrated, reflected beam and, necessarily, purifies the transmitted beam of this component. If enough plates are assembled, the transmitted beam approaches a linearly polarized condition. Pile-of-plates polarizers are especially helpful in those regions of the infrared and ultraviolet spectrum where dichroic sheet polarizers and calcite prisms are ineffective. Multilayer, thin film coatings—which can be prepared to show little absorption in the spectral region of interest—behave in a similar manner and can be used as polarization-sensitive reflectors and transmitters.

Another interesting and useful application of polarization by reflection is *Brewster's window*. The window (Figure 15-13) operates in the same way as a single plate of the pile-of-plates analyzer. Linearly polarized light (TM) incident at Brewster's angle is fully transmitted at the first surface. None is reflected. The angle of incidence θ_r at the second surface also satisfies Brewster's law for internal reflection, so that the light is again fully transmitted. The plate acts as a perfect window for TM polarized light.

One important application of the Brewster window occurs in the operation of some lasers.[4] The active medium of a gas laser is often bounded by two Brewster windows, forming the ends of the gas plasma tube. The light in the cavity makes repeated passes through the windows, on its way to and from cavity mirrors positioned beyond alternate ends of the gas tube. Upon each traversal, the TM mode is completely transmitted, whereas the TE mode is partially reflected (rejected). After many such traversals in the laser cavity, the beam is essentially free of TE vibrations and the emerging laser beam is polarized in the TM mode.

15-5 POLARIZATION BY SCATTERING

Before discussing the polarization of light that occurs by *scattering*, we introduce the general subject of scattering and point out some familiar consequences that are in themselves rather interesting. By the *scattering of light*, we mean the removal of energy from an incident light wave by a scattering medium and the re-emission of some portion of that energy in many directions. We can think of the elemental oscillator or scattering unit as an electronic charge bound to a nucleus (a *dipole oscillator*). The electron is set into forced oscillations at the same frequency as the alternating electric field of the incident light. The response of the

[4]This description presupposes the explanation of the elements of a laser, given in Chapter 12, "Lasers and the Eye."

electron to this driving force depends on the relationship between the driving frequency ω and the natural or resonant frequency ω_0 of the oscillator.[5] In most materials, resonant frequencies lie predominantly in the ultraviolet and the infrared, rather than in the visible region of the electromagnetic spectrum. Resonant frequencies in the ultraviolet are due to electronic oscillations, while those in the infrared are due to molecular vibrations. Because atomic masses involved in molecular vibrations are so much larger than the electron mass, amplitudes of induced molecular vibrations are small compared with those of electronic vibrations, and so can be neglected in this discussion. Thus, for light incident on most materials, the driving frequencies of the visible light are less than the resonant frequencies of the scattering medium. Since the higher frequencies (shorter wavelengths) of the visible spectrum are closer to the resonant frequencies of the medium, blue light is scattered more efficiently than red light.

Light scattering is most effective when the scattering centers are particles whose dimensions are small compared with the wavelength of the radiation, in which case we speak of *Rayleigh scattering*. The scattering of sunlight from oxygen and nitrogen molecules in the atmosphere, for example, is Rayleigh scattering, whereas the scattering of light from dense scattering centers—like the droplets of water in clouds and fog—is not. The radiated power in Rayleigh scattering can be shown to be inversely proportional to the fourth power of the wavelength of the incident radiation. Thus the oscillating dipoles radiate more energy in the shorter wavelength region of the visible spectrum than in the longer wavelength region. The radiated power for violet light of wavelength 400 nm is nearly 10 times as great as for red light of wavelength 700 nm. Rayleigh scattering explains why a clean atmosphere appears blue: Higher frequency blue light from the sun is scattered by the atmosphere down to the earth more than the lower frequency red light. On the other hand, when we are looking at the sunlight head-on at sunrise or sunset, after it has passed through a good deal of atmosphere, we see reddish or yellowish light, white light from which the blues have been preferentially removed by scattering.

In contrast to Rayleigh scattering, light scattered from larger particles such as those found in clouds, fog, and powdered materials like sugar, appears as white light. Here, "larger particles" refers to the size of the scattering particles relative to the wavelength of light. In this case, the scattering centers in the larger particles of condensed matter are closer together. The cooperative effect of many oscillators tends to cancel the radiation in all directions but the forward and backward directions. The scattering can now be understood in terms of the usual laws of refraction and reflection. The cooperative effect of the densely packed scatterers leads to a radiated power that is directly proportional to the fourth power of the wavelength, canceling the $1/\lambda^4$ dependence of isolated oscillators. Thus the scattered radiation is essentially wavelength independent, so that fog and clouds appear white by scattered light.

Of particular interest in the context of this chapter is the fact that scattered radiation may also be polarized. This phenomenon can be explained rather simply by the fact that light is a transverse, not a longitudinal, wave. As an example,

[5]The *resonant frequency* of an oscillator is just that frequency at which it oscillates freely after being set into motion. A simple pendulum has the single natural or resonant frequency given by $\dfrac{1}{2\pi}\sqrt{\dfrac{g}{\ell}}$. A more complicated oscillating system may have several resonant frequencies. When an oscillator is set into forced vibrations (by a *driving force*) at its resonant frequency, it responds with oscillations of maximum amplitude.

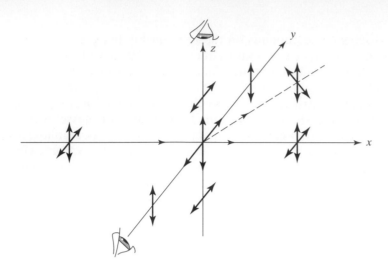

Figure 15-14 Polarization due to scattering. Unpolarized light incident from the left is scattered by a particle at the origin.

consider a vessel of water to which is added one or more drops of milk. The milk molecules quickly diffuse throughout the water and serve as effective scattering centers for a beam of light transmitted across the medium. In pure water the light does not scatter sideways but propagates only in the forward direction. The light scattered in various directions from the milk molecules, when examined with a polarizing filter, is found to be polarized, as shown in Figure 15-14. The perpendicular E components of the unpolarized light incident from the left sets the electronic oscillators of a scattering center into similar forced vibrations, producing radiation re-emitted in all directions. Light scattered in any direction can include only those identical E-vibrations executed by the oscillators, that is, along the y- and the z-directions. If scattered light is viewed along the $-y$-direction, it will be found to contain E-vibrations along the z-direction, but those along the y-direction are absent because they would represent longitudinal E-vibrations in an electromagnetic wave. Similarly, viewed along the z-direction, the z-vibrations are missing, and light is linearly polarized along the y-direction. Viewed along directions off the axes (dashed line), the light is partially polarized. The forward beam shows the same polarization as the incident light.

In the same way, when the sun is not directly overhead so that its light crosses the atmosphere above us, the light scattered down is found to be partially polarized. The effect is easily seen by viewing the clear sky through a rotating polarizing filter. The polarization is not complete, both because we see light that is multiply scattered into the eye and because not all electronic oscillators in molecules are free to oscillate in exactly the same direction as the incident E-vector of the light.

Ordinary polarization by scattering is generally weak and imperfect and so is not used as a practical means of artificially producing polarized light.

15-6 BIREFRINGENCE: POLARIZATION WITH TWO REFRACTIVE INDICES

Birefringent crystalline materials are so named because they are able to cause double refraction; that is, for a single incident beam, there appear two refracted beams due to the two different indices of refraction possessed by this material.

We have already seen that anisotropy (directional variation) in the binding forces affecting the electrons in a material can lead to a dependence on direction in the amplitudes of their oscillations in response to a stimulating electromagnetic wave and hence to a dependence on direction of absorption. Such a material exhibits dichroism. For this to occur, however, the stimulating optical frequencies must fall within the *absorption band* of the material.[6] Typically, the absorption band lies in the ultraviolet, above optical frequencies, so that the material is transparent to visible light. In this case, even with anisotropy of electron-binding forces, there is little or no effect on optical absorption, and the material does not appear dichroic. Still, the presence of anisotropic binding forces along the x- and y-directions leads to different strengths of interaction with the radiation and thus different indices of refraction n_x and n_y, respectively. Of course, different indices of refraction imply different velocities of propagation v_x and v_y through the crystal. The result is that such a crystal, while not appreciably dichroic, still manifests the property of birefringence. Recapitulating then, an ideal dichroic material is anisotropic in absorption but not in refractive index, whereas an ideal birefringent material is anisotropic in refractive index, but not in absorption. Both conditions require anisotropic crystalline structures and both conditions are frequency dependent. Calcite is birefringent in the visible spectrum, for example, and strongly dichroic in certain parts of the infrared spectrum. Other common materials, birefringent in the visible region, are quartz, ice, mica and cellophane.

The relationship of crystalline asymmetry to refractive index and the speed of light in the medium may be understood a bit more clearly by considering the case of calcite. The basic molecular unit of calcite is $CaCO_3$, which assumes a tetrahedral or pyramidal structure in the crystal. Figure 15-15a shows one of these molecules, assumed to be surrounded by identical structures that are similarly oriented. The carbon (C) and oxygen (O) atoms form the base (shaded) of the pyramid, as shown, with carbon lying in the center of the equilateral triangle of oxygen atoms. The calcium (Ca) atom is positioned at some distance above the carbon atom, at the apex of the pyramid. The direction through the crystal parallel to the line joining the calcium and carbon atoms is a direction of symmetry and is called the *optic axis* (OA) of the crystal. This symmetry is apparent in Figure 15-15b, which gives a view of a section of the atomic structure of calcite looking along its OA. Returning to Figure 15-15a, consider light entering the crystal from two different directions—from below and from the left—as shown. For light entering from below the triangular base plane, the perpendicular components of the electric vector are both perpendicular ($E_\perp$) to the OA. Therefore both components interact with the crystal in the same way. For light entering from the side, one of the two perpendicular components of the $\boldsymbol{E}$-vector is parallel ($E_\parallel$) to OA and the other is perpendicular ($E_\perp$). In this case the two components have dissimilar effects on the electrons in the base plane. The component $E_\parallel$ causes electrons in the base plane to oscillate along a direction perpendicular to the plane, whereas its orthogonal counterpart $E_\perp$ causes oscillations within the plane. Since the latter take place more easily, the interaction of $E_\perp$ is greater than the interaction of $E_\parallel$. Now greater interaction between light and crystal implies more slowing down of the light, or $v_\perp < v_\parallel$. This difference in the speed of light through the crystal for the two $\boldsymbol{E}$-components shows up as a difference in refractive indices, since $n = c/v$. Thus the crystal exhibits two refractive indices such that $n_\perp > n_\parallel$. The measured values for calcite are $n_\perp = 1.658$ and $n_\parallel = 1.486$ for light

---

[6]The absorption band is a continuum or semicontinuum of frequencies of radiation that is absorbed by the material. In other words, the material is transparent to radiation of frequencies outside its absorption band.

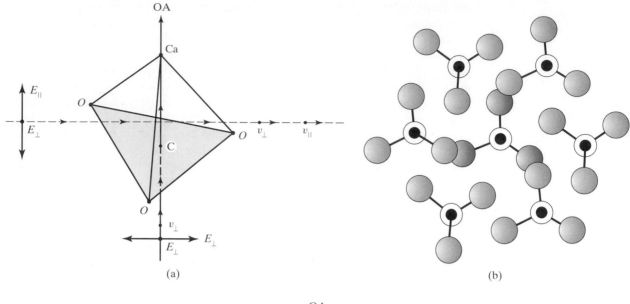

(a)

(b)

(c)

Figure 15-15 (a) Passage of light through a calcite crystal. Three oxygen atoms form the base of a tetrahedron. The optic axis is parallel to the line joining C and Ca atoms. (b) View of a portion of a calcite crystal along its optic axis. The calcium (clear) and carbon (dark) atoms (of different sizes) are aligned. The oxygen atoms forming the bases of the tetrahedrons are shown shaded. The structure has threefold symmetry of rotation about the optic axis. (c) Cleaved crystal of calcite in the shape of a rhombohedron. The optic axis direction through the crystal is parallel to a line passing symmetrically through a blunt corner where the three face angles equal 102°.

of wavelength $\lambda = 489.3$ nm. As Table 15-1 indicates, the inequality may be reversed in other materials. In materials that crystallize in the trigonal (like calcite), tetragonal, or hexagonal systems,[7] there is one unique direction through the crystal for which the atoms are arranged symmetrically. For example, the calcite atomic arrangement of Figure 15-15b shows a threefold rotational symmetry[8] about the optic axis. Such structures possessing a single optic axis are called *uni-*

[7]For a description of such crystalline systems, any textbook of solid state physics or crystallography may be consulted, for example, Charles Kittel, *Introduction to Solid State Physics* (New York: John Wiley & Sons, 1986).

[8]This means that if the tetrahedral (pyramidal) atomic structure of 3 oxygen, 1 carbon, and 1 calcium is rotated 120° about the OA axis, the resulting atomic orientation is identical with the original.

TABLE 15-1 REFRACTIVE INDICES FOR SEVERAL MATERIALS MEASURED AT SODIUM WAVELENGTH OF 589.3 nm

		n	
Isotropic (cubic)	Sodium chloride	1.544	
	Diamond	2.417	
	Fluorite	1.392	

		$n_\parallel$	$n_\perp$
Uniaxial (trigonal, tetragonal, hexagonal)	**Positive:**		
	Ice	1.313	1.309
	Quartz (SiO_2)	1.5534	1.5443
	Zircon ($ZrSiO_4$)	1.968	1.923
	Rutile (TiO_2)	2.903	2.616
	Negative:		
	Calcite ($CaCO_3$)	1.4864	1.6584
	Tourmaline	1.638	1.669
	Sodium Nitrate	1.3369	1.5854
	Beryl ($Be_3Al_2(SiO_3)_6$)	1.590	1.598

axial birefringent. Further, when $n_\parallel > n_\perp$, the crystals are said to be *uniaxial positive*, and when the inequality is reversed, they are *uniaxial negative*. Other crystal systems (for example, mica, in the so-called *orthorhombic system*) possess two such directions of symmetry or optic axes and are called *biaxial crystals*. Such materials then possess three distinct indices of refraction. Of course there are also cubic crystals such as salt (NaCl) or diamond (C) that are optically isotropic and possess the more common single index of refraction. This is also the case for materials that have no large-scale crystalline structure, such as glass or fluids. These materials are also optically isotropic, with a single index of refraction.

Naturally occurring calcite crystals are cleavable into rhombohedrons as a result of their crystallization into the trigonal lattice structure. The rhombohedron (Figure 15-15c) has only two corners where all three face angles (each 102°) are obtuse. These corners appear as the blunt corners of the crystal. The OA of calcite is directed through a blunt corner in such a way that it makes equal angles with the three faces there.

A birefringent crystal can be cut and polished to produce polarizing elements in which the OA may have any desired orientation relative to the direction of the incident light. Consider the cases represented in Figure 15-16. In (a), both representative E directions of the unpolarized light incident from the left are oriented perpendicular to the OA of the crystal. Both propagate at the same speed through the crystal with refractive index $n_\perp$. In (b) and (c), however, the

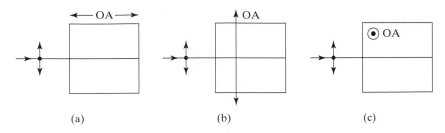

(a)   (b)   (c)

Figure 15-16 Light entering a birefringent plate with its optic axis in various orientations. (a) Light propagation along the optic axis, (b) and (c), two modes of light propagation perpendicular to the optic axis.

OA is parallel to one component of the E-field and perpendicular to the other. In this case each component propagates through the crystal with a different refractive index and speed. On emerging, the cumulative relative phase difference can be described in terms of the difference between optical paths for the two components. If the thickness of the crystal is d, the difference in optical paths is

$$\Delta = |n_\perp - n_\parallel|\, d$$

and the corresponding phase difference (or retardation) is

$$\Phi = 2\pi\left(\frac{\Delta}{\lambda_0}\right) = \left(\frac{2\pi}{\lambda_0}\right)|n_\perp - n_\parallel|\, d \tag{15-3}$$

where λ_0 is the vacuum wavelength. If the thickness of the plate is such as to make $\Phi = \pi/2$, it is a quarter-wave plate (QWP); if $\Phi = \pi$, we have a half-wave plate (HWP), and so on. These are called *zero-order* plates. Because such plates are extremely thin, it is more practical to make thicker QWPs of higher order, giving $\Phi = (2\pi)m + \pi/2$, where $m = 1, 2, 3, \ldots$. A thicker composite of two plates may also be joined, in which one plate compensates the retardance of all but the desired value of Φ of the other. In this way, we can fabricate optical elements that act as phase retarders. Mica and quartz are commonly used as *retardation plates*, usually in the form of thin, flat discs sandwiched between glass layers for added strength.

15-7 DOUBLE REFRACTION

In the cases depicted in Figure 15-16b and c, the light propagating through the crystal may develop a net phase difference between **E**-components perpendicular and parallel to the crystal's OA, but the beam remains a single beam of light. If now the OA is situated so that it makes an arbitrary angle with respect to the beam direction, as in Figure 15-17, the light experiences *double refraction*,[9] that is, two refracted beams emerge, labeled the *ordinary* and *extraordinary* rays. The extraordinary ray is so named because it does not obey Snell's law on refraction at the crystal surfaces. Thus if a calcite crystal is laid over a dark dot on a white piece of paper, or over an illuminated pinhole, two images are seen while looking into the top surface. If the crystal is rotated about the incident ray direction,

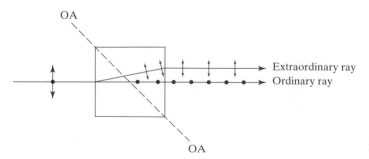

Figure 15-17   Double refraction.

[9]*Double refraction* is a term used to describe a manifestation of *birefringence* in materials, although it has literally the same meaning. Birefringence indicates the existence of two refractive indices, whereas double refraction refers to the splitting of a ray of light into ordinary and extraordinary parts.

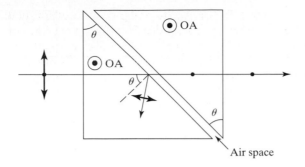

Air space

Figure 15-18 Glan-air prism. Refraction of the ray at the prism faces is not shown.

the extraordinary image is found to rotate around the ordinary image, which remains fixed in position. Furthermore, the two beams emerge linearly polarized in orthogonal orientations, as shown in Figure 15-17. Notice that the ordinary ray is polarized perpendicular to the OA and so propagates with a refractive index of $n_o = n_\perp = c/v_\perp$. The extraordinary ray emerges polarized in a direction perpendicular to the polarization of the ordinary ray. Inside the crystal, the extraordinary ray can be described in terms of components polarized in directions both perpendicular and parallel to the optic axis. The perpendicular component propagates with speed $v_\perp = c/n_\perp$, just as for the ordinary ray. The other component, however, propagates with a refractive index $n_e = n_\parallel = c/v_\parallel$. The net effect of the action of both components is to cause the unusual bending of the extraordinary ray shown in Figure 15-17.

If the two refracted rays, linearly polarized perpendicular to one another, can be well separated physically, then double refraction can be used to produce a linearly polarized beam of light. There are various devices that accomplish this. One of the most commonly used is the *Glan-air prism*, shown in Figure 15-18. Two calcite prisms with apex angle θ, as shown, are combined with their long faces opposed and separated by an air space. Their optic axes are parallel, with the orientation shown. At the point of refraction out of the first prism, the angle of incidence is equal to the apex angle θ of the prisms. The critical angle for refraction into air is given as usual by $\sin \theta_c = 1/n$ and so depends on the orientation of the $\boldsymbol{E}$-vibration relative to the OA. For $\boldsymbol{E} \parallel \text{OA}, n = 1.4864$ and $\theta_c = 42.3°$, while for $\boldsymbol{E} \perp \text{OA}, n = 1.6584$ and $\theta_c = 37.1°$. Thus by using prisms with apex angle intermediate between these values, the perpendicular component is totally internally reflected while the parallel component is transmitted. The second prism serves to reorient the transmitted ray along the original beam direction. The entire device constitutes a linear polarizer. When the air space between prisms is filled with a transparent material such as glycerin, the apex angle must be modified. Several other designs for polarizing prisms constructed from positive uniaxial material (quartz) are illustrated in Figure 15-19. Notice that in these cases, the ordinary and extraordinary rays are separated without the agency of total internal reflection. In each case, the OAs of the two prisms are perpendicular to one another, so that an $E_\perp$ component in the first prism, for instance, may become an $E_\parallel$ component in the second, with corresponding change in refractive index. Different relative indices for the two components result in different angles of refraction and separation into two polarized beams. We see that birefringent materials are useful in fabricating devices that behave as polarizers as well as in producing phase retarders such as QWPs, considered earlier in this chapter.

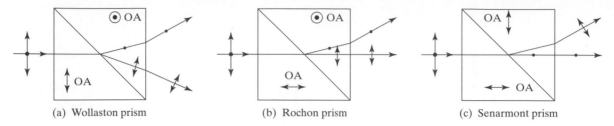

Figure 15-19 Polarizing prisms. (a) Wollaston prism, (b) Rochon prism, and (c) Senarmont prism.

15-8 OPTICAL ACTIVITY

Certain materials possess a property called *optical activity*. When linearly polarized light is incident on an optically active material, it emerges as linearly polarized light, but with its direction of vibration rotated from the original. Viewing the beam head-on, some materials produce a clockwise rotation (*dextrarotatory*) of the **E**-field, whereas others produce a counterclockwise rotation (*levorotatory*). Optically active materials include both solids (for example, quartz and sugar) and liquids (turpentine and sugar in solution). Some materials, such as crystalline quartz, produce either rotation, traceable to the existence of two forms of the crystalline structure that turn out to be mirror images (*enantiomorphs*) of one another. Optically active materials can modify the state of polarization of a beam of polarized light. The rotator mechanism involved in rotating the direction of vibration of linearly polarized light is distinct from the action of phase retarders, such as half-wave plates, which may produce the same result.

Optical activity is easily measured using two linear polarizers originally set for extinction, that is, with their TAs crossed in perpendicular orientations (Figure 15-20). When a certain thickness of optically active material is inserted between analyzer and polarizer, the condition of extinction no longer exists because the **E**-vector of the light is rotated by the optically active medium, say an amount β. The exact angle of rotation β can be measured by rotating the analyzer an amount β until extinction reoccurs, as shown. The rotation so measured depends on both the wavelength of the light and the thickness of the medium. The rotation (in degrees) produced by a 1-mm plate of optically active solid material is called its *specific rotation*. Table 15-2 gives the specific rotation ρ of quartz for a range of optical wavelengths. The amount of rotation caused by optically active liquids is much less by comparison. In the case of solutions, the specific rotation is defined as the rotation due to a 10-cm thickness and concentration of 1 gram of active solute per cubic centimeter of solution. The net angle of rotation β due to a light path L through a solution of d grams of active solute per cubic centimeter is then

$$\beta = \rho L d \tag{15-4}$$

where L is in decimeters and d is the concentration in grams per cubic centimeter. For example, 1 dm of turpentine rotates sodium light by $-37°$. The negative sign indicates that turpentine is levorotatory in its optical activity. Measurement of the optical rotation of sugar solutions is often used to determine concentration, via Eq. (15-4).[10]

---

[10]Ordinary corn syrup is often used in the optics lab to demonstrate optical activity.

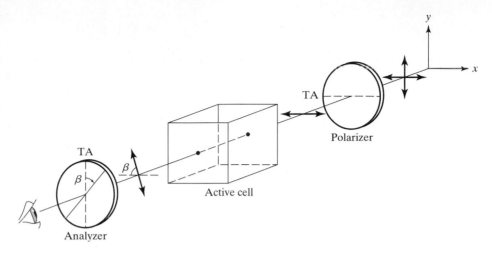

Figure 15-20 Measurement of optical activity. With the active material in place, the optical activity is measured by the angle β required to reestablish extinction.

❏ **Example**

Find the concentration of levorotatory glucose in a pure solution of the sugar, given that a 15-cm length of the solution rotates linearly polarized sodium light by 13.8°. The specific rotation of glucose for sodium light is −51.4°.

Solution From Eq. (15-4),

$$d = \frac{\beta}{\rho L} = \frac{13.8°}{(51.4° \text{ cm}^3/\text{g-dm})(1.5 \text{ dm})} = 0.179 \text{ g/cm}^3 \qquad ■$$

The dependence of specific rotation on wavelength means that if one views white light through an arrangement like that of Figure 15-20, each wavelength is rotated to a slightly different degree. This separation of colors is referred to as rotatory dispersion.

15-9 PHOTOELASTICITY

Consider the following experiment. Two polarizing filters acting as polarizer and analyzer are set up with a white-light source behind the pair. If the TAs of the filters are crossed, no light emerges from the pair, but if some birefringent material is inserted between them, light is generally transmitted in beautiful colors. To

TABLE 15-2 SPECIFIC ROTATION OF QUARTZ

λ (nm)	ρ (degrees/mm)
226.503	201.9
404.656	48.945
435.834	41.548
546.072	25.535
589.290	21.724
670.786	16.535

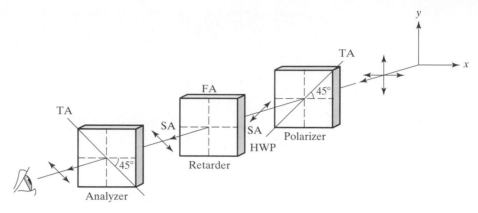

Figure 15-21 Light transmitted by crossed polarizers when a birefringent material acting as a half-wave plate is placed between them.

understand this unusual effect, consider Figure 15-21, where polarizer and analyzer TAs are crossed at 45° and −45°, respectively, relative to the x-axis. Suppose that the birefringent material introduced in the light beam constitutes a half-wave plate with its fast axis (FA) vertical, as shown. Its action on the incident linearly polarized light at +45° is to convert it to linearly polarized light perpendicular to the original direction, or at −45° inclination with the x-axis. This can be understood by resolving the incident light into equal orthogonal components along the FA and SA (slow axis) and with a 180° phase difference between them. As always, the effect of the HWP on linearly polarized light is to rotate it through 2α, or, in this case, 90° (refer to the Example at the end of Section 15-1). The light from the HWP is now polarized along a direction that is fully transmitted by the analyzer. If the retardation plate introduces phase differences other than 180°, the light is rendered elliptically polarized, and some portion of the light is still transmitted by the analyzer. Only if the phase difference is 360° or some multiple thereof, that is, if the retardation plate acts as a full-wave plate, will the character of the incident light be unmodified by the plate and the condition of extinction persist.

Now recall that the phase difference Φ introduced by a retardation plate is wavelength dependent, such that

$$\lambda_0\Phi = 2\pi d(n_\perp - n_\parallel) \tag{15-5}$$

where d is the thickness of the plate. For a given plate, the right side of Eq. (15-5) is constant throughout the optical region of the spectrum, if the small variation in the quantity $(n_\perp - n_\parallel)$ is neglected. It follows that the retardation is very nearly inversely proportional to the wavelength. Thus if the retardation plate acts as a HWP for red light, as in the arrangement of Figure 15-21, red light will be fully transmitted, whereas shorter visible wavelengths are only partially transmitted, giving the transmitted light a predominantly reddish hue. If the TA of the analyzer is now rotated by 90°, all components originally blocked are transmitted. Since the sum of the light transmitted under both conditions must be all the incident light, that is, white light, it follows that the colors observed under these two transmission conditions are complementary colors.

Sections of quartz or calcite and thin sheets of mica can be used to demonstrate the production of colors by polarization. Many ordinary materials also show birefringence, either under normal conditions or under stress, as in Figure 15-22. A crumpled piece of cellophane introduced between crossed polariz-

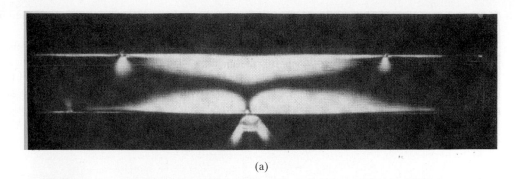

(a)

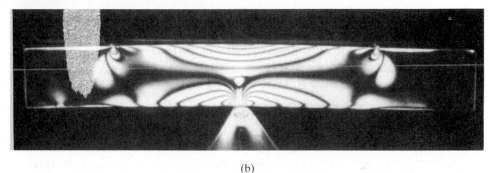

(b)

Figure 15-22 Photoelastic stress patterns for a beam resting on two supports (a) lightly loaded at the center, and (b) heavily loaded at the center. (From M. Cagnet, M. Francon, and J. C. Thrierr, *Atlas of Optical Phenomenon*, Plate 40, Berlin: Springer-Verlag, © 1962. Used by permission.)

ers shows a striking variety of colors, enhanced by the fact that light must pass through two or more thicknesses at certain points, so that Φ varies from point to point due to a change in thickness d. A similar effect is produced by wrapping glossy cellophane tape around a microscope slide, allowing for regions of overlap. Finally, Φ may also vary from point to point due to local variations in the quantity $(n_\perp - n_\parallel)$. Formed plastic pieces, such as a draftsman's drawing triangle or worker's safety glasses, often show such variations due to localized birefringent regions associated with strain. A pair of plastic safety goggles inserted between crossed polarizers shows a higher density of color changes in those regions under greater strain, because the difference in refractive index changes most rapidly in such regions. The birefringence induced by mechanical stress applied to normally isotropic substances such as plastic or glass is the basis for the method of stress analysis called *photoelasticity*. It is found that in such materials, an optic axis is induced in the direction of the stress, both in tension and in compression. Since the degree of birefringence induced is proportional to the strain, prototypes of mechanical parts may be fabricated from plastic and subjected to stress for analysis. Points of maximum strain are made visible by light transmitted through crossed polarizers when the stressed sample is positioned between the polarizers. Such polarized light patterns for a beam under light and heavy stress are shown in Figure 15-22.

PROBLEMS

15-1. Initially unpolarized light passes in turn through three linear polarizers with transmission axes at 0°, 30°, and 60°, respectively, relative to the horizontal. What is the irradiance of the product light, expressed as a percentage of the unpolarized light irradiance?

15-2. At what angles will light, externally and internally reflected from a diamond-air interface, be completely polarized? For diamond, $n = 2.42$.

15-3. How thick should a half-wave plate of mica be in an application where laser light of 632,8 nm is being used? Appropriate refractive indices for mica are 1.599 and 1.594.

15-4. Describe what happens to unpolarized light incident on birefringent material when the OA is oriented as shown. You will want to comment on the following considerations: Single or double refracted rays? Any phase retardation? Any polarization of refracted rays?

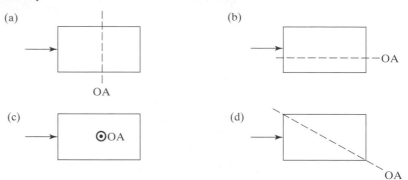

(a)

(b)

OA

OA

(c)

⊙OA

(d)

OA

(e) Which orientation(s) would you use to make a quarter-wave plate?

15-5. A number of dichroic polarizers are available, each of which can be assumed perfect; that is, each passes 50% of the incident unpolarized light. Let the irradiance of the incident light on the first polarizer be I_0.

(a) Using a sketch, show that if the polarizers have their transmission axes set at angle θ apart, the light transmitted by the pair is given by

$$I = \left(\frac{I_0}{2}\right) \cos^2 \theta$$

(b) What percentage of the incident light energy is transmitted by the pair when their transmission axes are set at 0° and 90°, respectively?

(c) Five additional polarizers of this type are placed between the two described above, with their transmission axes set at 15°, 30°, 45°, 60°, and 75°, in that order, with the 15° angle polarizer adjacent to the 0° polarizer, and so on. Now what percentage of the incident light energy is transmitted?

15-6. What minimum thickness should a piece of quartz have to act as a quarter-wave plate for a wavelength of 5893 Å in vacuum?

15-7. Determine the angle of deviation between the two emerging beams of a Wollaston prism constructed of calcite and with wedge angle of 45°. Assume sodium light.

15-8. A beam of linearly polarized light is changed into circularly polarized light by passing it through a slice of crystal 0.003 cm thick. Calculate the difference in the refractive indices for the two rays in the crystal, assuming this to be the minimum thickness showing the effect for a wavelength of 600 nm. Sketch the arrangement, showing the OA of the crystal, and explain why it occurs.

15-9. Light is incident on a water surface at such an angle that the reflected light is completely linearly polarized.

(a) What is the angle of incidence?

(b) The light refracted into the water is intercepted by the top, flat surface of a block of glass with index of 1.50. The light reflected from the glass is completely linearly polarized. What is the angle between the glass and water surfaces? Sketch the arrangement, showing the polarization of the light at each stage.

15-10. In each of the following cases, deduce the nature of the light that is consistent with the analysis performed. Assume a 100% efficient polarizer.

(a) When a polarizer is rotated in the path of the light, there is no intensity variation. With a QWP in front of (coming first) the rotating polarizer, one finds a variation in intensity but no angular position of the polarizer that gives zero intensity.

(b) When a polarizer is rotated in the path of the light, there is some intensity variation but no position of the polarizer giving zero intensity. The polarizer is set to give maximum intensity. A QWP is allowed to intercept the beam first with its OA parallel to the TA of the polarizer. Rotation of the polarizer now can produce zero intensity.

15-11. Light from a source immersed in oil of refractive index 1.62 is incident on the plane face of a diamond ($n = 2.42$), also immersed in the oil. Determine **(a)** the angle of incidence at which maximum polarization occurs and **(b)** the angle of refraction into the diamond.

15-12. The rotation of polarized light in an optically active medium is found to be approximately proportional to the inverse square of the wavelength.

(a) The specific rotation of glucose is 20.5°. A glucose solution of unknown concentration is contained in a 12-cm-long tube and is found to rotate linearly polarized light by 1.23°. What is the concentration of the solution?

(b) Upon passing through a 1-mm-thick quartz plate, red light is rotated about 15°. What rotation would you expect for violet light?

15-13. How much is linearly polarized, sodium light rotated when passed through a 20-cm length of sucrose solution prepared by dissolving 80.0 g of the sugar in water to make a volume of 1 liter? The specific rotation of sucrose, using sodium light, is 66.5°.

15-14. (a) A thin plate of calcite is cut with its OA parallel to the plane of the plate. What minimum thickness is required to produce a quarter-wave path difference for sodium light of 589 nm?

(b) What color will be transmitted by a zircon plate, 0.0182 mm thick, when placed in a 45° orientation between crossed polarizers?

15-15. The indices of refraction for the fast and slow axes of quartz with 546 nm light are 1.5462 and 1.5333, respectively.

(a) By what fraction of a wavelength is the e-ray retarded, relative to the o-ray, for every wavelength of travel in the quartz?

(b) What is the thickness of a zeroth-order QWP?

(c) If a multiple-order quartz plate 0.735 mm thick functions as a QWP, what is its order?

(d) Two quartz plates are optically contacted so that they produce opposing retardations. Sketch the orientation of the OA of the two plates. What should their difference in thickness be such that they function together like a zeroth-order QWP?

15-16. When a plastic triangle is viewed between crossed polarizers and with monochromatic light of 500 nm, a series of alternating transmission and extinction bands is observed. How much does ($n_\perp - n_\parallel$) vary between transmission bands to satisfy successive conditions for HWP retardation? The triangle is 1/16 inch thick.

15-17. A plane plate of beryl is cut with the optic axis in the plane of the surfaces. Plane polarized light is incident on the plate such that the E-field vibrations are at 45° to the optic axis. Determine the smallest thickness of the plate such that the emergent light is **(a)** plane polarized and **(b)** circularly polarized.

15-18. Find the angle at which a half-wave plate must be set to compensate for the rotation of a 1.15-mm levorotatory quartz plate using 546-nm wavelength light.

15-19. Among the so-called *Fresnel equations* is the equation giving the percent irradiance R reflected from a dielectric plane surface for the TE polarization mode:

$$R = \left(\frac{\cos\theta - \sqrt{n^2 - \sin^2\theta}}{\cos\theta + \sqrt{n - \sin^2\theta}} \right)^2$$

where θ is the angle of incidence and n is the ratio n_2/n_1.

(a) Calculate the percent reflectance for the TE mode when the light is incident from air onto glass of $n = 1.50$ at the polarizing angle.

(b) The reflectance calculated above is also valid for an internal reflection as light leaves the glass into air. This being the case, calculate the net percentage of the TE mode transmitted through a stack of 10 such plates relative to the incident irradiance I_0. Assume that the plates do not absorb light and that there are no multiple reflections within the plates.

(c) Calculate the *degree of polarization P* of the transmitted beam, given by

$$P = \frac{I_{TM} - I_{TE}}{I_{TM} + I_{TE}}$$

where I stands for the irradiance of either polarization mode.

15-20. A half-wave plate is placed between crossed polarizer and analyzer such that the angle between the polarizer TA and the FA of the HWP is θ. How does the emergent light vary as a function of θ?

REFERENCES

AZZAM, R. M., and N. M. BASHARA. 1987. *Ellipsometry and Polarized Light*. New York: Elsevier Science.

BENNETT, JEAN M., and HAROLD E. BENNETT. 1978. "Polarization." In *Handbook of Optics*, edited by Walter G. Driscoll and William Vaughan. New York: McGraw-Hill Book Company.

CHARNEY, ELLIOT. 1979. *The Molecular Basis of Optical Activity*. New York: Wiley-Interscience.

COLLETT, EDWARD. 1992. *Polarized Light Fundamentals and Applications*. New York: Marcel Dekker, Inc.

FEYNMAN, RICHARD P., ROBERT B. LEIGHTON, and MATTHEW SANDS. 1963. *The Feynman Lectures on Physics*, vol. 1. Reading, Mass.: Addison-Wesley Publishing Company. Ch. 32, 33.

HEAVENS, O. S, and R. W. DITCHBURN. 1991. *Insight into Optics*. New York: John Wiley & Sons.

KONNEN, G. P. 1985. *Polarized Light in Nature*. New York: Cambridge University Press.

LEWIS, JAMES W., DAVID S. KLIGER, and CORA E. RANDALL. 1990. *Polarized Light in Optics and Engineering*. New York: Academic Press, Inc.

MEYER-ARENDT, JURGEN R. 1994. *Introduction to Classical and Modern Optics*, 4th ed. Englewood Cliffs, N. J.: Prentice Hall.

WALDMAN, GARY. 1983. *Introduction to Light*. Englewood Cliffs, N. J.: Prentice Hall.

WATERMAN, TALBOT H. 1955. "Polarized Light and Animal Navigation." *Scientific American* (July): 88.

WEHNER, RUDINGER. 1976 "Polarized Light Navigation by Insects." *Scientific American* (July): 106.

WEISSKOPF, RICHARD F. 1969. "How Light Interacts with Matter." In *Lasers and Light*. San Francisco: W. H. Freeman and Company Publishers.

WELFORD, W. T. 1981. *Optics*, 2d ed. Oxford University Press.

16

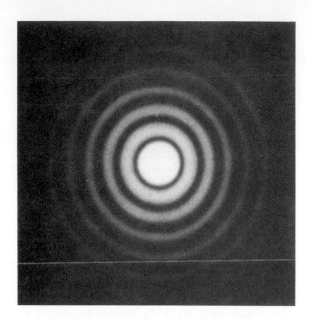

Fraunhofer Diffraction

INTRODUCTION

The wave character of light has been invoked to explain a number of phenomena, classified as "interference" effects in preceding chapters. In each case, two or more individual, coherent beams of light, originating from a single source and separated by amplitude or wave front division, were brought together again to interfere. Fundamentally, the same effect is involved in the *diffraction* of light, as we shall see.

In its simplest description, diffraction is any deviation from geometrical optics that results from the obstruction of a wave front of light. For example, an opaque screen with a round hole represents such an obstruction. On a viewing screen placed beyond the hole, the circle of light may show complex edge effects. This type of obstruction is typical in many optical instruments that utilize only the portion of a wave front passing through a round lens. Any obstruction, however, shows detailed structure in its own shadow that is quite unexpected on the basis of geometric optics.

Diffraction effects are a consequence of the wave character of light. Even if the obstacle is not opaque but simply causes local variations in the amplitude or phase of the wave front of the transmitted light, such effects are observed. Tiny bubbles or imperfections in a glass lens, for example, produce undesirable diffraction patterns when transmitting laser light. Because the edges of optical images are blurred by diffraction, the phenomenon leads to a fundamental limitation in instrument resolution, including resolution of vision, where the instrument is the eye. More often, though, the sharpness of optical images is more seriously degraded by optical aberrations due to the imaging components themselves, rather than to diffraction. When the quality of the image suffers only from the effects of diffraction, the image is said to be *diffraction-limited*. Diffraction-limited optics is good optics indeed.

The double slit studied previously constitutes an obstruction to a wave front in which light is blocked everywhere except at the two apertures. Recall that the irradiance of the resulting fringe pattern was calculated by treating the two openings as point sources, or long slits whose widths could be treated as points. The double slit interference pattern calculated earlier describes only the major features of the observed pattern. A more complete analysis of this experiment takes into account the finite size of the slits, treating each as a continuous array of point sources. When this is done, the problem is treated as a diffraction problem and the results correctly account for the actual details of the observed fringes.

Adequate agreement with experimental observations is possible through an application of the *Huygens-Fresnel principle*. According to Huygens, every point of a given wave front of light can be considered a source of secondary spherical wavelets. To this Fresnel added the assumption that the actual electric field at any point beyond the wave front is a superposition of all these wavelets, taking into account both their amplitudes and phases. Thus in calculating the diffraction pattern of the double slit at some point on a screen, one considers every point of the wave front emerging from each slit as a source of wavelets whose superposition produces the resultant field. This procedure then takes into account the continuous array of sources across both slits, rather than two isolated point sources, as in the interference calculation. Diffraction is often distinguished from interference on this basis: In diffraction phenomena, the interfering beams originate from a continuous distribution of sources; in interference phenomena the interfering beams originate from a discrete number of sources. Even though this does not represent a fundamental *physical* distinction, it demands fundamentally different mathematical techniques to carry out. The calculation of an interference pattern can be handled by algebra; the calculation of a diffraction pattern rigorously requires calculus. Nevertheless, the more predominant features of the diffraction pattern can be deduced, as they are in this chapter, without appealing to the methods of calculus.

A further classification of diffraction effects arises from certain mathematical approximations made when calculating the resultant fields. If both the source of light and observation screen are *effectively*[1] far enough from the diffraction aperture (for example, the two slits) so that wave fronts arriving both at the aperture and observation screen can be considered plane waves, we speak of *Fraunhofer*, or *far-field* diffraction, the type treated here. When this is not the case, and the curvature of the wave front cannot be ignored, we speak of *Fresnel* or *near-field* diffraction, which is not treated in this text. In the far-field approximation, the diffraction pattern changes uniformly in size only, as the viewing screen is moved further from the aperture. In the near-field approximation, the situation is more complicated: Both shape and size of the diffraction pattern depend on the distance between aperture and screen. As the screen is moved away from the aperture, the image of the aperture passes through the forms predicted, in turn, by geometrical optics, near-field diffraction, and far-field diffraction.

---

[1]Either actually distant or made to appear distant through the use of lenses.

16-1 DIFFRACTION FROM A SINGLE SLIT

The treatment of diffraction from a single, small opening or aperture regards the wave front incident on the aperture as a continuous array of even smaller areas that together fill the aperture. For the present, we imagine the aperture to be the rectangular slit shown in Figure 16-1a. In this drawing, the long dimension of the slit is aligned perpendicular to the page. The monochromatic wave front incident on the slit is a plane wave front, consistent with the requirements of Fraunhofer diffraction. According to Huygens' principle, we can consider each of the small, contiguous sections along the slit as a source of spherical wavelets, as shown for one representative section. When these wavelets reach a distant screen, they interfere to form a pattern of light and dark regions, the Fraunhofer diffraction pattern of a single slit. The requirement of a distant screen is most easily met by allowing the light emerging from the slit to be focused by a lens onto a screen positioned a distance from the lens equal to its focal length f. Parallel bundles of rays from the slit then all come to the same focus at the screen, as shown in Figure 16-1b. Rays parallel to the central axis focus at point O on the screen. Parallel rays leaving the slit at some other angle θ, like the representative set shown, focus at another point P on the screen. By considering all such bundles of rays from the slit, it is possible to make some conclusions about the nature of the resultant diffraction pattern that appears on the screen.

Before actually calculating the irradiance variation at the screen, we can get some idea of the pattern by considering the kind of interference that occurs at different points on the screen. It should be evident that the bundle of rays parallel to the axis that focus at point O are all in phase, having traveled the same

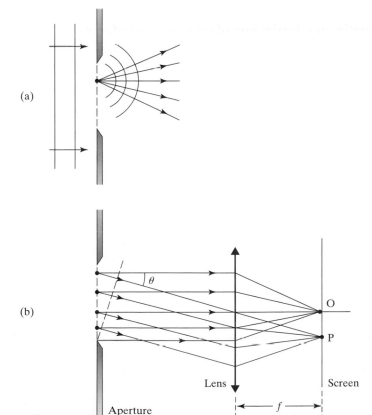

(a)

(b)

Figure 16-1 (a) Huygens' wavelets emanating from one interval of a wave front incident on a single-slit aperture. Corresponding rays diverge in all directions from the source. (b) Two sets of parallel rays, one parallel to the axis and the other making an angle θ with the axis. The rays are gathered together by a positive lens and focused on the screen a focal distance from the lens. All such bundles of rays form the diffraction pattern of a single slit.

number of wavelengths from the slit to the screen. Thus point O on the screen is a bright point of constructive interference. The same cannot be claimed for another point like P because each ray of the bundle at angle θ travels a different optical path length. Since these rays travel the same number of wavelengths from the perpendicular plane (shown dashed) to point P at the screen, they are generally out of step there due to their initial path differences from the slit to the perpendicular plane. We wish to show that there will in fact exist points like P on the screen where destructive interference between rays leads to a dark portion in the diffraction pattern. The argument proceeds as follows:

Consider Figure 16-2a, in which a single slit is positioned for Fraunhofer diffraction, with incoming monochromatic plane waves of wavelength λ, a con-

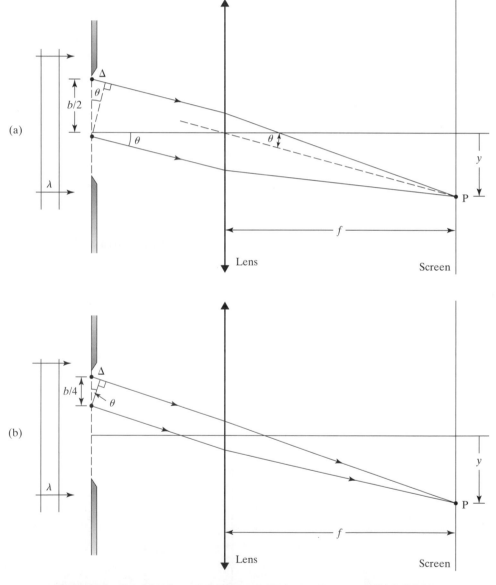

Figure 16-2   Condition for zero-irradiance in single slit diffraction. In (a) the angle θ is such that rays from the top and center of the slit are paired off to produce destructive interference at P. In (b), rays from the top and $\frac{3}{4}$ slit position are similarly paired off. In both cases, consecutive pairs of rays, moving down the slit, cancel in turn to account for zero irradiance at P for the entire slit.

verging lens of focal length f, and a screen on which to view the diffraction pattern. The slit width has the dimension b, as shown. Let the width b be divided into many very small sections—as small as you please. Consider the two (solid) parallel light rays, both making an angle θ with the axis, and originating from two separated portions of the slit—the very top section and the section just below the slit's midpoint. A line drawn from the origin of the lower ray, perpendicular to the upper ray, cuts off the line segment shown as Δ, which is the path difference between the two rays. This path difference is related to slit width b and angle θ by the right triangle shown, such that $\sin \theta = \dfrac{\Delta}{b/2}$. When the size of Δ is half the wavelength, the two rays interfere destructively at P. The same can be said of the rays leaving the two sections of the slit immediately below those just considered, and so on down the slit, until the entire width of the slit has been accounted for. Thus a dark point on the screen occurs when

$$\sin \theta = \frac{\lambda}{b}$$

The distance y from the central axis to the dark point on the screen is simply related to the angle θ by geometry: $y = f \tan \theta$.

The argument used above can now be extended by dividing the slit into four equal parts, as indicated in Figure 16-2b. In this case the path difference between rays exiting the top sections of adjacent quarters of the slit is related to the angle θ and the slit width b as follows:

$$\sin \theta = \frac{\Delta}{b/4} = \frac{\lambda/2}{b/4} \quad \text{or} \quad \sin \theta = \frac{2\lambda}{b}$$

Continuing this approach, we can generalize by dividing the slit into an even number of $2m$ equal parts, where $m = 1, 2, 3, \ldots$, so that $\sin \theta = \dfrac{\Delta}{b/2m} = \dfrac{\lambda/2}{b/2m}$ and the condition for a dark spot is

$$\sin \theta = \frac{m\lambda}{b} \quad \text{where } m = 1, 2, 3, \ldots \tag{16-1}$$

The bright peaks or maxima, which are not so simply located, fall—as it turns out—approximately midway between the minima. It should be evident that dividing the slit into an odd number of parts would not allow complete cancellation of all sections of the slit, the feature that makes the above approach successful.

❑ Example

A slit 0.120 mm wide is illuminated by monochromatic light. On a screen, placed 2 m away from the slit, the two, second-order minima in the diffraction pattern are measured to be 4.00 cm apart. Determine the wavelength.

Solution The distance L of the screen from the slit is large enough to justify the assumption of Fraunhofer diffraction, without the use of a focusing lens.

The minima in the diffraction pattern are $y = +2$ and -2 cm relative to the central axis. Using Eq. (16-1), we have, for small angles,

$$\lambda = \frac{b \sin \theta}{m} \cong \frac{b\,y}{m\,L} = \frac{(0.012 \text{ cm})(2.0 \text{ cm})}{(2)(200 \text{ cm})} = 6 \times 10^{-5} \text{ cm} = 600 \text{ nm} \quad \blacksquare$$

The technique adopted above is useful in showing that the Fraunhofer diffraction pattern of a single slit possesses alternating light and dark regions and pinpoints the locations of the zero irradiance points. However, it does not describe the irradiance pattern in detail. We now use another approach to accomplish this goal. We will see that the location of minima in the pattern just deduced also follows from the more general irradiance formula we are about to derive.

Single Slit Diffraction by Phasor Addition.

We now consider the resultant irradiance at a point on the screen as the superposition of many parallel light waves originating from the wave front at the slit, which itself has been divided into a large number of smaller sections or "sources." The light wave from each tiny section differs in phase from that of its neighbor because of the path differences already noted above. This is illustrated in Figure 16-3, where the path differences of adjoining sections is shown as Δ. If the number of intervals is N, then the total path difference for light emanating from the edges of the slit is $\Delta_T = N\Delta$, as shown. From the right triangle so formed, one can read off the relationship we shall soon make use of,

$$\sin \theta = \frac{\Delta_T}{b} = \frac{N\Delta}{b} \tag{16-2}$$

The lens of focal length f provides the customary service of collecting parallel bundles of rays making angle θ with the central axis and focusing them in a Fraunhofer diffraction pattern on the screen.

Mathematically, the problem is to divide the slit into a large number N of sections and then to find a way to add the individual waves, all making the same angle θ with the axis, all of the same amplitude E, but differing in phase ϕ each time we move from one tiny section to the next. The method we shall employ has

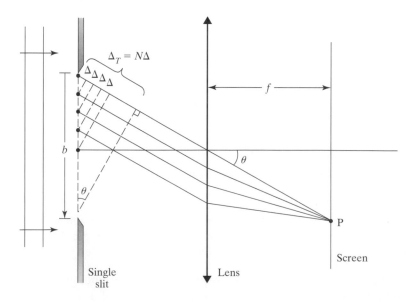

Figure 16-3 Analysis of diffraction from a single slit by addition of phasors. The slit is divided into N small intervals having the same consecutive path difference Δ. Parallel rays at inclination θ with the axis are focused at P, where the sum of their phasors produces the resultant irradiance on the screen.

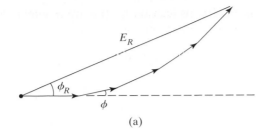

(a)

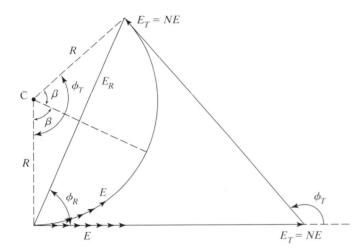

Figure 16-4 (a) Illustration of phasor addition. Five phasors of equal amplitude and phase differences add to give the resultant phasor of amplitude E_R with phase angle ϕ_R relative to the first phasor. (b) Construction used to determine the diffraction of a single slit by phasor addition. A large number N of phasors of equal amplitude and phase difference add along a circular arc to produce the resultant phasor amplitude E_R with resultant phase angle ϕ_R.

already been presented in Chapter 11, where we discussed phasor addition. In Figure 16-4a, we recall this procedure for the example of five waves of equal amplitude and differing in phase by a step-wise increase of ϕ. The result of the addition, similar to adding vectors, is a light wave of resultant amplitude E_R and phase ϕ_R. If drawn to scale, these quantities could be found by measurement on the diagram. Trigonometry can be used to calculate the result with more accuracy.

Notice that as the slit is divided into a larger number N of tiny sections, each section represents a smaller portion of the wave front and so produces a light ray of smaller amplitude. In addition, the phase differences between successive rays become smaller. The effect on a phasor diagram like that of Figure 16-4a would be to smooth out the individual phasors into the arc of a circle. This is shown in Figure 16-4b, where the resultant amplitude of the N phasors is E_R and the resultant phase is ϕ_R. Also shown are the N phasors laid out along a horizontal, all of the same phase. These phasors represent the phasor diagram for all light rays leaving the slit parallel to the central axis. The sum of their amplitudes is clearly $E_T = NE$, which must also be the length of the arc of the circle. Let the circular arc have a center at point C, and a radius of length R. The angle subtended by the arc at C, which extends from the tail of the first phasor to the head of the last phasor, must be the sum of the individual phases, the total phase $\phi_T = N\phi$. Now let the angle ϕ_T be bisected to form two equal right triangles of angle β, as shown.

From either right triangle, one has the relationship $\sin \beta = \dfrac{E_R/2}{R}$, or

$$E_R = 2R \sin \beta$$

Finally, the angle $\phi_T = 2\beta$ in radians is equal to the arc length E_T divided by the radius R of the circle, or

$$E_T = (2\beta)R$$

Dividing the last two Equations to eliminate R, we get

$$E_R = E_T \frac{\sin \beta}{\beta} \qquad (16\text{-}3)$$

Since the irradiance is proportional to the square of the amplitude, we can write, squaring both sides and lumping together constants into I_0,

$$I_R = I_0 \left(\frac{\sin \beta}{\beta} \right)^2 \qquad (16\text{-}4)$$

The ratio $(\sin \beta/\beta)$ is a well-known ratio (sinc β) that approaches a value of 1 the closer the angle β approaches zero. Thus, at $\beta = 0$, $I_R = I_0$. To relate this to the pattern on the screen, we recall the relationship between path difference Δ and corresponding phase difference ϕ, that is,

$$\phi_T = \frac{2\pi}{\lambda} \Delta_T$$

Then, with the help of Eq. (16-2), we can write

$$2\beta = \phi_T = \frac{2\pi}{\lambda}\Delta_T = \frac{2\pi}{\lambda} b \sin \theta$$

or

$$\beta = \frac{\pi b}{\lambda} \sin \theta \qquad (16\text{-}5)$$

Thus, when $\beta = 0$, $\theta = 0$, and we are looking at the center of the diffraction pattern where the irradiance is I_0. As θ increases, β increases appropriately.

Plots of $\sin \beta/\beta$ (solid line) and the single slit irradiance (dashed line) given by Eq. (16-4) are both shown in Figure 16-5. The irradiance pattern is characterized by a strong central maximum with secondary maxima and minima symmet-

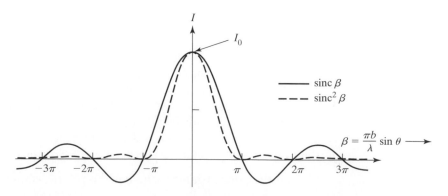

Figure 16-5 The function $\sin \beta/\beta$ (solid line) and its square, the irradiance function (dashed line), for single-slit Fraunhofer diffraction.

rically placed on either side. The maximum irradiance of the nearest secondary peak is only 4.7% that of the central peak. Although the plots shown have β as the independent variable, Eq. (16-5) permits interpretation in terms of $\sin \theta$, (proportional to β), where θ locates the position of the pattern on the screen. For example, the plot—as well as Eq. (16-4)—shows that the irradiance falls to zero for $\beta = m\pi$, where $m = \pm 1, \pm 2, \dots$. Substituting into Eq. (16-5), this gives

$$m\pi = \frac{\pi b}{\lambda} \sin \theta$$

so that the minima in the pattern satisfy

$$\text{minima:} \qquad m\lambda = b \sin \theta \quad m = \pm 1, \pm 2, \dots \qquad (16\text{-}6)$$

identical with the results found previously, Eq. (16-1). Since $\sin \theta \cong y/f$ for the central features of the irradiance pattern when θ is small, the minima fall at vertical distances y from the central axis given by

$$y \cong \frac{m\lambda f}{b} \qquad (16\text{-}7)$$

The secondary maxima fall very nearly—but not exactly—halfway between the minima.[2]

The central maximum represents essentially the image of the slit on a distant screen. We observe that the edges of the image are not sharp but reveal a series of maxima and minima that tail off into the shadow surrounding the image. These effects are typical of the blurring of images due to diffraction and will be seen again in other cases of diffraction to be considered. The angular width of the central maximum is defined as the angle $\Delta\theta$ between the first minima on either side, as illustrated in Figure 16-6. Using Eq. (16-6) with $m = \pm 1$ and approximating $\sin \theta$ by θ (in radians), we get

$$\Delta\theta = \frac{2\lambda}{b} \qquad (16\text{-}8)$$

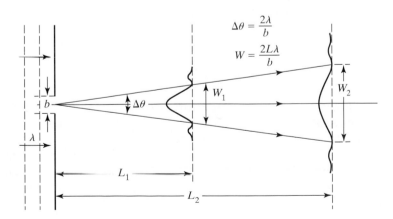

Figure 16-6   Far-field diffraction pattern of a single slit, showing the angular width of the principal maximum.

[2]It can be shown that the maxima fall precisely at values of β that satisfy the relation $\beta = \tan \beta$.

From Eq. (16-8) it follows that the central maximum of angular width $\Delta\theta$ spreads as the slit width b is narrowed. For this reason—since the length of the slit is very large compared with its width—the diffraction pattern due to points of the wave front along the length of the slit has a very small angular width and is not prominent on the screen. Of course, the dimensions of the diffraction pattern also depend on the wavelength, as indicated in Eq. (16-8).

16-2 RECTANGULAR AND CIRCULAR APERTURES

We have been describing diffraction from a slit having a width b much smaller than its length, as illustrated in Figure 16-7a. When both dimensions of the slit are comparable and small, each produces appreciable spreading, as illustrated in Figure 16-7b. For the aperture dimension a we write, analogously, for the irradiance, as in Eq. (16-4),

$$I = I_0\left(\frac{\sin \alpha}{\alpha}\right)^2 \quad \text{where} \quad \alpha \equiv \left(\frac{k}{2}\right)a \sin \theta \tag{16-9}$$

The two-dimensional pattern now exhibits zero irradiance for points x, y satisfied by either

$$y = \frac{m\lambda f}{b} \quad \text{or} \quad x = \frac{n\lambda f}{a} \tag{16-10}$$

where both m and n represent nonzero integral values. The irradiance over the screen turns out to be just a product of the irradiance functions in each dimension, or

$$I = I_0\left(\frac{\sin \beta}{\beta}\right)^2\left(\frac{\sin \alpha}{\alpha}\right)^2 \tag{16-11}$$

Photographs of single aperture diffraction patterns for rectangular and square apertures are shown in Figures 16-7c and d.

When the aperture is circular, the irradiation pattern looks like that of Figure 16-8, a pattern with rotational symmetry about the optical axis. Analysis shows that the irradiance of the central maximum now falls to zero when

$$D \sin \theta = 1.22\lambda \tag{16-12}$$

where D is the diameter of the circular aperture. Equation (16-12) should be compared with the analogous equation for the narrow rectangular slit, $b \sin \theta = m\lambda$. We see that $m = 1$ for the first minimum in the slit pattern is replaced by the number 1.22 in the case of the circular aperture. The central maximum is essentially a circle of light without a sharp boundary, the diffracted "image" of the circular aperture, and is called the *Airy disc*. Note that the far-field angular *radius* of the Airy disc, according to Eq. (16-12), is very nearly

$$\Delta\theta = \frac{1.22\lambda}{D} \tag{16-13}$$

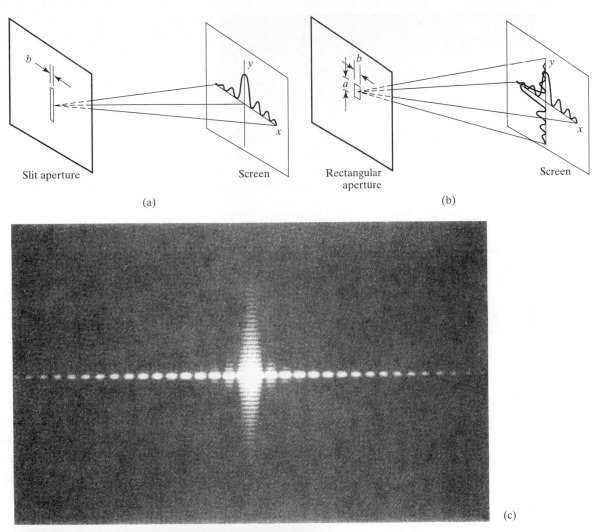

(a)

(b)

Slit aperture   Screen   Rectangular aperture   Screen

(c)

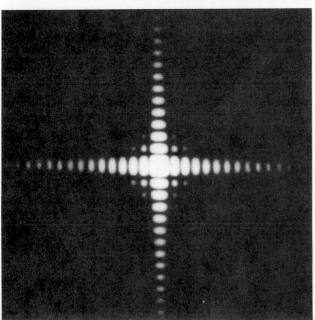

Figure 16-7 (a) Single-slit diffraction from a narrow, rectangular slit. Only the small dimension b of the slit causes appreciable spreading of the light along the x-direction on the screen. (b) Single-slit diffraction of a rectangular aperture. Both dimensions of the aperture are small and a two-dimensional diffraction pattern appears on the screen. (c) Diffraction image of a single slit, as in the representation (a). (d) Diffraction image of a single, square aperture. (Photos (c) and (d) are from M. Cagnet, M. Francon, and J. C. Thrierr, *Atlas of Optical Phenomenon*, Plate 17, Berlin: Springer-Verlag, © 1962. Used by permission.)

(d)

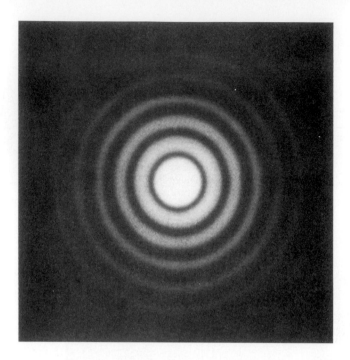

Figure 16-8 Photo of the irradiance pattern of a circular aperture. By far the largest amount of light energy is diffracted into the central maximum. The circle of light at the center corresponds to the zeroth order of diffraction and is known as the *Airy disk*. (From M. Cagnet, M. Francon, and J. C. Thrierr, *Atlas of Optical Phenomenon*, Plate 16, Berlin: Springer-Verlag, © 1962. Used by permission.)

❏ **Example**

Determine the radius of the Airy disc in the diffraction pattern formed by visible light of 632.8 nm when passing through a circular aperture of radius 20 μm. The Fraunhofer diffraction pattern is focused on a screen a distance of 1 m from a positive lens.

Solution   The radius of the Airy disc extends from its center to the circle that defines the first minimum in the diffraction pattern. The angular width of the Airy disc is given by Eq. (16-13):

$$\Delta\theta = \frac{(1.22)(632.8 \times 10^{-9})}{20 \times 10^{-6}} = 3.86 \times 10^{-2} \text{ rad}$$

If d is the diameter of the Airy disc, then

$$d = f \sin(\Delta\theta) = (1) \sin(3.86 \times 10^{-2}) = 3.86 \times 10^{-2} \text{ m} = 3.86 \text{ cm}$$

Thus, the radius of the Airy disc, half its diameter, is 1.93 cm.   ∎

16-3 RESOLUTION

In forming the Fraunhofer diffraction pattern of a single slit, as in Figure 16-1b, we notice that the distance between slit and lens is not crucial to the details of the pattern. The lens merely intercepts a larger solid angle of light when the distance is small. If this distance is allowed to go to zero, aperture and lens coincide, as in the objective of a telescope. Thus the image formed by a telescope with a round objective is subject to diffraction effects for a circular aperture, as in Figure 16-8. The sharpness of the image of a distant point object—a star, for example—is then limited by diffraction. The image occupies essentially the region of the Airy disc. An eyepiece viewing the primary image and providing further magnification merely enlarges the details of the diffraction pattern formed by the lens. The limit of *resolution* is already set in the primary image. The inevitable blur that diffraction produces in the image restricts the resolution of the instrument, that is, its ability to provide distinct images for distinct object points, either physically close together (as in a microscope), or separated by a small angle at the lens (as in a telescope). Figure 16-9a illustrates the diffraction of two point objects formed by a single lens. The point objects and the centers of their Airy discs are both separated by the angle θ. If the angle is large enough, two distinct images are clearly seen, as shown in the photograph of Figure 16-9b. Imagine now that the objects S_1 and S_2 are brought closer together. When their image patterns begin to overlap substantially, it becomes more difficult to discern the patterns as distinct, that is, to resolve them as belonging to distinct object points. A photograph of the two images at the limit of resolution is shown in Figure 16-9c.

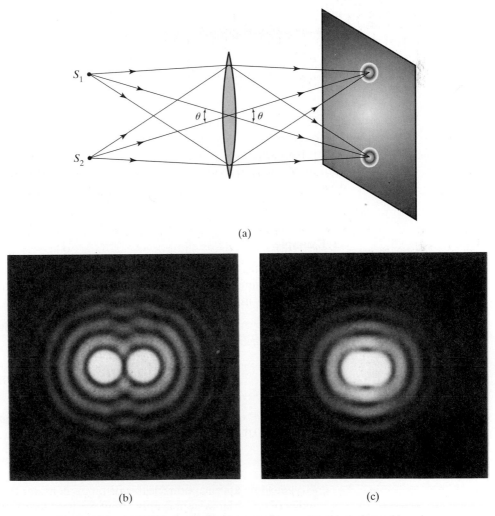

(a)

(b)   (c)

Figure 16-9 (a) Diffraction-limited images of two point objects formed by a lens. As long as the Airy disks are well separated, the images are well resolved. (b) Separated images of two incoherent point sources. In this diffraction pattern, the two images are well resolved. (c) Image of a pair of incoherent point sources at the limit of resolution. (Photos from M. Cagnet, M. Francon, and J. C. Thrierr, *Atlas of Optical Phenomenon*, Plate 16, Berlin: Springer-Verlag, © 1962. Used by permission.)

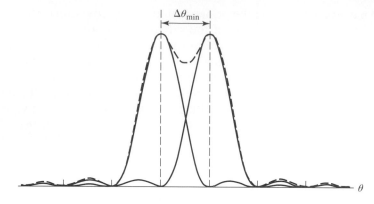

Figure 16-10 Rayleigh's criterion for just-resolvable diffraction patterns. The dashed curve is the observed sum of independent diffraction peaks.

Rayleigh's criterion for just-resolvable images—a somewhat arbitrary but useful criterion—requires that the centers of the image patterns be no nearer than the angular radius of either Airy disc, as in Figure 16-10. In this condition the maximum of one pattern falls directly over the first minimum of the other. Thus for *limit of resolution*, we have, as in Eq. (16-13),

$$(\Delta\theta)_{min} = \frac{1.22\lambda}{D} \tag{16-14}$$

where D is now the diameter of the lens. In accordance with this result, the minimum resolvable angular separation of two object points may be reduced (the resolution improved) by increasing the lens diameter and decreasing the wavelength. We consider several applications of Eq. (16-14), beginning with the following example.

❏ **Example**

Suppose that each lens on a pair of binoculars has a diameter of 35 mm. How far apart must two stars be before they are theoretically resolvable by either of the lenses in the binoculars?

Solution   According to Eq. (16-14), using $\lambda = 550$ nm as the "center" of the visible spectrum,

$$(\Delta\theta)_{min} = \frac{1.22\lambda}{D} = \frac{1.22(550 \times 10^{-9})}{35 \times 10^{-3}} = 1.92 \times 10^{-5} \text{ rad}$$

or about 4″ of arc. If the stars are near the center of our galaxy, a distance d of around 30,000 light years, then their actual separation s is approximately

$$s = d\Delta\theta_{min} = (30{,}000)(1.92 \times 10^{-5}) = 0.58 \text{ light years}$$

To get some appreciation for this distance, consider that the planet Pluto at the edge of our solar system is only about 5.5 light *hours* distant. If the stars are being detected by their long-wavelength radio waves—the lenses being replaced by dish antennas—the minimum resolution must, by Eq. (16-14), be much less. ∎

If the lens is the objective of a microscope, as indicated in Figure (16-11), the problem of resolving objects close to one another is basically the same. Mak-

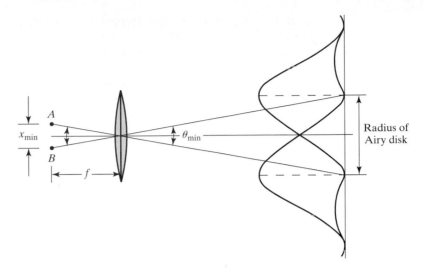

Figure 16-11 Minimum angular resolution of a microscope.

ing only rough estimates, we shall ignore the fact that the wave fronts striking the lens from nearby object points A and B are not plane, as required in far-field diffraction equations. The minimum separation x_{min} of two just-resolved objects near the focal plane of the lens is then given by

$$x_{\text{min}} = f\theta_{\text{min}} = f\left(\frac{1.22\lambda}{D}\right)$$

The ratio D/f is the *numerical aperture*, with a typical value of 1.2 for a good oil-immersion objective. Thus

$$x_{\text{min}} \cong \lambda$$

The resolution of a microscope is roughly equal to the wavelength of light used, a fact that explains the advantage of ultraviolet, X-ray, and electron microscopes in high-resolution applications.

The limits of resolution due to diffraction also affect the human eye, which may be approximated by a circular aperture (pupil), a lens, and a screen (retina), as in Figure 16-12. Night vision, which takes place with large, adapted pupils of around 8 mm, is capable of higher resolution than daylight vision. Unfortunately there is not enough light to take advantage of this situation! On a bright day the pupil diameter may be 2 mm. Under these conditions Eq. (16-14) gives $(\Delta\theta)_{\text{min}} = 33.6 \times 10^{-5}$ radian, for an average wavelength of 550 nm. Experimentally, one finds that two points separated by 1 mm at a distance of about 2 m are just barely resolvable, giving $(\Delta\theta)_{\text{min}} = 50 \times 10^{-5}$, about 1.5 times the theoretical

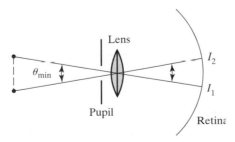

Figure 16-12 Diffraction by the eye with pupil as aperture limits the resolution of objects subtending angle θ_{min}.

limit. One's own resolution (*visual acuity*) can easily be tested by viewing two lines drawn 1 mm apart at increasing distances until they can no longer be seen as distinct. It is interesting to note that the theoretical resolution just determined for a 2-mm diameter pupil is consistent with the value of 1′ of arc (29×10^{-5} radian) used by Snellen to characterize normal visual acuity (see Chapter 10).

16-4 MULTIPLE SLIT DIFFRACTION

We generalize much of the previous discussion of diffraction from a single slit and interference from a double slit (Section 13-2) by treating the problem of N slits, where N can be a large number. It is, in fact, when N is large that we deal with the *diffraction grating*, certainly one of the most useful tools in spectral analysis. The problem we set for ourselves is that of finding the irradiance pattern due to interference from N discrete sources or slits whose width is—for the present—considered negligible. Then we shall see that the individual slits, because they do have finite width, modify the interference pattern by their own diffraction envelope.

Irradiance peaks in the diffraction pattern due to N slits are easily found, as illustrated in Figure 16-13 . Plane waves of monochromatic light impinge on an aperture of N slits spaced a distance a apart. Several of the slits are shown near the central axis. Regarding each slit as having negligible width, Huygens' wavelets emanate from each. The lens, placed at its focal length f from the screen, images sets of parallel rays from the slits, all making an angle θ with the central axis. When the path difference $\Delta = a \sin \theta$ for rays from successive slits is an integral number m of wavelengths, all rays arrive at the screen in phase and their

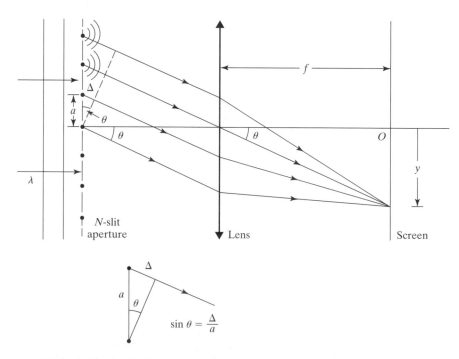

Figure 16-13 Representative grating slits illuminated by collimated monochromatic light. The construction illustrates the formation of the principal maxima due to interference between rays of light from N slits, each of negligible width.

constructive interference there produces a maximum or *peak* in the irradiance. Thus the condition for an irradiance peak is $\Delta = m\lambda$, or

$$m\lambda = a \sin \theta \qquad (16\text{-}15)$$

which is called the *diffraction grating equation*. Notice that when $\Delta = (m + \frac{1}{2})\lambda$, rays are out of phase and interfere destructively to give a point of zero irradiance. When $\theta = 0°$ for rays parallel to the central axis, a central peak corresponding to $m = 0$ is observed. Other peaks occur on either side of the central peak, corresponding to values of $m = \pm1, \pm2, \ldots$.

❏ **Example**

Where do first order diffraction peaks occur on the screen for a grating of 5000 slits/cm if light of 546 nm is directed onto the grating along the central axis, and the lens has a focal length of 1 m?

Solution The spacing a between the slits is

$$a = \frac{1 \text{ cm}}{5000} = 2 \times 10^{-4} \text{ cm} = 2 \times 10^{-6} \text{ m}$$

The distance y on the screen is measured from the central axis and is, using the geometry in Figure 16-13 (with θ small so that $\sin \theta \cong \tan \theta$) together with the grating equation,

$$y = f \sin \theta = f \frac{m\lambda}{a} = (1) \frac{(\pm 1)(546 \times 10^{-9})}{2 \times 10^{-6}} = \pm 0.273 \text{ m}$$

Thus, first order irradiance peaks occur on the screen at 27.3 cm above and below the axis point O. The minima or zero-irradiance points then fall halfway between successive peaks. ∎

The argument used above allows us to locate the interference peaks in the ideal pattern, but gives no information on other details of the irradiance function. This is remedied by the following approach, which appeals again to the technique of phasor addition.

Multiple Slit Diffraction by Phasor Addition. Refer now to Figure 16-14. Let each ray of light from the wave front at the grating have amplitude E and successive phase difference ϕ. For all rays parallel to the axis, the rays are in phase and the resultant phasor is $E_T = NE$, the total length of the phasors added head to tail along the horizontal. When successive rays have the phase difference ϕ, the phasor diagram situates the individual phasors along the arc of a circle whose length is also E_T. The arc lies on a circle whose center is at point C, and whose radius is R. The total phase difference between the first and last phasors is $\phi_T = N\phi$, as shown. The resultant phasor has the amplitude E_R, drawn from the first to the last phasor, and the phase ϕ_R. The key relationships are found in the triangles OCA and COB, which are individually sketched apart from the main diagram. We see that

$$\text{From } \Delta \text{ OCA} \qquad \sin\left(\frac{\phi}{2}\right) = \frac{E/2}{R} \qquad \text{or} \qquad E = 2R \sin\left(\frac{\phi}{2}\right)$$

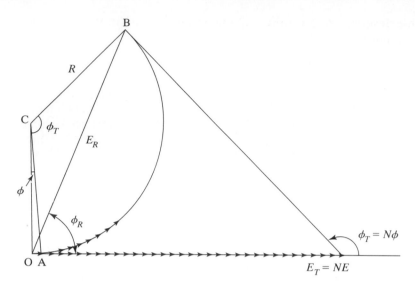

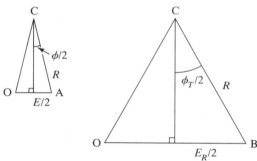

Figure 16-14 Phasor addition diagram for interference from an aperture with N slits of negligible width. The resultant is shown as the amplitude E_R with phase angle ϕ_R.

$$\text{From } \Delta \text{ COB} \quad \sin\left(\frac{\phi_T}{2}\right) = \frac{E_R/2}{R} \quad \text{or} \quad E_R = 2R \sin\left(\frac{N\phi}{2}\right)$$

Dividing these equations for the two triangles to eliminate R, we have

$$\frac{E_R}{E} = \frac{\sin(N\phi/2)}{\sin(\phi/2)} \tag{16-16}$$

Squaring both sides to get irradiance,

$$I_R = I_0 \frac{\sin^2(N\phi/2)}{\sin^2(\phi/2)} \tag{16-17}$$

where we have collected all constants in the proportionality between irradiance and square of the E-field amplitude into the factor I_0. It is customary, and a bit neater, to introduce the phase angle $\alpha \equiv \phi/2$ so that Eq. (16-17) becomes

$$I_R = I_0 \frac{\sin^2(N\alpha)}{\sin^2 \alpha} \tag{16-18}$$

The physical meaning of ϕ is the phase difference between rays arriving at the screen from successive slits. Thus,

$$\phi = 2\pi \frac{\Delta}{\lambda} = 2\pi \frac{a \sin \theta}{\lambda}$$

so that $\alpha = \phi/2$ is simply half that phase difference, or

$$\alpha = \frac{\phi}{2} = \frac{\pi}{\lambda} a \sin \theta \qquad (16\text{-}19)$$

Equation (16-18) is the irradiance on screen due to *interference* between N slits. Of course, every slit must have a finite width, but so far that width has been ignored. As we have seen, each slit produces its own diffraction pattern, given by Eq. (16-4). A rigorous calculation of the irradiance of N *finite-width* slits yields

$$I = I_0 \left(\frac{\sin \beta}{\beta} \right)^2 \left(\frac{\sin N\alpha}{\sin \alpha} \right)^2 \qquad (16\text{-}20)$$

where I_0 includes all constants. Recall that $\beta = (\pi/\lambda)b \sin \theta$ where b is the width of a single slit. Examining the factors in the irradiance of Eq. (16-20), we see that it is just a product of the interference function for N slits times the diffraction function for a single slit.

For the cases $N = 1$ and $N = 2$, Eq. (16-20) reduces to the results obtained previously for single-slit and double-slit diffraction. We have already examined the factor in β (Figure 16-5). Let us examine the factor in α, which describes the interference between the slits. As α approaches a value equal to some multiple of π, the ratio $(\sin N\alpha/\sin \alpha)^2$ approaches a value of N^2, so that $I = N^2 I_0$. Figure 16-15 shows such interference peaks for a double slit ($N = 2$) where $a = 6b$, that is, where the slit spacing a is six times the slit width b. In Figure 16-15a, the interference peaks called *principal maxima* at $\alpha = \pm m\pi$ are shown as closely spaced peaks. The height of these peaks is $N^2 I_0 = 4I_0$. However, superimposed on these peaks is the diffraction function $\sin^2 \beta/\beta^2$, shown by a dashed line. Since the irradiance is a product of the two functions, the diffraction function serves as an envelope, limiting the irradiance of the interference peaks, as shown in Figure 16-15b and in the corresponding photo, Figure 16-15d. Figure 16-15c is a photo of the diffraction envelope alone.

In Figure 16-16 are shown similar plots for the case of eight slits, for which $a = 3b$. Notice that the principal maxima are separated by $(N - 2)$ small, *secondary maxima*, and $(N - 1)$ minima, or points of zero irradiance. When N is large, therefore, these variations become small in amplitude and are not observed. Actual photos for diffraction from 2, 3, 4, and 5 slits are shown in Figure 16-17.

The practical device that makes use of such multiple slit diffraction is the *diffraction grating*. For large N, its principal maxima are bright, distinct, and spatially well-separated. For these maxima, the *diffraction grating equation* is

$$m\lambda = a \sin \theta \qquad m = 0, \pm 1, \pm 2, \ldots$$

where m is called the *order* of the diffraction. This follows from the fact that $\alpha = \frac{\pi a}{\lambda} \sin \theta$, as in Eq. (16-19), and because principal maxima occur when $\alpha = m\pi$. Together, these relations lead to the diffraction grating equation in agreement with that found earlier in Eq. (16-15).

According to the grating equation, the zeroth order of interference, $m = 0$, occurs at $\theta = 0°$—the direction of the incident light—for all wavelengths. Thus

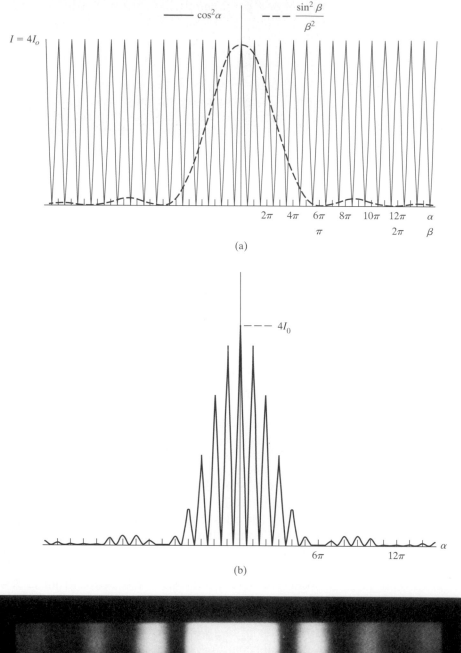

$$I = 4I_o$$

$—\cos^2\alpha$ $---\dfrac{\sin^2\beta}{\beta^2}$

2π 4π 6π 8π 10π 12π α

π 2π β

(a)

$---4I_0$

6π 12π α

(b)

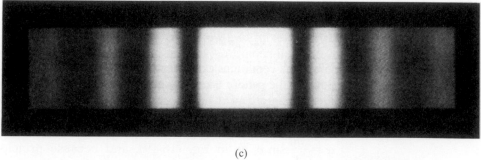

(c)

Figure 16-15 (a) Interference (solid line) and diffraction (dashed line) functions plotted for double-slit Fraunhofer diffraction when the slit separation is six times the slit width ($a = 6b$). (b) Irradiance for the double slit of (a). The curve represents the product of the interference and diffraction factors. (c) Diffraction pattern due to a single slit.

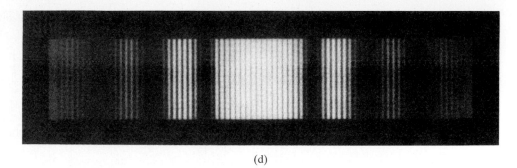

(d)

Figure 16-15 (continued) (d) Diffraction pattern due to a double-slit aperture, with each slit like the one that produced (c). (Photos from M. Cagnet, M. Francon, and J. C. Thrierr, *Atlas of Optical Phenomenon*, Plate 18, Berlin: Springer-Verlag, © 1962. Used by permission.)

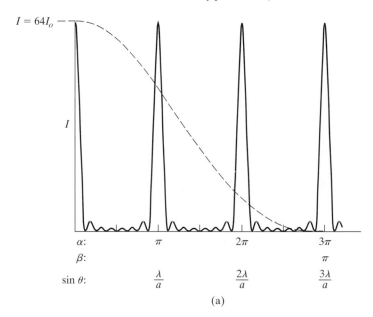

(a)

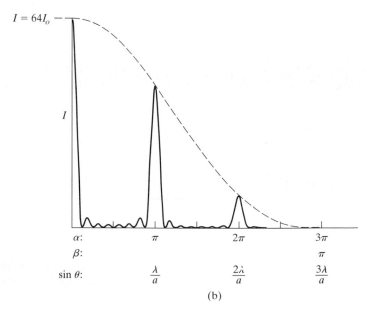

(b)

Figure 16-16 (a) Interference (solid line) and diffraction (dashed line) functions plotted for the multiple-slit Fraunhofer diffraction, when $N = 8$ and $a = 3b$. (b) Irradiance function for the multiple slit of (a). The irradiance is limited by the diffraction envelope (dashed line).

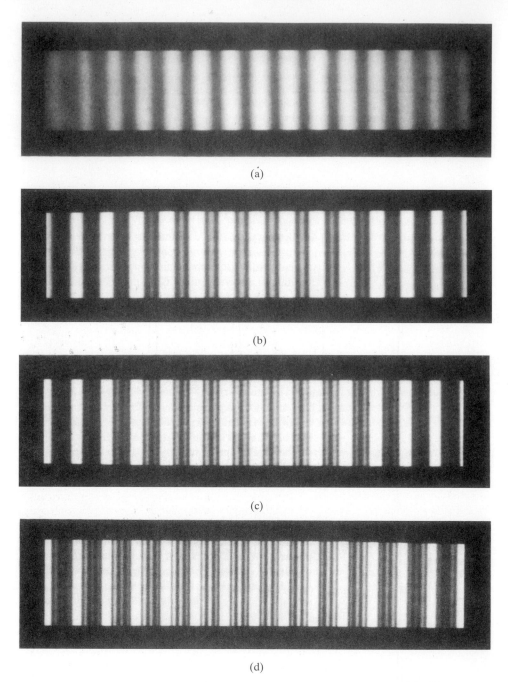

Figure 16-17 Diffraction fringes produced in turn by two, three, four, and five slits. (From M. Cagnet, M. Francon, J. C. Thrierr, *Atlas of Optical Phenomenon*, Plate 19, Berlin: Springer-Verlag, © 1962. Used by permission.)

light of all wavelengths appears in the central or zeroth-order peak of the diffraction pattern. Higher orders, both plus and minus, produce other peaks or *spectral lines*, appearing on either side of the zeroth-order peak. The direction θ of each principal maximum varies with wavelength. For $m \neq 0$, therefore, the grating separates different wavelengths of the light present in the incident beam, a feature that explains its usefulness in wavelength measurement and spectral analysis. As a dispersing element, the grating is superior to a prism in several

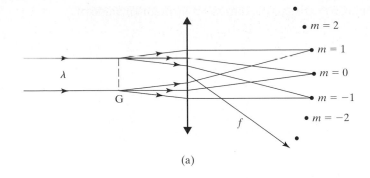

(a)

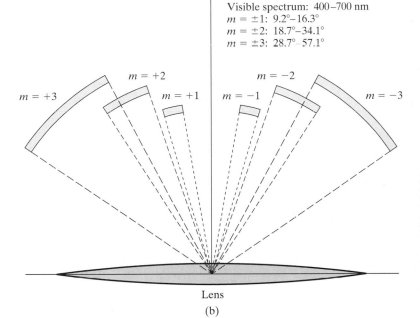

Visible spectrum: 400–700 nm
$m = \pm1$: 9.2°–16.3°
$m = \pm2$: 18.7°–34.1°
$m = \pm3$: 28.7°–57.1°

$m = +3$ $m = +2$ $m = +1$ $m = -1$ $m = -2$ $m = -3$

Lens

(b)

Figure 16-18 (a) Formation of the orders of principal maxima for monochromatic light incident normally on grating G. The grating can replace the prism in a spectroscope. Focused images have the shape of the collimator slit (not shown). (b) Angular spread of the first three orders of the visible spectrum for a diffraction grating with 400 grooves/mm. Orders are shown at different distances from the lens for clarity. In each order, the red end of the spectrum is deviated most. Normal incidence is assumed.

ways. Figure 16-18a illustrates the formation of the spectral orders of diffraction for monochromatic light. Figure 16-18b shows the angular spread of the continuous spectrum of visible light for a particular grating. Note that second and third orders, in this case, overlap. Unlike the prism, a grating produces greater deviation from the zeroth-order point for *longer* wavelengths.

❏ **Example**

Light of wavelength 500 nm is incident normally on a grating with 400 grooves/mm. The diffraction pattern is focused on a screen using a converging lens of 50 cm focal length. (a) Determine the position on the screen of the first- and second-order principal maxima. (b) Also find the total angular spread of the visible spectrum (400–700 nm) in first and second orders when white light is diffracted by the grating (refer to Figure 16-18b).

Solution   The groove separation or *grating constant* is

$$a = \frac{1}{400 \text{ gr/mm}} = 2.5 \times 10^{-3} \text{ mm/gr} = 2.5 \times 10^{-6} \text{ m}$$

(a) The monochromatic light is diffracted through an angle of

$$\text{For } m = 1, \qquad \sin \theta = \frac{\lambda}{a} = \frac{500 \times 10^{-9}}{2.5 \times 10^{-6}} = 0.200 \quad \text{or} \quad \theta_1 = 11.5°$$

$$\text{For } m = 2, \qquad \sin \theta = \frac{2\lambda}{a} = 0.400 \quad \text{or} \quad \theta_2 = 23.6°$$

Positions on the screen are found from

$$y_1 = f \tan \theta_1 = 50 \tan 11.5° = 10.2 \text{ cm}$$

$$y_2 = f \tan \theta_2 = 50 \tan 23.6° = 21.8 \text{ cm}$$

(b) For the short wavelength end of the visible spectrum at 400 nm,

$$\text{For } m = 1, \qquad \sin \theta = \frac{\lambda}{a} = \frac{400 \times 10^{-9}}{2.5 \times 10^{-6}} = 0.160 \quad \text{or} \quad \theta_1 = 9.21°$$

$$\text{For } m = 2, \qquad \sin \theta = \frac{2\lambda}{a} = 0.320 \quad \text{or} \quad \theta_2 = 18.7°$$

For the long wavelength end of the visible spectrum at 700 nm,

$$\text{For } m = 1, \qquad \sin \theta = \frac{\lambda}{a} = \frac{700 \times 10^{-9}}{2.5 \times 10^{-6}} = 0.280 \quad \text{or} \quad \theta_1 = 16.3°$$

$$\text{For } m = 2, \qquad \sin \theta = \frac{2\lambda}{a} = 0.560 \quad \text{or} \quad \theta_2 = 34.1°$$

Thus the visible spectrum encompasses an angular width of $16.3° - 9.2° = 7.1°$ in first order, and $34.1° - 18.7° = 15.4°$ in second order. It follows that the diffracted images or spectral lines of closely spaced wavelengths are better separated in higher-order grating diffraction. ■

PROBLEMS

16-1. A collimated beam of mercury green light at 546.1 nm is normally incident on a slit 0.015 cm wide. A lens of focal length 60 cm is placed behind the slit. A diffraction pattern is formed on a screen placed in the focal plane of the lens. Determine the distance between **(a)** the central maximum and first minimum, and **(b)** the first and second minima.

16-2. Call the irradiance at the center of the central Fraunhofer diffraction maximum of a single slit I_0 and the irradiance at some other point in the pattern I. Obtain the ratio I/I_0 for a point on the screen that is $\frac{3}{4}$ wavelength farther from one edge of the slit than the other.

16-3. The width of a rectangular slit is measured in the laboratory by means of its diffraction pattern at a distance of 2 m from the slit. When illuminated normally with a parallel beam of laser light at 632.8 nm, the distance between third minima on either side of the principal maximum is measured. An average of several tries gives 5.625 cm. Assuming Fraunhofer diffraction, what is the slit width?

16-4. In viewing the far-field diffraction pattern of a single slit illuminated by a discrete-spectrum source, with the help of absorption filters, one finds that the fifth minimum of one wavelength component coincides exactly with the fourth minimum of the pattern due to a wavelength of 620 nm. What is the other wavelength?

16-5. Calculate the rectangular slit width that will produce a central maximum in its far-field diffraction pattern having an angular breadth of $30°, 45°, 90°,$ and $180°$. Assume a wavelength of 550 nm.

16-6. Consider the far-field diffraction pattern of a single slit of width 2.125 μm, when illuminated normally by a collimated beam of 550 nm light. Determine **(a)** the angular radius of its central peak, and **(b)** the ratio I/I_0 at points making an angle of $\theta = 5°, 10°, 15°,$ and $22.5°$ with the axis.

16-7. A telescope objective is 12 cm in diameter and has a focal length of 150 cm. Light of mean wavelength 550 nm from a distant star enters the scope as a nearly collimated beam. Compute the radius of the central disk of light forming the image of the star on the focal plane of the lens.

16-8. Suppose that a CO_2 gas laser emits a diffraction-limited beam at wavelength 10.6 mm, power 2 kW, and diameter 1 mm. Assume that (by *multimoding*) the laser beam has essentially uniform irradiance over its cross section. approximately how large a spot would be produced on the surface of the moon, a distance of 376,000 km away from such a device, neglecting any scattering by Earth's atmosphere? What will be the irradiance at the lunar surface?

16-9. Assume a 2-mm diameter laser beam (632.8 nm) is diffraction limited and has a constant irradiance over its cross section. On the basis of spreading due to diffraction alone, how far must it travel to double in diameter?

16-10. Two headlights on an automobile are 45 in. apart. How far away will the lights appear to be just resolvable to a person whose nocturnal pupils are just 5 mm in diameter? Assume an average wavelength of 550 nm.

16-11. Assume the range of pupil variation during adaptation of a normal eye is from 2 to 7 mm. What is the corresponding range of distances over which it can detect the separation of objects 1 in. apart?

16-12. A double-slit diffraction pattern is formed using mercury green light at 546.1 nm. Each slit has a width of 0.100 mm. The pattern reveals that the fourth-order interference maxima are missing from the pattern.
(a) What is the slit separation?
(b) What is the irradiance of the first three orders of interference fringes, relative to the zeroth-order maximum?

16-13. (a) Show that the number of bright fringes seen under the central diffraction peak in a Fraunhofer double-slit pattern is given by $2(a/b) - 1$, where a/b is the ratio of slit separation to slit width.
(b) If 13 bright fringes are seen in the central diffraction peak when the slit width is 0.30 mm, determine the slit separation.

16-14. (a) Show that in a double-slit Fraunhofer diffraction pattern, the ratio of widths of the central diffraction peak to the central interference fringe is $2(a/b)$, where a/b is the ratio of slit separation to slit width. Notice the result is independent of wavelength.
(b) Determine the peak-to-fringe ratio in particular when $a = 10b$.

16-15. Make a rough sketch for the irradiance pattern due to seven, equally-spaced slits having a separation-to-width ratio of 4. Label points on the x-axis with corresponding values of α and β.

16-16. A 10-slit aperture, with slit spacing five times the slit width of 1×10^{-4} cm, is used to produce a Fraunhofer diffraction pattern with light of 435.8 nm. Determine the irradiance of the principal interference maxima of order 1, 2, 3, 4, and 5, relative to the central fringe of zeroth order.

16-17. A rectangular aperture of dimensions 0.100 mm along the x-axis and 0.200 mm along the y-axis is illuminated by coherent light of wavelength 546 nm. A 1-m focal length lens intercepts the light diffracted by the aperture and projects the diffraction pattern on a screen in its focal plane.
(a) What is the distribution of irradiance on the screen near the pattern center, as a function of x and y (in mm) and I_0, the irradiance of the pattern center?

(b) How far from the pattern center are the first minima along the x and y directions?

(c) What fraction of the I_0 irradiance occurs at 1 mm from pattern center along the x and y directions?

(d) What is the irradiance at the point ($x = 2, y = 3$) mm?

16-18. What is the angular half-width (from central maximum to first minimum) of a diffracted beam for a slit width of (a) λ; (b) 5λ; (c) 10λ?

16-19. What is the angular separation in second order between light of wavelengths 400 nm and 600 nm when diffracted by a grating of 5000 grooves/cm?

16-20. A grating, used in second order, diffracts light of 546.1 nm wavelength through an angle of 30°. How many lines per mm are ruled on the grating?

16-21. A grating is ruled with 10,000 grooves across a width of 2 cm. The two, third-order maxima in the Fraunhofer diffraction of monochromatic light appear 90° apart. What is the wavelength of the light?

16-22. Two closely-spaced yellow lines of the mercury spectrum occur at wavelengths of 577.0 nm and 579.1 nm. Light from a mercury arc source is filtered to exclude other spectral lines and allowed to fall normally on a plane transmission grating having 2500 lines per cm. The diffracted light is then focused on a flat screen by a lens of 120.0 cm focal length. Determine the separation on the screen of the two spectral lines in **(a)** first, and **(b)** second orders.

REFERENCES

BALL, C. J. 1971. *Introduction to the Theory of Diffraction*. New York: Elsevier Science.

COWLEY, JOHN M. 1995. *Diffraction Physics*, 3rd. ed. New York: Elsevier Science.

FEYNMAN, RICHARD P., ROBERT B. LEIGHTON, and MATTHEW SANDS. 1975. *The Feynman Lectures in Physics*, vol. 1. Reading, Mass.: Addison-Wesley Publishing Company. Ch. 30.

HEAVENS, O. S., and R. W. DITCHBURN. 1991. *Insight into Optics*. New York: John Wiley & Sons.

HUTLEY, M. C. 1982. *Diffraction Gratings*. New York: Academic Press, Inc.

INGALLS, ALBERT G. 1952. "Ruling Engines." *Scientific American* (June): 45.

WELFORD, W. T. 1981. *Optics*, 2d ed. Oxford University Press.

17

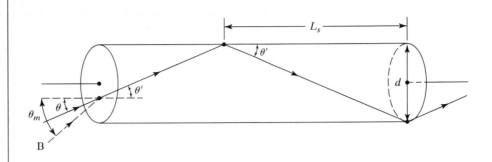

Fiber Optics

INTRODUCTION

The channeling of light through a transparent conduit has taken on great importance in recent times, as any review of laser applications would confirm. This is specially true because of its application in computers, communications, and laser medicine. As long as a transparent, solid cylinder, such as a glass fiber, has a refractive index greater than that of its surrounding medium, some of the light launched into one end succeeds in emerging from the other end, though somewhat attenuated, due to the large number of total internal reflections. A comprehensive treatment of fiber optics requires a wave approach in which the fundamental laws of electricity and magnetism (Maxwell's equations) are solved in a dielectric, subject to the physical boundary conditions[1] at the fiber walls. In this chapter we adopt a simpler and more intuitive approach, describing the propagating wavefronts by their rays, the normals to the wavefronts.

---

[1] A boundary condition is a fixed value of some variable at a specific boundary location that the solution must satisfy.

We consider now the manner in which light propagates through an optical fiber. The conditions for successful propagation are developed mainly from the point of view of geometrical optics. In addition, we consider only the meridional rays, which intersect the fiber's central axis.[2]

Consider a short section of straight fiber, pictured in Figure 17-1a. The fiber itself has a refractive index n_1, the encasing medium (called *cladding*) has index n_2, and the end faces are exposed to a medium of index n_0. Ray A, entering the left face of the fiber, is refracted there and transmitted to point C on the fiber surface, where it is partially refracted out of the fiber and partially reflected internally. The internal ray continues to D, then to E, and so on. After multiple reflections the ray will have lost a large part of its energy. Ray A does not meet the conditions for total internal reflection, that is, it strikes the fiber surface at points C, D, E, ... such that its angle of incidence φ is less than the critical angle φ_c, or

$$\varphi < \varphi_c = \sin^{-1}\left(\frac{n_2}{n_1}\right) \qquad (17\text{-}1)$$

Ray B on the other hand, which enters at a smaller angle θ_m with respect to the axis, strikes the fiber surface at F in such a way that it is refracted parallel to the fiber surface. Other rays, as in Figure 17-1b, incident at angles $\theta < \theta_m$, experience total internal reflection at the fiber surface. Such rays are propagated along the

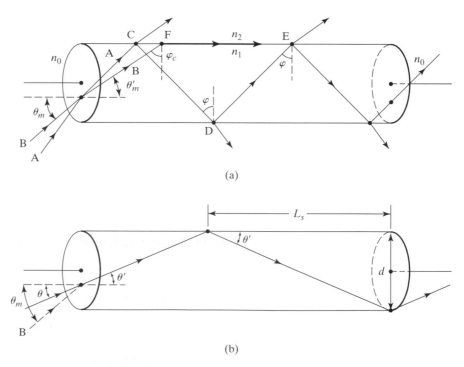

(a)

(b)

Figure 17-1 (a) Propagation of light rays through an optical fiber. Ray B defines the maximum input cone of rays satisfying total internal reflection at the walls of the fiber. (b) Propagation of a typical light ray through an optical fiber.

[2]Other rays, the *skew* rays, do not lie in a plane containing the central fiber axis. These rays take a piecewise spiral path through the fiber.

fiber by a succession of such reflections, without loss of energy due to refraction out of the cylinder. However, depending upon the fiber's transparency to the light, some attenuation occurs by absorption.

Ray B thus represents an extreme ray that defines a cone of rays, all of which satisfy the condition for total internal reflection within the fiber. The maximum half-angle θ_m of this cone is evidently related to the critical angle of reflection φ_c and can be determined. At the input face,

$$n_0 \sin \theta_m = n_1 \sin \theta_m'$$

and at a point like F,

$$\sin \varphi_c = \frac{n_2}{n_1}$$

Using the geometrical fact $\theta_m' = 90° - \varphi_c$ and the trigonometric identity $\sin^2 \varphi_c + \cos^2 \varphi_c = 1$, these relations can be combined to give the *numerical aperture*

$$\text{N.A.} \equiv n_0 \sin \theta_m = n_1 \cos \varphi_c = \sqrt{n_1^2 - n_2^2}$$

If $n_0 = 1$, the numerical aperture is simply the sine of the half-angle of the largest cone of meridional rays (i.e., rays coplanar with the fiber axis) that is propagated through the fiber by a series of total internal reflections. The numerical aperture clearly cannot be greater than unity, unless $n_0 > 1$. A numerical aperture of 0.6, for example, corresponds to an acceptance cone of 74°. The light-gathering ability of an optical fiber increases with its numerical aperture.

Also from Figure 17-1b, the *skip distance L_s* between two successive reflections of a ray of light propagating in the fiber is given by

$$L_s = d \cot \theta'$$

where d is the fiber diameter. By relating θ' to the entrance angle θ by Snell's law,

$$L_s = d\sqrt{\left(\frac{n_1}{n_0 \sin \theta}\right)^2 - 1} \tag{17-3}$$

❏ **Example**

Determine the number of internal ray reflections per meter for a fiber in air when the incident ray enters the fiber end face at an angle of 30°. The fiber's refractive index is 1.60 and its diameter is 50 μm.

Solution Substitute the given values $d = 50\ \mu\text{m}, n_0 = 1, n_1 = 1.60$, and $\theta = 30°$ into Eq. (17-5) to find the skip distance:

$$L_s = 50\sqrt{\left(\frac{1.60}{(1) \sin 30}\right)^2 - 1} = 152\ \mu\text{m} \qquad ■$$

Thus in 1 m of fiber, there are $1/L_s$, or approximately 6580 reflections. Table 17-1 lists various core and cladding possibilities, for which the critical angle, numerical aperture, and skip distances have been calculated. With so many reflections occurring, the condition for total internal reflection must be met accurately over the entire length of the fiber. Surface scratches or irregularities as

TABLE 17-1 CHARACTERIZATION OF SEVERAL OPTICAL FIBERS

Core/cladding	n_0	n_1	n_2	φ_c	θ_{max}	N.A	$1/L_s$
Glass/air	1	1.50	1.0	41.8°	90.0°	1	8944
Plastic/plastic	1	1.49	1.39	68.9°	32.5°	0.54	3866
Glass/plastic	1	1.46	1.40	73.5°	24.5°	0.41	2962
Glass/glass	1	1.48	1.46	80.6°	14.0°	0.24	1657

Note: The reciprocal of the skip distance ($1/L_s$ or skips per meter) is calculated for a fiber of diameter 100 μm and at $\theta = \theta_{max}$.

well as surface dust, moisture, or grease become sources of loss that rapidly diminish light energy. If only 0.1% of the light is lost (or 99.9% survives) at each reflection over a length of 1 m, this attenuation—for the example above— would reduce the energy by a factor of $1/(0.999)^{6580}$ or about 720. Therefore, to protect the optical quality of the fibers, it is essential that they be coated with a layer of plastic or glass, which is called the *cladding*. Cladding material need not be highly transparent but must be compatible with the fiber core in terms of, for example, expansion coefficients. The index of refraction n_2 of the cladding, where $n_2 < n_1$, influences the critical angle and numerical aperture of the fiber.

The cladding around the fiber cores has another important function, which is to prevent what is called *frustrated total internal reflection*. When the process of total internal reflection is treated as the interaction of a wave disturbance with the electron oscillators that compose the medium, it becomes apparent that there is some short-range penetration of the wave beyond the boundary. Although the wave amplitude decreases rapidly beyond the boundary, a second medium introduced into this region can couple into the wave and provide a means of carrying away energy that otherwise would have returned into the first medium. Thus, if bare optical fibers are packed closely together in a bundle, there will be some leakage between fibers, a phenomenon called *crosstalk* in communications applications. The presence of cladding of sufficient thickness prevents leakage, or, to put it obliquely, negates the frustration of total internal reflection.

The optical fiber cores are assumed above to be homogeneous in composition, characterized by a single index of refraction n_1. Light is propagated through them by multiple total internal reflections. Such fibers are called *step-index* fibers because the refractive index changes abruptly between core and cladding. See Figure 17-2a. They are *multimode* fibers if they permit a discrete number of *modes* to propagate. A mode is just one of the zigzag paths that light can take through the fiber. It can be shown that not all angles of propagation lead to self-sustaining guided modes in the fibers. Successful modes of propagation are those that satisfy a coherent phase condition: rays belonging to the same propagating wave front must remain in step despite the phase changes that accumulate in reflection and in traversing different optical paths. When the fiber is thin enough so that only one mode (a ray in the axial direction) satisfies this condition, it is said to be a *single mode* fiber.

Another type of fiber, the *graded-index* or GRIN fiber, is produced with an index of refraction that decreases continuously from the core axis as a function of radius, as shown in Figure 17-2b. A process of continuous refraction then bends rays of light entering the fiber, as illustrated in Figure 17-2c. Notice that at any point of the path, Snell's law is obeyed on a microscopic scale. Ray containment

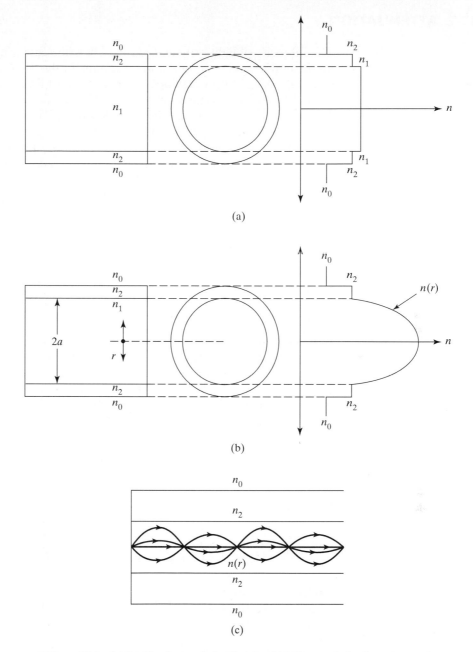

Figure 17-2 (a) Profile of a step-index fiber, in which the core index is constant and slightly greater than that of the cladding. (b) Profile of a graded-index (GRIN) fiber, showing a parabolic variation of the refractive index within the core. (c) Several ray paths within a GRIN fiber, showing their self-confinement due to continuous refraction.

now occurs by continuous refraction rather than total reflection. By Fermat's principle, we conclude that any two rays, such as those pictured in Figure 17-2c, are isochronous, a property of graded-index fiber that is important in reducing distortion of wave forms propagated through the fiber. (We shall return to this topic later in this chapter.) Like ordinary fibers, graded-index fibers are also cladded for protection.

The intensity of light propagated through a fiber invariably attenuates due to dimensional irregularities and imperfections in the fiber. If the fiber must be bent, some of the light rays may not be able to maintain the condition for total internal reflection, and some energy will escape from the core. If the bending is not too severe, that is, if the bending radius is much greater than the fiber radius, this loss may not be significant. Other extrinsic losses occur as light is coupled into and out of the fiber. In addition, defects and inhomogeneities in the core index, which inevitably develop during production, result in some intrinsic loss by scattering. Such losses are due mainly to the presence of defects in the form of impurity oxides and metallic ions, as well as localized variations in the density and refractive index of the core material. Because the dimensions of such scattering centers are much smaller than the wavelength of the light, losses are described by *Rayleigh scattering*. According to this theory,[3] the amount of scattering is inversely proportional to the fourth power of the wavelength. Thus longer wavelengths are subject to less scattering. For example, an optical fiber transmitting at the wavelength of 1.3 μm, rather than 800 nm, represents a $(1.3/0.8)^4$ or sevenfold reduction in Rayleigh scattering losses.

Until recently, by far the greatest loss in fiber-optic propagation has been by material absorption. In certain wavelength regions, absorption by impurity atoms and atomic defects present in the fiber may be much greater than that due to the fiber material itself. Due to improved methods of production, modern fibers made from fused silica exhibit cumulative absorption losses over 1 km of length, comparable to losses once measured over only 1 m of length.

Absorption losses over a length L of fiber can be described by the usual exponential law for light irradiance I,

$$I = I_0 e^{-a/L}$$

where a is an *attenuation* or *absorption coefficient* for the fiber, a function of wavelength.[4] For optical fibers, the defining equation for the absorption coefficient in *decibels per kilometer* (db/km)

$$\alpha z \equiv 10 \log_{10}\left(\frac{P_1}{P_2}\right) \tag{17-4}$$

where P_1 and P_2 refer to power levels of the light at cross sections 1 and 2, illustrated in Figure 17-3. The distance z is usually chosen to be 1 km. Dramatic ad-

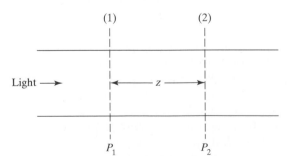

Figure 17-3 Schematic used to define the absorption coefficient for a glass fiber.

[3]Rayleigh scattering was discussed in greater detail in Section 15-5.

[4]Since rays that strike the fiber wall at smaller angles of incidence travel a greater distance through the same axial length L of the absorbing medium, a is also a function of the angle of incidence.

vances have been made in reducing the absorption of fused silica so that today fibers rated at 0.2 db/km (operating at 1.55 μm) are readily available. Plastic fibers are less expensive but not nearly as transparent. Their overall attenuation is at least an order of magnitude higher than for glass. Glass fibers are therefore preferable in long-distance applications. Glass fibers are also better transmitters in the near-infrared region.

A relatively low-loss fiber might have a loss rating of 10 db/km, which means that only 10% of the light energy launched into a km-long fiber at one end will arrive at the other end. Fused silica fibers having attenuations of 3 to 5 db/km are readily available today. At transmission wavelengths greater than or equal to 1.2 μm, fibers can be produced having attenuations as low as 0.5 db/km.

❏ **Example**

Suppose that at an attenuation of 5 db/km, a particular communications fiber can transmit laser pulses a distance of 14 km before any amplification is necessary. What is the net attenuation of the signal before amplification, in both db and fraction of the original intensity?

Solution   The total attenuation is calculated from the product,

$$\alpha z = 5 \text{ db/km} \times 14 \text{ km} = 70 \text{ db}$$

Thus

$$10 \log_{10} \frac{P_1}{P_2} = 70 \text{ db}$$

$$\log_{10} \frac{P_1}{P_2} = 7 \quad \text{or} \quad \frac{P_1}{P_2} = 10^7$$

so that $P_2 = 10^{-7} P_1$. The net attenuation of 70 db means that the energy has been reduced to 10^{-7} of its original value at input.   ∎

17-3 DISTORTION

Light transmitted by a fiber may not only lose power by the mechanisms just mentioned, it may also lose information through *pulse broadening*. When input light is modulated it can convey information, as in the case of radio waves (amplitude or frequency modulation: AM or FM). If the input is a pattern of square-wave pulses, digital information can be sent over the fibers. Consider a single square wave input into a fiber. The output pulse detected at the other end suffers, in general, from both attenuation and distortion. To be detected as a single pulse, the output pulse must not spread to the extent of significant overlap with neighboring pulses. This requirement places a limitation on the frequency of input pulses or the rate at which bits of information may be sent. The major causes of pulse broadening due to distortion include *modal distortion* and *material dispersion*, described in the following sections.

Modal Distortion.   Figure 17-4 indicates schematically the input of a square wave (a digital signal) into a fiber. Modal distortion occurs because propagating rays (fiber modes) travel different distances before arriving at the output. Consequently, these rays arrive at different times, broadening the square

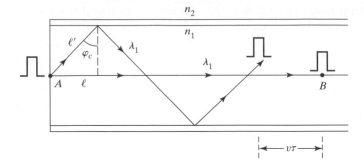

Figure 17-4 Symbolic representation of modal distortion. A square wave input arrives at the fiber end at different times, depending on the path taken. Shown are the extreme paths: an axial ray and one propagating at the critical angle. Their separation after any time interval τ is given by $v\tau$.

wave. The shortest distance L from A to B is taken by the axial ray; the longest distance L' from A to B is taken by the steepest propagating ray that reflects at the critical angle φ_c. The sine of the critical angle is equal both to the ratio of refractive indices and to the distances ℓ and ℓ' shown in the right triangle of Figure 17-4. Thus

$$\sin \varphi_c = \frac{n_2}{n_1} = \frac{\ell}{\ell'} = \frac{L}{L'} \tag{17-5}$$

The last member of the equation follows because the ratio of the total paths of the two rays after any number of reflections must be the same as their ratio over one reflection. The transit time for each ray is its distance L divided by its velocity v, where $v = c/n_1$. Thus the difference in transit times τ is given by

$$\tau = \frac{L'}{v} - \frac{L}{v} = \frac{n_1}{c} (L' - L) \tag{17-6}$$

❏ **Example**

Find the (a) length and (b) difference in transit times τ for the longest and shortest trajectories in a step-index plastic fiber of 1 km that has a core index of 1.49 and a cladding index of 1.39.

Solution   (a) The shortest trajectory L occurs for an axial ray and is thus equivalent to the fiber length: $L = 1$ km. The longest trajectory L' occurs for a ray incident at just greater than the critical angle and is found from Eq. (17-5):

$$\frac{L'}{L} = \frac{n_1}{n_2}$$

so that

$$L' = \frac{n_1}{n_2}L = \frac{1.49}{1.39} (1 \text{ km}) = 1.072 \text{ km}$$

(b) The transit time difference τ is then, using Eq. (17-6),

$$\tau = \frac{n_1}{c} (L' - L) = \frac{1.49}{3 \times 10^8 \text{ m/s}} (1.072 - 1) \times 10^3 \text{ m} = 3.58 \times 10^{-7} \text{ s} = 358 \text{ ns} \quad \blacksquare$$

Modal dispersion can be reduced by using fibers of very small diameter (typically around one optical wavelength) that eliminate all but the fastest modes from propagation. Whereas multimode fibers may have diameters in the approximate range of 25 to 50 μm, fibers of core diameter less than about 15 μm can function as single mode when used with appropriate cladding. However, such small fibers are expensive and more difficult to handle when they must be spliced and coupled together. The graded-index (GRIN) fibers mentioned previously are easier to manage, with core diameters in the 50 to 100 μm range. The fact that the various rays are isochronous in this case should make clear the advantages of such fibers in reducing modal dispersion. Equalizing the time delays of the propagating modes in a fiber reduces pulse broadening and, consequently, increases the pulse-rate capability or the *bandwidth* of the fiber. Thus bandwidths of graded-index multimode fibers are greater than those of step-index multimode fibers, and the bandwidths of single-mode fibers are greater still.

Material Dispersion. Even if modal distortion is absent, some pulse broadening still occurs because the refractive index is a function of wavelength. The dependence of refractive index on wavelength is generally referred to as *dispersion*. Dispersion for a silica fiber is shown in Figure 17-5. Since no light source can be precisely monochromatic, the light propagating in the fiber is characterized by a spread of wavelengths determined by the purity of the light source. Each wavelength component has a different refractive index and therefore a different speed through the fiber. Pulse broadening occurs because each component arrives at a slightly different time. The more monochromatic the light, the less the distortion due to material dispersion (also called *chromatic dispersion*). The best monochromatic sources are semiconductor lasers, but light-emitting diodes can also be used in many applications.

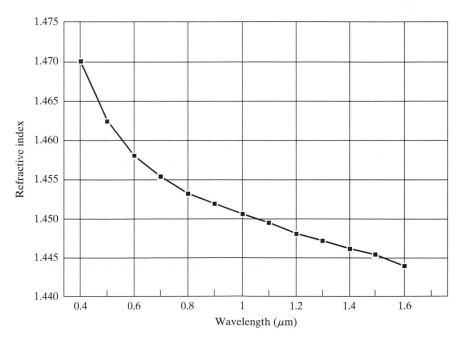

Figure 17-5 Dispersion in fused quartz.

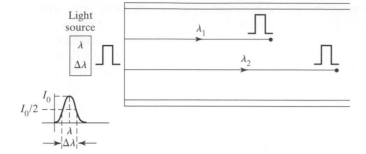

Figure 17-6 Symbolic representation of material dispersion. A square wave input arrives at the fiber end at different times, depending on wavelength. The spectral output of the light source is characterized both by a central wavelength λ and a spectral width $\Delta\lambda$.

Pulse broadening due to material dispersion is negligible at wavelengths near an inflection point[5] in the ordinary dispersion curve (n versus λ). For silica glass fibers, this occurs near 1.3 μm.

Figure 17-6 illustrates material dispersion by showing the progress of two square pulses (initially coincident) in a fiber at wavelengths λ_1 and λ_2. If the corresponding refractive indices are n_1 and n_2, the figure implies that $n_1 > n_2$. These wavelengths are but two in a continuum described by the spectral width $\Delta\lambda$ of the source, usually chosen as the width of the source's spectral output at half-maximum, as shown in Figure 17-6. It can be shown that the pulse broadening due to *material dispersion*, which we symbolize by the Greek delta δ, is proportional to the spectral width $\Delta\lambda$ of the light source,

$$\delta = -M\,\Delta\lambda \tag{17-7}$$

where M is a proportionality factor that depends on λ. For pure silica the value of M can be found from Figure 17-7, where it is plotted versus the peak wavelength of the spectral output of the light source. Common units of δ are ps/km, so the factor M should be used in units of ps/nm-km to agree with Eq. (17-7).[6]

❑ **Example**

Using Figure 17-7, calculate the pulse spread due to material dispersion in pure silica for both LED (light-emitting diode) and LD (laser diode) light sources. Consider the source wavelength to be 0.82 μm with a spectral width of 20 nm for the LED and 1 nm for the more monochromatic LD.

Solution   At 0.82 μm, Figure 17-7 gives a value for M near 100 ps/nm-km or 0.10 ns/nm-km. Using Eq. (17-7) then, we have

LED: $\delta = (0.10\text{ ns/nm-km})(20\text{ nm}) = 2.0\text{ ns/km}$

LD: $\delta = (0.10\text{ ns/nm-km})(1\text{ nm}) = 0.10\text{ ns/km}$

At 0.82 μm, the LD is 20 times better than the LED, as a direct result of its superior monochromaticity.   ∎

Notice also that pulse broadening due to material dispersion is much smaller than that due to modal distortion. Material dispersion therefore becomes

---

[5]An inflection point in a curve is the point where the slope is changing from increasing to decreasing, or vice versa.

[6]A picosecond (ps) is 10^{-12} second or 10^{-3} nanosecond (ns).

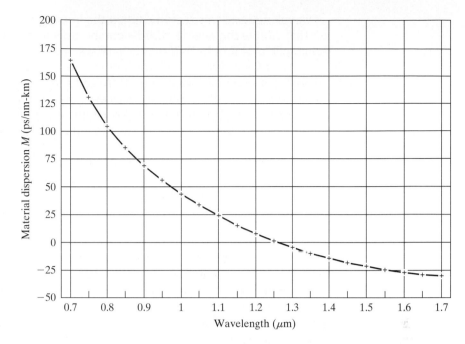

Figure 17-7 Material dispersion in pure silica. The quantity M, representing the pulse broadening in picoseconds (ps) per unit of spectral width in nanometers (nm), per unit of fiber length in kilometers (km), is plotted against the wavelength. Pulse broadening becomes zero at 1.27 μm and is negative as the wavelength increases further.

significant only when modal distortion is greatly reduced, as in single mode fibers. Consequently, in the presence of modal distortion, the advantage of superior monochromaticity of a LD over a LED is lost. In applications where fiber lengths are short enough, plastic fibers and LED sources may well represent the best compromise between performance and cost. Finally, notice from Figure 17-7 that M actually passes through zero at around 1.27 μm, so that material dispersion can also be reduced by finding light sources that operate in this spectral region.

17-4 SOURCES AND DETECTORS

In addition to the development of low-loss fibers and appropriate cladding materials, the practical implementation of the basic principles of fiber optics has been achieved through the simultaneous development of appropriate semiconductor light sources and detectors. The most widely used light source in fiber-optic systems is a solid-state device, the light-emitting diode (LED). For example, solid solutions of gallium arsenide-gallium phosphide (GaAs-GaP) permit light emission somewhere in the range of 550 to 910 nm, depending on the relative concentrations of the two materials. For fiber-optic applications, the wavelength is selected that best matches the spectral transmission characteristics of the fiber. In most glass fibers, longer wavelengths in the approximate range of 1.2 to 1.4 μm are preferred because they correspond to a region of low material absorption and of minimal material dispersion. Longer wavelengths also lead to a reduction in losses due to Rayleigh scattering, as we have seen. The advantages of wavelengths in this region explain the interest in LED semiconductor materials that emit in the near infrared—compounds such as GaAs and InP, ternary systems such as AlGaAs, and quaternary systems such as InGaAsP.

The plastic dome of the LED device can serve as a lens. The dome is shaped so that it projects a radiation pattern optimized to match precisely the acceptance cone of the fiber. Such devices are rugged, dependable, and consume little power. Their spectral output, however, typically spans an interval of about 35 nm, compared with about 2 nm for a semiconductor injection laser diode. The semiconductor laser is ideally suited as a source that can couple maximum radiation into the optical fiber. It is very small, with an emitting area that is well matched to the input face of the fiber. It provides an intense, coherent, and well-collimated output beam. Because of the semiconductor laser's higher cost, shorter lifetime, and more elaborate power supply, as well as the need to provide cooling, however, the LED very often remains the compromise choice.

Semiconductors that serve as detectors are also useful at the output end. Those photoconducting or photovoltaic cells, whose response is matched to the spectral output of the source, are most often employed. The two most commonly used photodiode detectors are the *pin* and the *avalanche* diodes, designed for greater sensitivity and efficiency of response.

17-5 APPLICATIONS

The simplest use of optical fibers, either singly or in bundles, is as light pipes. For example, a flexible bundle of fibers might be used to transport light from inside a vacuum system to the outside, where it can be more easily analyzed. Interestingly, the rods and cones of the human eye have been shown to act as light pipes, transmitting light along their lengths, as in optical fibers. For such nonimaging applications the fibers can be randomly distributed within the cable. When imaging is required, however, the fiber ends at input must be coordinated with the fiber ends at output. To maintain this coordination, fibers at either end are bonded together. The fiberscope, a bundle of such fibers, end-equipped with objective lens and eyepiece, is routinely used by physicians, for example, to examine regions of the stomach, lungs, and duodenum. Some of the fibers function as light pipes, transporting light from an external source to illuminate inaccessible areas internally. Other fibers return the image.

Fibers can be bound rigidly by fusing their cladding. In this way, for example, fiber optic faceplates are made for use as windows in cathode ray tubes. Further, when such fused-fiber bundles are tapered by heating and stretching, images can be magnified or diminished in size, depending on the relative areas of input or output faces.

The resolving power of imaging fibers depends on the accuracy of fiber alignment and, as might be expected, on the individual fiber diameter d. A conservative estimate (Siegmund, 1978) of fiber resolving power RP is given by

$$\text{RP (lines/mm)} = \frac{500}{d(\mu\text{m})}$$

Thus a 5-μm fiber, for example, can produce a high resolution of about 100 lines/mm.

Flexible optical fibers also find increasing application as sensing devices. Because the light carried by the optical fiber is very sensitive to changing ambient conditions, such as strain, vibration, temperature, and pressure, fibers are used as sensors to make measurements without the need for electrical connections. Fiber optic systems used to monitor the environment of a particular location can be sensitive to a change in pressure of 0.01 psi or a change in position of one Angstrom unit (10^{-8} m). Any external influence that affects the fiber length causes a phase change in the light signal, which is detected and analyzed.

The most far-reaching applications of optical fibers lie in the area of voice or video communications and data transmission. The replacement of microwaves and radio waves by light waves is especially attractive, since the information-carrying capacity of the carrier wave increases directly with the width of the frequency band available. Optical frequencies are some five orders of magnitude higher than, say, microwave frequencies corresponding to wavelengths of several centimeters. This means, for example, that a 24-fiber cable has the capacity to handle over 8000 two-way conversations, compared with only 144 if the fibers are replaced by the usual copper conductors and the carriers are microwaves. Replacement of copper coaxial cable by fiber optic cable thus offers greater communication capacity in a smaller space. Additionally, in contrast with metallic conduction techniques, communication by light offers the possibility of complete electrical isolation, immunity to electromagnetic interference, and freedom from signal leakage. The latter is especially important where security of information is vital, as in computer networks that handle confidential data. An effective way to transmit information accurately, given the distortion that occurs due to modal and spectral broadening, is to encode the signal into digital form before transmitting it. In this way, retrieval of the signal at some distance down the line depends only on the recognition of either the presence or the absence of a pulse representing a binary digit. Minor distortions and noise can therefore be tolerated as long as pulses can be detected and regenerated, free of distortion. At the receiving end, the signal is decoded to reproduce the original input.

Figure 17-8 illustrates such a fiber-optic communications system, from message source to message output. At the input end of the fiber-optic cable, the information to be transmitted is converted by some type of transducer from an electrical signal to an optical one; at the ouput end it is reconverted from an optical to an electrical signal. The fiber serves as an *optical waveguide* to propagate

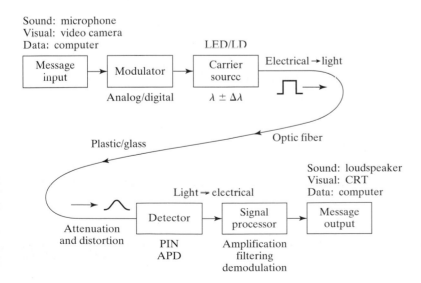

Figure 17-8 Overview of a fiber-optics communication system.

the information with as little distortion and loss of power as possible, over a distance that can range from meters to thousands of kilometers.

Looking at Figure 17-8, we can follow a signal from input to output. The message source might be audio, providing an analog electrical signal from a microphone; it might be visual, providing an analog signal from a video camera; or it might be digitally-encoded information, like computer data in the form of a train of pulses. Analog and digital formats are convertible into one another so that the choice of format for transmission through the fiber is always available, regardless of the original nature of the signal.

The purpose of the *modulator* is to perform this conversion when desired and to impress this signal onto the carrier wave generated by the *carrier source*. The carrier wave can be modulated to contain the signal information in various ways, usually by amplitude modulation (AM), frequency modulation (FM) or digital modulation. In fiber optics systems, the carrier source is either a light-emitting diode (LED) or a laser diode (LD). The ideal carrier wave is a single-frequency wave of adequate power to propagate long distances through the fiber. However, no real light source can propagate radiation of a single frequency. The LD and LED are characterized by a frequency band with central wavelength λ and spectral width $\Delta\lambda$. As we have seen, the LD is preferable because it has a narrower spectral width, approaching the ideal single frequency more closely than the LED; however, other considerations make the LED acceptable in some systems, as we have seen.

In Figure 17-8, the carrier source output into the optic fiber is represented by a single square pulse. As this pulse propagates through the fiber, it suffers both attenuation (loss of amplitude) and distortion (change in shape) due to several mechanisms discussed earlier. If the fiber is very long, it may be necessary to intercept the signal at one or more intermediate points with a *repeater* that amplifies and restores the original pulse shape.[7] At the remote end of the fiber, the light signal is coupled into a detector that changes the optical signal back into an electrical signal. This service is performed by a semiconductor device, most commonly a *PIN diode*, an *avalanche diode*, or a *photomultiplier* (Section 2-7). Of course, the response of the detector should be well matched to the optical frequency of the signal received. The detector output is then handled by a signal processor, whose function is to recapture the original electrical signal from the carrier, a process that involves filtering, amplification, and possibly, a digital-to-analog conversion. The message output may then be communicated by loudspeaker (audio), by cathode-ray tube (video), or by computer input (digital).

Within the last two decades, many intercity networks and some transoceanic networks have been outfitted with fiber optic cable. The aim of continuing research efforts is to achieve fiber-optic systems that can handle large-capacity data transmission (measured in bits per second) and that permit large distances between stations (*repeater spacing*), where signal restoration and amplification must occur. Repeater spacing decreases rapidly with the demand for higher data rates. For example, a glass-fiber network connecting the states of Virginia and Massachusetts uses fiber cables containing around 50 fiber pairs. Light pulses are transported along the fibers at a rate of around 90 million bits per second (90 Mbps), the equivalent of approximately 1350 telephone circuits per pair

---

[7]In some applications, the fiber is able to overcome its losses and reduce or eliminate the need for repeater stations. For example, lightly erbium-doped fibers are used, pumped from the transmitter end so that the fiber itself serves as a distributed amplifier. See Urquhart and Whitley (1990) and Desurvire (1992).

of fibers. Pulses are reamplified every 6.2 miles. At the present time, using single-mode fibers and a laser source at 1.55 μm, data rates of around 400 Mbps with repeater spacings greater than 71 miles appear to be achievable—a significant improvement.

PROBLEMS

17-1. **(a)** Show that, for a guided ray traveling at the steepest angle relative to the fiber axis, the skip distance L_s can be expressed by

$$L_s = \frac{n_2 d}{\sqrt{n_1^2 - n_2^2}}$$

(b) How many reflections occur per meter for such a ray in a step-index fiber with $n_1 = 1.460$, $n_2 = 1.457$, and $d = 50$ μm?

`17-2. Refractive indices for a step-index fiber are 1.52 for the core and 1.41 for the cladding. Determine **(a)** the critical angle; **(b)** the numerical aperture; **(c)** the maximum incidence angle θ_m for light that is totally internally reflected.

17-3. A step-index fiber 0.0025 inch in diameter has a core of index 1.53 and a cladding of index 1.39. Determine **(a)** the numerical aperture of the fiber; **(b)** the *acceptance angle* (or maximum entrance cone angle); **(c)** the number of reflections in 3 ft of fiber for a ray at the maximum entrance angle, and for one at half this angle.

17-4. **(a)** Show that the actual distance x_s a ray travels during one skip distance is given by

$$x_s = \frac{n_1 d}{\sin \theta}$$

where θ is the entrance angle and the fiber is used in air.

(b) Show that the actual total distance x_t a ray with entrance angle θ travels over a total length L of fiber is given by

$$x_t = \frac{n_1 L}{(n_1^2 - \sin^2 \theta)^{1/2}}$$

(c) Determine x_s, L_s, and x_t for a 10-m-long fiber of diameter 50 μm, core index of 1.50, and a ray entrance angle of 10°.

17-5. A flux of 5 μW exists just inside the entrance of a fiber 100 m long. The flux just inside the fiber exit is only 1 μW. What is the absorption coefficient of the fiber in db/km?

17-6. An optic fiber cable 3 km long is made up of three 1-km lengths, spliced together. Each length has a 5-db loss and each splice contributes a 1-db loss. If the input power is 4 mW, what is the output power?

17-7. The attenuation of a 1-km length of RG-19/U coaxial cable is about 12 db at 50 MHz. Suppose the input power to the cable is 10 mW and the receiver sensitivity is 1 μW. How long can the coaxial cable be under these conditions? If optical fiber is used instead, with a loss rated at 4 db/km, how long can the transmission line be?

17-8. A germanium-doped silica fiber has an attenuation loss of 1.2 db/km due to Rayleigh scattering alone, when light of wavelength 0.90 μm is used. Determine the attenuation loss at 1.55 μm.

17-9. **(a)** Show that the attenuation db / km is given by

$$\alpha = 10 \log_{10} (1 - f)$$

where f is the overall fractional power loss from input to output over a 1-km long fiber.

(b) Determine the attenuation in db / km for fibers having an overall fractional power loss of 25%, 75%, 90%, and 99%.

17-10. Determine **(a)** the length, and **(b)** transit time for the longest and shortest trajectories in a step-index fiber of 1 km, having a core index of 1.46 and a cladding index of 1.45.

17-11. Evaluate modal distortion in a fiber by calculating the difference in transit time through a 1-km fiber required by an axial ray and a ray entering at the maximum entrance angle of 35°. Assume a fused silica core index of 1.446. What is the maximum frequency of input pulses that produce nonoverlappping pulses on output for this case of modal dispersion?

17-12. Calculate the time delay between an axial ray and one that enters a 1-km-long fiber at an angle of 15°. The core index is 1.48.

17-13. Calculate the delay between the fastest and slowest modes in a 1-km-long step index fiber with $n_1 = 1.46$ and a *relative index difference*

$$\Delta \equiv \frac{n_1 - n_2}{n_2} = 0.003$$

using a light source at wavelength 0.9 μm.

17-14. Determine the material dispersion in a 1-km length of fused silica fiber when the light source is **(a)** a LED centered at 820 nm with a spectral width of 40 nm, and **(b)** a LD centered at 820 nm with a spectral width of 4 nm.

17-15. The total delay time $\Delta\tau$ due to *both* modal distortion and material dispersion is given by

$$(\Delta\tau)^2 = (\Delta\tau_{\text{mod}})^2 + (\Delta\tau_{\text{mat}})^2$$

Determine the total delay time in a 1-km fiber for which $n_1 = 1.46, \Delta = 1\%, \lambda = 820$ nm, and $\Delta\lambda = 40$ nm.

17-16. Determine the theoretical limit to the number of TV station channels that could transmit on a single optical beam of 1 μm wavelength. The bandwidth of a single TV channel is 6 MHz.

REFERENCES

BARNOSKI, MICHAEL K., ed. 1976. *Fundamentals of Optical Fiber Communications.* New York: Academic Press.

BOYD, WALDO T. 1982. *Fiber Optics Communications, Experiments and Projects.* Indianapolis: Howard W. Sams and Co.

CHEO, PETER K. 1989. *Fiber Optics and Optoelectronics.* Englewood Cliffs, N.J.: Prentice Hall.

COOK, J. S. 1973. "Communication by Optical Fiber." *Scientific American* (Nov.): 28.

DESURVIRE, EMMANUEL. 1992. "Lightwave Communications: The Fifth Generation." *Scientific American*: 114.

GHATAK, A. K., and K. THYAGARAJAN. 1989. *Optical Electronics.* New York: Cambridge University Press.

GOWAR, JOHN. 1984. *Optical Communication Systems.* Englewood Cliffs, N.J.: Prentice Hall International.

HEAVENS, O. S., and R. W. DITCHBURN. 1991. *Insight into Optics.* New York: John Wiley & Sons.

KARIM, MOHAMMAD A. 1990. *Electro-Optical Devices and Systems.* Boston: PWS-Kent Publishing Company. Ch. 9.

KATZIR, ABRAHAM. 1989. "Optical Fibers in Medicine." *Scientific American* (May): 120.

NEROU, JEAN PIERRE. 1988. *Introduction to Fiber Optics.* Sainte-Foy, Quebec: Les Editions Le Griffon D'Argile.

PALAIS, JOSEPH C. 1992. *Fiber Optic Communications*, 3rd ed. Englewood Cliffs, N.J.: Prentice Hall.

SEIPPEL, ROBERT G. 1984. *Fiber Optics*. Englewood Cliffs, N.J.: Prentice Hall.

SHOTWELL, R. ALLEN. 1996. *An Introduction to Fiber Optics*. Englewood Cliffs, N.J.: Prentice Hall.

SIEGMUND, WALTER P. 1978. "Fiber Optics." In Handbook of Optics, edited by Walter G. Driscoll and William Vaughan. New York: McGraw-Hill Book Company.

URQUHART, PAUL, and TIMOTHY J. WHITLEY. 1990. "Long Span Fiber Amplifiers." *Applied Optics* 29 (24): 3503.

WELFORD, W. T. 1981. *Optics,* 2d. ed. New York: Oxford University Press.

YARIV, AMNON. 1979. "Guided Wave Optics." *Scientific American* (Jan.): 64.

YARIV, AMNON. 1995. *Optical Electronics,* 4th ed. New York: Oxford University Press, Inc.

ZANGER, HENRY. 1990. *Fiber Optics*. Englewood Cliffs, N.J.: Prentice Hall.

Answers to Selected Problems

Chapter 1

1-2. (a) 6.6×10^{-34} m   (b) 3.9 Å

1-3. 3.6×10^{-17} W

1-4. 3.10 eV and 1.77 eV

1-5. 3.75×10^{17}

Chapter 2

2-1. $3.9 - 7.9 \times 10^{14}$ Hz

2-2. (a) 2055 lm   (b) 39.8 W/sr   (c) 10^5 W/m^2; 4.1×10^5 lm/m^2
(d) 9.95 W/m^2; 40.9 lx   (e) 0.0195 W; 0.0803 lm

2-3. (a) He-Cd appears about $1.3 \times$ brighter.   (b) about 2.4 mW

2-4. (a) 900 cd   (b) 85.4 lm/m^2 or lx

2-5. 1.055 : 1

2-6. 320 lx

2-7. (a) 1.7×10^9 cd/m^2 (b) $2\pi L$

2-8. (a) $I_{x2}/I_{x1} \cong 5$ (b) $I_{x1} \cong 35.6$ cd, $I_{x2} \cong 180$ cd

2-9. (a) 2.25×10^{-3} lm   (b) 28.6×10^6 lm/m^2 or 7.16 lm/mm^2

2-10. 78.4 cd

2-11. (a) 100%   (b) 1.31 lm

2-12. 5800 K

2-13. (a) 0.4830 μm   (b) 0.0756 W

2-14. 6266 K; 462.5 μm

2-15. 6105 K

Chapter 3

3-1. $t = (\Sigma_i n_i x_i)/c$

3-2. 4.00 mm

3-3. 3 ft, with top edge of mirror at a height half-way between the person's eye level and the top of the person's head.

3-4. The ray emerges from the bottom at 45°.

3-5. Reflection from the bottom surface; 1.60

3-6. 1.55

3-7. 1.153 cm

3-8. 8 cm

3-9. Light from the bubble is refracted through the plane surface both directly and after reflection from the spherical mirror; 3.33 cm and 10 cm.

3-10. 12.5 cm; 75 cm

3-11. 10 cm behind the rear surface; 3×

3-12. (a) $f = n_1 r/(n_2 - n_1)$ (b) $r > 0$ (convex) and $r < 0$ (concave), respectively

3-13. (a) center, 4/3 actual size   (b) 6.4 cm behind the glass, 8/7 actual size

3-14. Virtual, inverted, 15 cm from the window, twice the object size

3-15. 13.0 cm

3-16. +20 cm or −20 cm

3-17. 22.5 cm behind the lens; 1.50 times the actual size

3-18. (a) −6.7 cm   (b) −10 cm or −60 cm

3-19. −50 cm

3-20. 3.33 mm in front of the objective; erect and magnified

3-21. Final image between lens and mirror at 21/34 f from lens, virtual, inverted, and 1/17 the original size

3-22. (a) 33.3 cm, 2×   (b) 86.67 cm, 2×   (c) 7.37 cm, −0.316×

3-23. 1.63

3-24. 150 cm and 600 cm; inverted

3-25. (a) 10, 5, −2.5 diopters; 12.5 diopters   (b) −8.33 D, 4.17 D; 24 cm

3-26. (a) −16.67 D   (b) −22.17 D   (c) −18.75 cm

3-27. (a) +16.67 D   (b) −20.00 D   (c) −30.0 cm, to right of mirror, virtual, erect, 6×

3-28. 12.5 cm beyond the lens

3-31. $f_1/f_2 = n(n - 1)$

3-32. (b) $\Sigma_i (t_i \tan \theta_i)$

3-33. Incident on plane side: 8 cm beyond lens; on curved side: 5.33 cm beyond the lens

3-34. +40 cm, +30 cm; −30 cm, −40 cm

3-35. 25,000 ft

Chapter 4

4-1. $f_1 = -62.05$ cm; $f_2 = 46.66$ cm; $a = 2.91$ cm; $b = -0.98$ cm

4-2. (a) $f_1 = 14.06$ cm $= -f_2$; $a = 1.17$ cm; $b = -0.73$ cm   (b) 8.92 cm left of lens center
(c) 9.78 cm left of lens center, with 9.6% error

4-3. Erect, virtual image at 6.67 cm left of second vertex, 0.556 in. high

4-4. $f_1 = -11.51$ cm; $f_2 = 15.31$ cm; $a = 0.400$ cm; $b = -1.16$ cm; $j = 4.20$ cm; $k = 2.64$ cm;
$v = 18.9$ cm from H_2; $h_i = -1.18$ cm

4-5. (a) $f_1 = -20.15$ cm $= -f_2$; $a = 10$ cm $= -b$; $j = 10$ cm $= -k$
(b) Image is inverted, real, 61.38 cm from sphere center, and $m = -2.05$

4-6. $f_1 = -3.33$ cm, $f_2 = 4.33$ cm, $a = -0.83$ cm, $b = -2.17$ cm, $j = 0.17$ cm, $k = -1.17$ cm

4-7. $f_1 = -16.7$ cm, $f_2 = 23.33$ cm, $a = -1.67$ cm, $b = -4.67$ cm, $j = 5.00$ cm, $k = 2.00$ cm

4-8. (a) $f_1 = -6$ in., $f_2 = +6$ in., $a = +4$ in., $b = -4$ in., $j = a, k = b$
(b) 2 in. beyond ball (or right of V_2)

4-9. $a = b = +10$ cm; $f_2 = +150$ cm $= -f_1$; $j = k = +10$ cm

Chapter 5

5-1. (a) 15.62 mm   (b) 1.05 mm beyond the retina; 0.91 mm in front of the retina
(c) +4.0 D; −4.0 D

5-2. (a) +60.60 D   (b) 22.7 cm in front of eye   (c) −4.40 D

5-3. No. His range of clear vision is from 9.52 to 22.2 cm in front of the eye.

5-4. The range is now from 14.3 to 33.3 cm.

5-5. (a) $u_{fp} = -10.0$ cm; $u_{np} = -5.55$ cm; $P_c = -10.0$ D; $P_s = -11.76$ D
(b) $u_{fp} = +40.0$ cm; $u_{np} = -16.7$ cm; $P_c = +2.50$ D; $P_s = +2.41$ D
(c) $u_{fp} = -53.76$ cm; $u_{np} = -22.94$ cm; $P_c = -1.86$ D; $P_s = -1.91$ D

5-6. +15.0 D; −30.0 D

5-7. −3.75 D; −2.73 D

5-8. +3.75 D; +2.73 D

5-9. −15.0 D; +30.0 D

5-10. +10.0 D; −20.0 D

5-11. −2.50 D; −1.82 D

5-12. $P_s = +10.50$ D

5-13. $P_c = -10.27$ D

5-14. +4.0 D

5-15. +2.33 D

5-16. 93.9 μm

Chapter 6

6-1. The line image is real, 18.75 cm past the lens and 15.75 cm long.

6-2. The line image is virtual at a distance of 65.2 cm on the object side of the lens and the line length is 5.65 cm.

6-3. $v = -11.11$ cm; length = 2.80 cm

6-4. The line image is virtual at 13.33 cm from the lens and is 0.67 cm long.

6-5. The line image is now virtual, 21.4 cm from the lens, and 17.97 cm long.

6-7. The Sturm interval extends from 15.0 cm to 60.0 cm. The circle of least confusion lies at 24.0 cm from the lens, with diameter of 0.60 cm.

6-8. The diameter of the circle of least confusion is 4.57 mm at 28.6 cm from the lens. The interval of Sturm extends from 5.82 mm to 21.3 mm.

6-9. Spherical +6.0/cylindrical -4.0 × 90; Spherical +2.0/cylindrical +4.0 × 180

6-10. Spherical −5.0/cylindrical +10.0 × 90; Spherical +5.0/cylindrical −10.0 × 180

6-11. 11.0 × 30/3.0 × 120; 11.8 to 200 cm

6-12. 18.0 × 90 / 14.0 × 180; 5.55 to 11.11 cm

6-13. Spherical +2.87/cylindrical +1.78 × 90; Spherical +4.65/cylindrical −1.78 × 180

6-14. Spherical −4.25/cylindrical +6.19 × 90; Spherical +1.94/cylindrical −6.19 × 180

6-15. With SMA, the vertical bar is broadened; with SHA, the horizontal bar is broadened

Chapter 7

7-2. For $\sigma = 0.70$, radii are +17.65 cm and −100 cm; for $\sigma = 3.00$, radii are +7.50 cm and +15.0 cm

7-3. For $\sigma = 0$: +20 cm and −20 cm; for $\sigma = 1$: +10 cm and infinite; for $\sigma = 2$: +6.67 cm and +20 cm

7-4. For $\sigma = 2$: −8.67 cm and −26.0 cm; for $\sigma = 4$: −5.20 cm and −8.67 cm

7-6. $r_1 = 18.62$ cm; $r_2 = -33.75$ cm

7-7. $r_1 = +20.0$ cm $= -r_2$

7-8. $\sigma = +0.867$, closer to $\sigma = +1$ for incident spherical surface than to $\sigma = -1$ for incident plane surface

7-9. (a) $\sigma = +0.714$　　(b) $r_1 = +17.5$ cm, $r_2 = -105$ cm (c) $\sigma = -0.714$, $r_1 = +105$ cm, $r_2 = -17.5$ cm

7-10. (a) $\sigma = +0.80$ (b) $r_1 = +16.7$ cm, $r_2 = -150$ cm (c) $\sigma = -0.80$, $r_1 = +150$ cm, $r_2 = -16.7$ cm

7-11. ±20 cm

7-12. ±20 cm

7-13. −17.71 cm

7-14. (a) 15.7 cm　　(b) −3.476 cm

7-15. $r_{11} = 8.5168$ cm, $r_{22} = -434.89$ cm; $f_D = 20.0000$ cm, $f_C = 20.0096$ cm $= f_F$

7-16. (a) $r_{11} = 3.4535$ cm, $r_{22} = -12.6576$ cm　　(b) $f_D = 5.0000$ cm, $f_C = 5.0026$ cm $= f_D$ (c) $P_{1D} = 0.3695$ cm^{-1}, $P_{2D} = -0.1695$ cm^{-1}　　(d) yes

7-17. (a) $r_{11} = -5.2415$ cm, $r_{22} = 53.1840$ cm　　(b) $f_{1D} = -4.5770$ cm, $f_{2D} = 8.4399$ cm (c) $f_D = -10.0000$ cm, $f_C = -10.0050$ cm $= f_F$

Chapter 8

8-1. (b) Any position farther than 15 cm from the first lens (c) Any position nearer than 15 cm from the first lens

8-2. (a) Lens is the aperture stop.
(b) Entrance and exit pupils coincide with the lens; same size
8-3. (a) Aperture is the aperture stop.
(b) Entrance pupil coincides with aperture; diameter = 5 cm
(c) Exit pupil is 60 cm to the right of lens; diameter = 20 cm
8-4. (a) Aperture is the aperture stop.
(b) Entrance pupil is 24 cm to the right of the lens; diameter = 15 cm
(c) Exit pupil coincides with the aperture; diameter = 5 cm
8-5. (a) Lens is the aperture stop.
(b) Entrance pupil coincides with the lens; diameter = 5 cm
(c) Exit pupil coincides with the lens; diameter = 5 cm
8-6. (a) Image is virtual, erect, 4 cm high, 12 cm to the left of lens, in focal plane.
(c) $\Delta\omega = 0.19$ sr
8-7. Entrance pupil is the aperture stop; exit pupil is 3.33 cm in front of the lens, with an aperture of 3.33 cm; image is 10 cm behind the lens, inverted and 2 cm long.
8-8. Exit pupil is the aperture stop; entrance pupil is 4.29 cm behind the lens, with an aperture of 3.43 cm; image is 10.5 cm behind the lens, inverted, and 3 cm long.
8-9. Entrance pupil is the aperture stop; exit pupil is 12 cm in front of the lens with an aperture of 6 cm; image is 10.5 cm behind the lens, inverted and 1.5 cm long.
8-10. (b) 10 cm right of L_2 (c) both at L_1
(d) 6.25 cm right of L_2 and 1.5 cm in diameter
(e) field stop at A; entrance window in object plane with a 4-cm diameter; exit window in image plane with a 2-cm diameter
8-11. (b) Entrance pupil is a virtual image, slightly magnified, 3.05 mm to the right of the corneal surface or 0.55 mm to the left of the real pupil.
(c) Exit pupil is a virtual image, slightly magnified, located 3.52 mm to the left of the rightmost surface of the lens or 0.08 mm to the right of the real pupil.

Chapter 9

9-2. 53′
9-3. (a) Left lens, $\alpha \cong 2.2°$; right lens, $\alpha \cong 3.3°$ (b) Left, base down; right, base up
9-4. (a) 5 pd   (b) 5.1°   (c) base-in prism, each lens
9-5. (a) crown: $A = 1.511, B = 4240$ nm^2, $n_D = 1.523$;
flint: $A = 1.677, B = 13190$ nm^2, $n_D = 1.715$
(b) crown: -4.146×10^{-5} nm^{-1}; flint: -1.290×10^{-4} nm^{-1}
9-6. (a) 50.0°   (b) 1/55.5   (c) $A = 1.6205, B = 6073.7$ nm^2; 4.297×10^{-5} nm^{-1}
9-7. 0.01909
9-8. 5.99°, 2.16′
9-9. 4.82°, 4.37°
9-10. (a) $M_e = 1 \times 10^4$ W/m^2, $I_e = 3.98$ W/sr, $L_e = 1.59 \times 10^3$ W/m^2-sr
(b) $\Phi_F = 7.81 \times 10^{-5}$ W   (c) $E_e = 18$ W/m^2
9-12. 5.7 cm
9-13. $f = 53.3$ cm, 13.33 cm, 1.86 cm
9-14. 5.3 to 7.0 ft
9-15. 1.3×10^7 lx
9-16. (a) 0.90 cm   (b) 5.45 cm, 3×
9-17. (a) 27.8 mm   (b) $f/3.1, f/5.4, f/9.4$ (c) 16.0, 9.26, 5.35 mm
(d) 0.03, 0.09, 0.27 s
9-19. (a) 2.8 cm   (b) 10×
9-20. (a) 320×   (b) 0.516 cm
9-21. (a) 46.7×   (b) 8.68 cm
9-22. 5 cm
9-23. 14.9 cm
9-24. (a) 7×   (b) 2 cm   (c) 5 mm   (d) 2.3 cm   (e) 337 ft
9-25. (b) 7.50×; 8.70×
9-26. 1.05 cm
9-27. (a) 8 cm, 3×   (b) 7.38 cm, 2.6×

9-28. 1.25 cm farther from objective

9-29. (a) 12.5×   (b) 15×   (c) 0.13 cm, 3 mm   (d) 3.8°

9-31. 2.5 ft; −180×

Chapter 10

10-1. +41.6 D

10-2. Block-letter sizes are 1.309 in. for 20/300; 0.436 in. for 20/100; 0.262 in. for 20/60; 0.087 in. for 20/20; 0.065 in. for 20/15. Letter details are 1/5 block letter size in each case.

10-3. (a) 8.33 mm; +120 D   (b) 42.86 mm; +23.33 D
(c) $f_2 = -f_1 = 42.35$ mm; $P = 23.6$ D

10-4. (a) 22.33 mm from the cornea   (b) 21.59 mm from the cornea

10-5. (a) 24.17 mm from the cornea   (b) 24.05 mm from the cornea

10-6. Focal point F_1 is 15.31 mm in front of cornea; focal point F_2 is 24.17 mm behind the cornea (at the retinal position of the Gullstrand eye)

10-7. (a) $f_1 = -18.98$ mm; $f_2 = +25.36$ mm; $a = 0.63$ mm; $b = -6.37$ mm, so focal point F_1 is 18.35 mm to the left of the cornea; focal point F_2 is 18.99 mm to the right of rear surface of lens; principal point H_1 is 0.63 mm to the right of cornea; principal point H_2 is 0.83 mm to the right of the cornea
(c) For this model, F_2 falls a distance $26.19 - 24.17 = 2.02$ mm further from the cornea.

10-8. Focal point F_1 is 16.67 mm to the left of cornea; focal point F_2 is 22.22 mm to the right of cornea (at the retinal surface)

10-9. +4.0 D

10-10. E_NP is virtual, 9.06 mm in diameter; E_XP is virtual, 8.27 mm in diameter.

10-11. (a) +3.2 D   (b) yes

10-12. −2.0 D; −21.4 cm

10-13. (a) myopia; astigmatism   (b) myopia   (c) hyperopia
(d) hyperopia; astigmatism

10-14. (a) At $P = -19.00$ D, $f = -5.26$ cm, diverging lens;
at $P = +17.00$ D, $f = 5.88$ cm, converging lens
(b) At $P = -19.00$ D, $\Delta f = -0.070$ cm; at $P = -5.00$ D, $\Delta f = -1.00$ cm;
at $P = +8.00$ D, $\Delta f = 0.39$ cm; at $P = +17.00$ D, $\Delta f = 0.087$ cm
(c) a plane piece of glass

10-15. (a) 15×
(b) The angular magnification of retinal detail is 15 times larger than if the same retinal detail were removed from the eye and examined directly at the distance of 25 cm.

Chapter 11

11-1. (a) $y = 2 \sin 2\pi(z/5 + t/3)$ (b) $y = 2 \sin (2\pi/5)(z + 5t/3)$

11-2. (a) $y = 5 \sin (\pi x/25)$ (b) $y = 5 \sin (\pi/25)(x + 8)$

11-3. (a) 0.01 cm   (b) 1000 Hz   (c) 628.3 cm^{-1} (d) 6283 s^{-1} (e) 1 ms
(f) 10 cm/s   (g) 10 cm

11-4. $y = 15 \sin (kx + \pi/3)$

11-5. (b) 90°, 60°, 0°, −90°, 108°   (c) Subtract 90° from each

11-6. $E = 1028$ V/m; $B = 3.43 \times 10^{-6}$ T

11-7. 5×10^{-7} T

11-8. (a) 1.01×10^3 V/m, 3.37×10^{-6} T   (b) 4.76×10^{21} / m^2-s
(c) $E = 1010 \sin 2\pi(1.43 \times 10^6 r + 4.28 \times 10^{14} t)$, r in m, t in s

11-9. (a) 8.75×10^{-3} W/m^2, 2.57 V/m   (b) 2×10^{13} W/m^2, 1.23×10^8 V/m, 0.410 T

11-10. $v = 0.168$ c

11-11. $v = -0.917$c

11-12. (b) $E_R = 8.53 \sin (\omega t + 0.2\pi)$

11-13. $E_R = 6.08 \sin (2\pi t + 1.13)$

11-14. $y = 11.6 \sin (\omega t + 0.402\pi)$

11-15. $E = 1.71 \sin (\omega t + 4.57)$

11-16. $f(t) = 2.48 \sin (20t - 0.2\pi)$

11-17. 14 cm; 1.57 cm; 0.785 cm; 0 cm/s; T seconds

11-18. (a) 1.5 cm; 25 Hz; 20 cm; 5 m/s; opposite directions   (b) 10 cm   (c) −3 cm

11-19. (a) 0.9 cm   (b) 30 cm   (c) 10.8 km

12-1. $A_{21}/B_{21} \propto 1/\lambda^3$ and $\lambda_{uv} < \lambda_{ir}$, so $(A_{21}/B_{21})_{uv} > (A_{21}/B_{21})_{ir}$; larger values of the ratio A_{21}/B_{21} will favor spontaneous emission over stimulated emission, thereby making a population inversion more difficult to achieve.

12-2. 0.00138; spontaneous emission is about 314 times larger than stimulated emission. This is to be expected since the ratio calculated is greater than 1 whenever $hf \gg kT$.

12-3. (a) See Table 12-1   (b) Linewidth is about one ten-millionth as wide.

12-4. 0.13 ms; 40 km

12-5. 0.80 mrad

12-7. (b) 2.5×10^5 W/m²-sr-Hz

12-8. (a) 10^{-4} rad   (b) 10 μm   (c) 1.27×10^7 W/m²

12-9. (a) 0.81 μm; 0.75 μm; 0.585 μm; 0.525 μm   (b) 76%; 70%; 55%; 49%

12-10. (a) 31.5 W   (b) 1.26%

12-11. (a) Argon ion; freq-doubled YAG; Nd:YAG; Ti:sapphire
(b) CO_2; Erbium:YAG

12-12. (a) Glass absorbs 10.6 μm radiation   (b) 4.4×10^{-4} rad   (c) 14.5 μm
(d) 3×10^6 W/cm²

12-13. (a) 10^6 W   (b) 5 μm   (c) 5.1×10^{12} W/cm²

12-14. (a) 1.04×10^{-2} W/cm²   (b) 783 W/cm²   (c) ~75,000

12-15. (a) 0.21 J/cm²   (b) 4.6

12-16. Nd:YAG; tunable dye; GaAs; alexandrite; Ti:sapphire

12-17. (a) 1×10^4 W   (b) $10\times$
(c) selection of suitable optical materials to transmit at 2.94 μm

12-18. (a) liquid dye laser at 1.0 μm (1.5 db/km); Nd:YAG laser at 1.06 μm (1.3 db/km);
Ti:sapphire laser at 1.08 μm (1.2 db/km)
(b) 0.66   (c) 81%

13-1. (a) 11,945 and 21,235 W/m²   (b) 18,723 W/m²   (c) 51,903 W/m²   (d) 0.96

13-2. 0.86, 0

13-3. 0.8; 3.73/1

13-4. (b) 1.78, 2.55, 4.00, 13.9

13-5. Lloyd's mirror interference fringes are produced, aligned parallel to the slit, and separated by 0.273 mm. The irradiance of the pattern is given by $I = 4I_0 \sin^2 (115y)$, with y measured in cm from the mirror surface.

13-6. 509 nm

13-7. 514.5 nm

13-8. To acquire coherent beams; 560 nm

13-9. (a) 83.3 cm   (b) 83.3 fringes   (c) 150 nm

13-10. 9.09×10^{-5} cm; orders 4 and 3, respectively

13-11. 498 nm

13-12. 1.33; 103 nm

13-13. (a) 2.78%   (b) 89.3 nm   (c) 1%

13-14. Soap film becomes wedge-shaped under gravity; the angle of the wedge is $1'14''$.

13-15. 15

13-16. 1.16×10^{-3} cm

13-17. 1.09 mm; 184

14-1. 3 m

14-2. (a) 436 nm   (b) 1.09 μm

14-4. 603.5 nm; 2.39 mm; 2.87×10^{-4} cm

14-5. 928 nm

14-6. 436 nm

14-7. One mirror makes a wedge angle of 0.0172° with the image of the other, reflected through the beam splitter. Fizeau fringes result.

14-8. 23.75 μm

14-9. (a) 80,000   (b) 79,994

14-10. (a) $n = 1 + N\lambda/2L$ (b) 153

14-11. (a) 11.2°   (b) 45.9°
14-12. 79.1 nm or $\lambda/8$
14-13. 63.3 m/s
14-14. 1.8×10^{10} bits

Chapter 15

15-1. 28.1%
15-2. 67.5°; 22.5°
15-3. 0.0633 mm
15-4. (a) single refraction, phase retardation, any polarization possible
   (b) single refraction, no phase retardation, unpolarized
   (c) same as (a)
   (d) double refraction, no phase retardation in each separated beam, each beam linearly polarized
   (e) cases (a) and (c)
15-5. (b) 0%   (c) 33%
15-6. 0.0162 mm
15-7. 20°
15-8. 0.005
15-9. (a) 53.12°   (b) 11.5°
15-10. (a) mixture of unpolarized and circularly polarized light
   (b) elliptically polarized light
15-11. (a) 56.2°   (b) 33.8°
15-12. (a) 0.05 g/ml   (b) about 46°
15-13. 10.64°
15-14. (a) 8.57×10^{-5} cm   (b) green
15-15. (a) 0.0091   (b) 15 μm   (c) 12   (d) 15 μm
15-16. 3.15×10^{-4}
15-17. (a) 36.8 μm   (b) 18.4 μm
15-18. 14.7°
15-19. (a) 14.8%   (b) 2.03% of I_0 (c) 0.92
15-20. $I_0 \sin^2 (2\theta)$

Chapter 16

16-1. (a) 0.218 cm   (b) 0.218 cm
16-2. 0.090
16-3. 0.135 mm
16-4. 496 nm
16-5. 2.125, 1.44, 0.778, and 0.55 μm
16-6. (a) 15°   (b) 0.678, 0.166, 0, 0.0461
16-7. 8.4×10^{-4} cm
16-8. 9725 km in diameter; 2.69×10^{-11} W/m^2
16-9. 5.2 m
16-10. 5.3 miles
16-11. 75.7 to 265 m
16-12. (a) 0.400 mm   (b) 0.8106, 0.4053, 0.09006
16-13. (b) 2.10 mm
16-14. (b) 20
16-16. 0.875, 0.573, 0.255, 0.0547, 0
16-17. (a) $I = I_0 \dfrac{\sin^2 (1.151y) \sin^2 (0.575x)}{0.438x^2y^2}$
   (b) 5.46 mm along x; 2.73 mm along y
   (c) 0.895 at (1,0); 0.629 at (0,1)   (d) 0.005 at (2,3)
16-18. (a) 90°   (b) 11.5°   (c) 5.7°
16-19. 13°18′
16-20. 458 lines/mm
16-21. 471.4 nm
16-22. (a) 0.63 mm   (b) 1.26 mm

17-1. (b) 1284

17-2. (a) 68.1°   (b) 0.567   (c) 34.5°

17-3. (a) 0.64   (b) 79.5°   (c) 3281

17-4. (c) 432 μm; 429 μm; 10.07 m

17-5. -70 db/km

17-6. 0.080 mW

17-7. 3.33 km; 10 km

17-8. 0.136 db/km

17-9. (b) -1.25 db, -6.02 db, -10 db, -20 db

17-10. (a) 1.0069 km; 1 km   (b) 4.900 μs; 4.867 μs

17-11. 431 ns; 2.32 MHz

17-12. 77.2 ns

17-13. 14.6 ns/km

17-14. (a) 4 ns (b) 0.4 ns

17-15. 48.9 ns

17-16. 50 million

Index